# Chemical Triggering

## Reactions of Potential Utility in Industrial Processes

# TOPICS IN APPLIED CHEMISTRY

**Series Editors: Alan R. Katritzky, FRS**
*Kenan Professor of Chemistry*
*University of Florida, Gainesville, Florida*

**Gebran J. Sabongi**
*Laboratory Manager, Encapsulation Technology Center*
*3M, St. Paul, Minnesota*

**CHEMICAL TRIGGERING**
**Reactions of Potential Utility in Industrial Processes**
Gebran J. Sabongi

**STRUCTURAL ADHESIVES**
Edited by S. R. Hartshorn

A Continuation Order Plan is available for this series. A continuation order will bring delivery of each new volume immediately upon publication. Volumes are billed only upon actual shipment. For further information please contact the publisher.

# Chemical Triggering

## Reactions of Potential Utility in Industrial Processes

Gebran J. Sabongi

*Laboratory Manager*
*Encapsulation Technology Center*
*3M*
*St. Paul, Minnesota*

**Plenum Press • New York and London**

Library of Congress Cataloging in Publication Data

Sabongi, G. J. (Gebran J.)
 Chemical triggering.

 (Topics in applied chemistry)
 Includes bibliographical references and index.
 1. Chemical processes. 2. Chemical reactions—Industrial applications. I. Title. II.
Series.
 TP155.7.S23   1987                        660.2′844                        87-25742

 ISBN-13: 978-1-4612-8239-6        e-ISBN-13: 978-1-4613-0907-9
 DOI: 10.1007/978-1-4613-0907-9

© 1987 Plenum Press, New York
Softcover reprint of the 1st edition 1987

A Division of Plenum Publishing Corporation
233 Spring Street, New York, N.Y. 10013

For

Susan, Katie,
Odette, and Farid

with all my love

# Preface

Chemical reactions which can, on demand, be switched on and off are valuable for industrial applications.

In order to make the best use of these reactions, it is essential to have them readily available for a research chemist. The chemical literature, in general, has not yet identified or grouped such reactions. However, their existence is relatively abundant.

This book is meant as a survey of those reactions which have potential utility in industrially useful processes. These reactions are grouped under the title of chemical release reactions which can be triggered by heat, light, electric current, etc., to release a specific compound from, or change in the physical or chemical properties of, a unimolecular reactant.

The book is divided into chapters covering ways to trigger the release of certain chemicals. Each chapter is further divided into sections, each beginning with a brief introduction of analogies of the discussed reactions and of how they were used in reported industrial processes.

This survey is not meant to be absolute or exhaustive but rather to be directive, to be as complete as possible, and to provide food for further thought.

G. J. Sabongi

*St. Paul, Minnesota*

# Contents

## 2. Triggered Release of Acids, Bases, Radicals, Nitrenes, and Carbenes

# 5. Triggered Isomerization and Color Change

# Introduction

The idea of putting together a monograph which could be used as a guide and reference for chemists in industry was born when I left academic life and moved into industry and applied research. The idea was also catalyzed by suggestions, interest, help, and unlimited support from Professor Alan R. Katritzky, a friend and my research advisor during my doctorate studies.

In applied industrial research, a number of processes are based on a few chemical reactions commonly used in academic research, which are viewed from a different angle. This difference is the concept behind this book. Could we put together a list of chemical reactions which are available in the chemical literature and would be valuable for industrial applications if viewed from a different perspective?

To do this, we had to define which of these reactions to include, the selection of their possible applications, and the conditions under which they would be utilized. After the survey of the common needs and a variety of industrially used reactions, it was apparent that chemical release reactions which can be triggered by a variety of triggering energies should be the main topic. The selection of their possible areas of application was a more difficult choice but it had to be narrowed to those which would utilize fine chemicals, such as image reproduction, information recording, and polymer applications, as opposed to bulk chemical applications.

The triggering conditions and nature of the triggering agent under which the chemical reaction would be used are determined by the nature of the application. Ideally, such reactions would be dormant in the absence of the triggering agent (energy) but would be triggered to completion in a short time under the action of the triggering energy.

Finally, the idea crystallized into a plan, which encompasses a survey of chemical reactions from the literature that, under specific triggering conditions, would release chemical reagents, including small molecules, acids, bases, and gases. These released chemicals could then be used in

1

industrially useful processes. No specific areas of potential use for such reactions will be discussed and it is left to the reader to determine their potential value. This approach was chosen in order to leave an element of inventiveness in any possible application.

The compilation of this survey is not meant to be absolute or exhaustive but directive and as complete as possible. The book is divided into chapters covering ways to trigger the release of certain chemicals. Each chapter is further divided into sections beginning with a brief introduction of analogies of the discussed reactions and how they were used in reported industrial applications. Such an introduction is meant to give the reader a quick and brief overview of areas of published applications. The rest of the sections of the chapter comprise a group of reactions which have in common the triggering agent (energy). The subsections include reactions which have in common the type of released molecule.

In order to make the best use of the included information, let us first define what is meant by triggered release. It is a chemical process which is switched on by the action of an energy, such as heat, light, or electric current, that causes the release of a specific chemical compound or the manifestation of a new physical property. The life of the reaction, and thus the release process, is limited to the time the triggering energy is switched on and ceases when the energy is switched off.

The thermally triggered release reactions surveyed are limited to those which proceed at temperatures between 70 and 200°C. These thermal limits eliminate processes which are unstable at room temperatures and those which require extreme temperatures in order to proceed. The ideal reaction is one which proceeds to completion in the shortest time and in high conversion yields. The common source for thermal energy is via conduction using a heated element, a liquid, or infrared radiation.

The reaction efficiency is described by the yield of the formed product expressed as a percentage and calculated as follows:

$$\text{Percent yield} = \frac{\text{Number of moles of product formed}}{\text{Number of moles of product theoretically possible}} \times 100$$

The photochemical processes have about the same requirements as the thermal ones. In general, most of the photochemical reactions are sensitive to ultraviolet radiation, and a few are sensitive to the energy from visible light. Some ultraviolet-sensitive reactions can be made sensitive to the energy of visible light by dye sensitization.

The efficiency of the reaction can be described by the quantum yield, which relates the amount of product released to the amount of energy absorbed. Mathematically the quantum yield is described as

$$\phi = \frac{\text{No. of moles of product formed (or reactants consumed)}}{\text{No. of Einsteins (moles) of light absorbed}}$$

If $\phi = 0$, all the absorbed energy is lost in physical processes and none is involved in the chemical reaction. When $\phi = 1$, the absorbed quanta are transformed stoichiometrically into product. When $\phi > 1$, a photo-induced chain reaction is occurring with an amplification factor.

The generation of the required photo energy can be accomplished by the use of a variety of sources. The most common are those which can provide a wide emission spectrum and thus have broad applicability. Examples are carbon arc, tungsten, mercury arc, pulsed xenon, and metal halide sources, whose emission spectra are graphically represented in Figure 1.

The use of optical filters in order to irradiate selectively at specific wavelengths is a well known technique which allows triggering at these wavelengths.

The low-pressure mercury lamp commonly used for photochemical release reactions has been recently studied by Hammond and Gallo[1] and its emission line at 253 nm examined as a function of mercury pressure, tube radius, and operating current.

Seliger and McElroy[2] and Calvert and Pitts[3] have published a range of chemical and other optical filters useful in selectively irradiated photochemical reactions.

Developments in photochemical instrumentation and techniques have been reviewed by West[4] and by McLaren and Shuger.[5] The latter review focuses on protein and nucleic acid photochemistry. Schonberg[6] has reviewed several light sources useful in photochemical reactions, their properties, manufacturers, emission wavelengths and corresponding energies.

Table 1 describes the emission energy of the mercury arc light source.

The recent commercialization of monochromatic lasers and laser diodes has allowed their use as another source of light energy. A laser beam can be described as a monochromatic, parallel, and coherent light beam which can deliver high light intensities by focusing the light output on a very small area. Seliger and McElroy[2] and Turro[8] have reviewed different aspects of lasers and their properties. Applications of lasers to chemistry have been periodically reviewed in the *Photochemistry: A Specialist Periodical Report* series of reviews[9] published by the Chemical Society of Great Britain.

The most commonly used lasers can be grouped into seven families:

1. Carbon dioxide lasers have the same active medium but are produced in four different configurations.
2. Chemical lasers have the same hardware which is used to obtain laser action from hydrogen fluoride or from its isotopic variants with emission at various wavelengths.

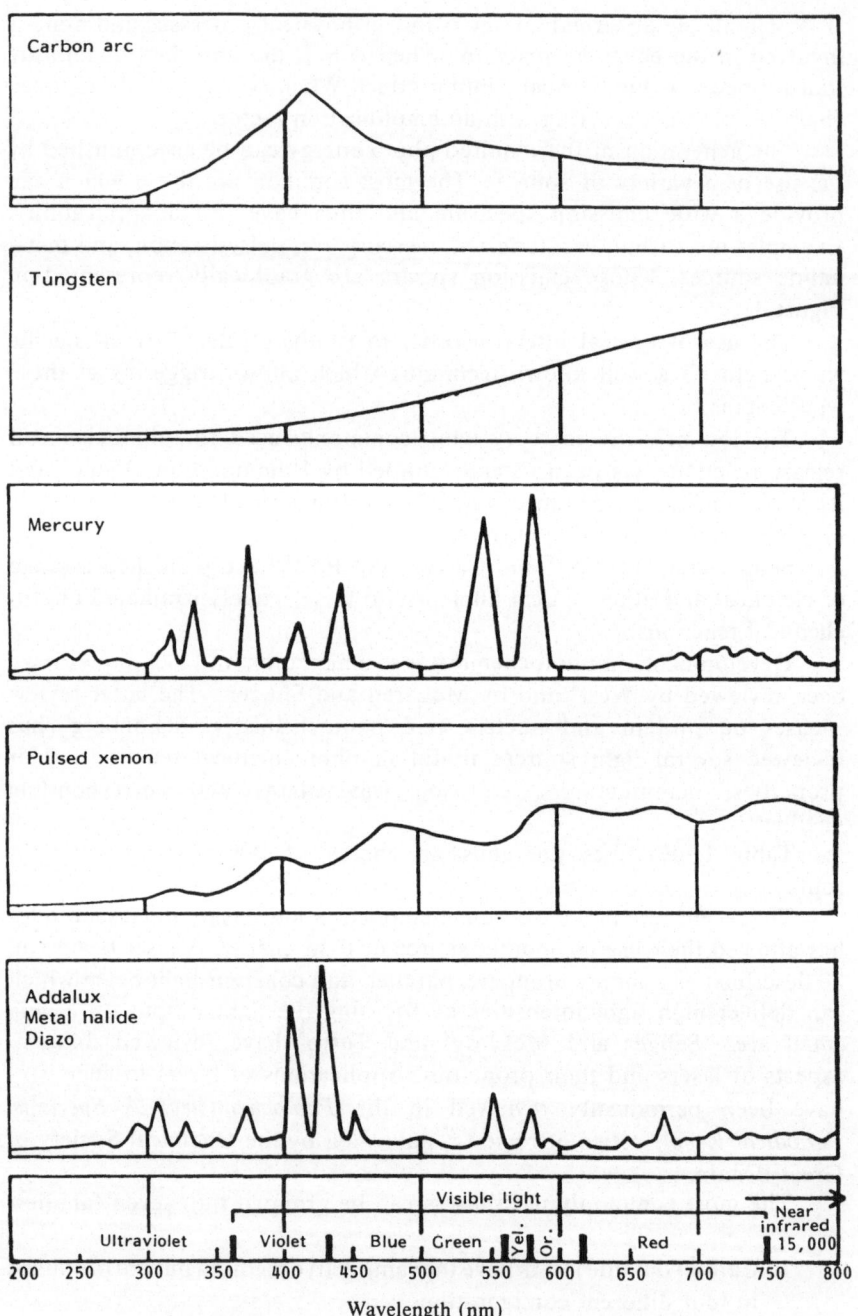

**Figure 1.** Spectral energy distribution of various light sources.

Table 1. Emission Energy from a Hanovia
679Å Lamp[a]

| Mercury line (nm) | Radiated wattage |
|---|---|
| 248.2 | 2.3 |
| 253.7 | 5.8 |
| 265.2 | 4.0 |
| 280.4 | 2.4 |
| 296.7 | 4.3 |
| 302.5 | 7.2 |
| 313.0 | 13.2 |
| 334.1 | 2.4 |
| 366.0 | 25.6 |
| 404.5 | 11.0 |
| 435.8 | 20.2 |
| 546.1 | 24.5 |
| 578.0 | 20.0 |

[a] Ref. 7.

3. Dye lasers rely on some organic dye, the active medium, in a liquid solvent and derive their energy from light emitted by different types of lasers or in some cases flash lamps. Their characteristics depend on the light source used.

4. Excimer lasers utilize different gases in the same hardware to produce light of different wavelengths.

5. Ion lasers emit light from ionized argon and/or krypton molecules; the hardware used to build them and mixed gas lasers is the same.

6. Semiconductor laser light is emitted at a $p-n$ junction in a semiconductor diode; the wavelength depends on the semiconductor composition.

7. Solid-state lasers utilize light from a flash lamp or arc lamp to excite laser emission from atoms in a crystalline or glass host.

Table 2 contains further information on the typers of lasers available.

A number of currently available reviews provide excellent references in areas related to the subject matter of this volume, as well as detailed and complementary background information. In the areas of photo- and thermochemistry, reviews edited/authored by Bryce-Smith,[9] by Turro et al.,[10] by Srinivasan and Roberts,[11] by Henderson and Marsden,[12] by Schonberg,[6] by Brown,[13] by Zewail,[14] by Pappas,[15a] by Allen,[15b] by Hurd,[16] by Harrison and Harrison,[17] by Hodge and Sherrington,[18] by Greene,[19] and by McOmie[20] are useful.

Table 2. Lasers, Their Power and Wavelengths

| Laser type | Output power | $\lambda_{max}^{emission}$ (nm) |
|---|---|---|
| *Excimer* | | |
| Argon fluoride | 10.0 W | 193 |
| Krypton fluoride | 25.0 W | 249 |
| Xenon chloride | 8.0 W | 308 |
| Xenon fluoride | 7.0 W | 351 |
| *Dye laser pumped by* | | |
| Nitrogen | | |
| Excimer | } 0.05–2.0 W | } Tunable (300–1000) |
| Neodynium-YAG | | |
| Flash lamp | 0.5–50.0 W | 340–940 |
| Ion laser | 20–800 mW | 400–900 |
| Nitrogen | 1–330 mW | 337 |
| *Ion* | | |
| Argon | 5.0 mW–20.0 W | Main lines, 486 and 514 |
| Krypton | 5.0 mW–20.0 W | Strongest, 647 |
| Argon-krypton | 0.5–6.0 W | 450–670 |
| Helium-cadmium | 2–40 mW; 1.5–50 mW | 442; 325 |
| Helium-neon | 0.1–50 mW | 633–1152 |
| *Semiconductor diode* | | |
| Ruby | 0.03–100 J | 694 |
| GaAs/GaAlAs | 1–40 mW | 780–905 |
| Neodynium-YAG (pulsed) | 400 | 1064 |
| Neodynium-doped glass | 0.15–100 J | 1064 |
| Neodynium-YAG (continuous) | 0.04–600 W | 1064 |
| InGaAsP | 1–7 mW | 1100–1600 |
| *Chemical lasers* | | |
| Hydrogen fluoride | 0.01–150 W | $26.3 \times 10^3$ |
| Deuterium fluoride | 0.01–100 W | $36.4 \times 10^3$ |
| *Carbon dioxide* | | |
| Flow gas | 50–15,000 W | $9–11 \times 10^3$ |
| Sealed tube | 3–100 kJ | $10.6 \times 10^3$ |
| Pulsed TEA | 0.03–75 J | $9–11 \times 10^3$ |
| Waveguide | 0.1–40 W | $9–11 \times 10^3$ |

Finally, errors of interpretation and fact are almost inevitable in a book of this nature. The responsibility for these is mine and I would appreciate having them brought to my attention.

The preparation of this book would not have been possible without the help of several people. I would like to express my appreciation to the 3M Company and its officers for their support, Professor Alan R. Katritzky whose encouragement, constructive suggestions, and review of the included material have been invaluable, and Dr. Melville R. Sahyun of the Science

Research Laboratory, Dr. Spencer F. Silver of the Specialty Chemicals Division, and Dr. Terry L. Davis of the Encapsulation Technology Center in 3M for their review of the manuscript prior to submission to the publishers. Last and by no means least, I would like to thank my wife for her help in the typing of the manuscript and both her and my daughter for their understanding and encouragement during the years this manuscript was in preparation.

## REFERENCES

1(a) T. J. Hammond and C. F. Gallo, *Appl. Optics 15*, 64 (1976); (b) T. J. Hammond and C. F. Gallo, *Appl. Optics 15*, 308 (1976).
2. H. H. Seliger and W. D. McElroy, *Light, Physical and Biological Action*, pp. 349-357 (lasers), 358-360 (filters), Academic Press, New York (1965).
3. J. G. Calvert and J. N. Pitts, *Photochemistry*, p. 148, Wiley, New York (1966).
4. M. A. West, in: *Photochemistry: A Specialist Periodical Report* (D. Bryce-Smith, ed.), The Chemical Society, London, Vol. 3, p. 170 (1972); Vol. 4, p. 85 (1973); Vol. 5, p. 80 (1974); Vol. 8, p. 3 (1977); Vol. 10, p. 3 (1979); Vol. 13, p. 3 (1983).
5. A. D. McLaren and D. Shuger, *Photochemistry of Proteins & Nucleic Acids*, pp. 358-387, Pergamon Press, Oxford.
6. A. Schonberg, *Preparative Organic Photochemistry*, pp. 472-474, Springer-Verlag, New York (1968).
7. T. R. Evans, in: *Techniques in Organic Chemistry* (P. A. Leermakers and A. Weisberger, eds.), Vol. 14, p. 297, Interscience, New York (1969).
8. N. J. Turro, *Molecular Photochemistry*, Benjamin, New York (1965).
9. D. Bryce-Smith, ed., *Photochemistry: A Specialist Periodical Report*, The Chemical Society, London (1970-1983).
10. N. J. Turro, G. J. Hammond, J. N. Pitts, D. Valentine, A. D. Broadbent, W. B. Hammond, and E. Whittle, *Annual Survey of Photochemistry*, Wiley, London.
11. R. Srinivasan and T. D. Roberts, *Organic Photochemical Syntheses*, Wiley-Interscience, New York (1971).
12. S. T. Henderson and A. M. Marsden, *Lamps and Lighting*, Crane, Russak and Co., New York (1972).
13. G. H. Brown, ed., *Techniques of Chemistry: Photochromism*, Vol. 3, Wiley-Interscience, New York (1971).
14. A. H. Zewail, ed., *Advances in Laser Chemistry*, Springer-Verlag, Berlin (1978).
15. (a) S. P. Pappas, ed., *UV Curing: Science and Technology*, A Technology Marketing Publication, Connecticut (1978); (b) N. S. Allen, ed., *Developments in Polymer Photochemistry*, Vols. 1-3, Applied Science Publishers, London (1980-82).
16. C. D. Hurd, *The Pyrolysis of Carbon Compounds*, The Chemical Catalog Company Inc., New York (1929).
17. I. T. Harrison and S. Harrison, *Compendium of Organic Synthetic Methods*, Vols. 1 and 2, Wiley-Interscience, New York (1974).
18. P. Hodge and D. C. Sherrington, eds., *Polymer Supported Reactions in Organic Synthesis*, J. Wiley and Sons, Chichester (1980).
19. T. W. Greene, *Protective Groups in Organic Synthesis*, Wiley-Interscience, New York (1981).
20. J. F. W. McOmie, ed., *Protective Groups in Organic Chemistry*, Plenum Pub. Corp., London (1983).

# 1

# Triggered Release of Gases

## 1.1. PROCESSES UTILIZING THE TRIGGERED RELEASE OF GASES

The triggered release of gases has found various industrial applications. Most of these are in the fields of imaging and polymer science.

Numerous gases have been used, both inert and reactive. The latter include acidic and basic gases. The examples included in this section will cover those most commonly cited in the literature.

Hydrogen halides and ammonia will be omitted from this chapter due to their inclusion in Chapter 2, which deals with triggered release of acids and bases.

### 1.1.1. Applications within Imaging Science

Uses of released gases in imaging science are mainly of two types. The first is the use of the triggered released gas as an agent in recording an image or information. In this case the triggering agent and the energy for recording the information are one and the same.

The second application is one which uses the triggered release of the gas as a stabilizing, amplifying, modifying, or erasing agent for the recorded information. In this case the triggering agent is most likely of a different nature or intensity from the recording energy. Examples of these two applications are numerous. The following are highlights of the most common.

#### 1.1.1.1. Forming Images

Vesicular imaging is a well-known technique for recording information using heat or light as imaging energies. Vesicules are created imagewise by the triggered release of a gas within a polymeric film matrix. This latent image is subsequently amplified by thermal expansion to form a visible opaque image. The vesicular images can appear as white or black depending on whether viewing is by transmitted or reflected light.

The binder must have certain specific properties, such as proper diffusivity to the released gas, correct permeability, which allows the gas to escape into the atmosphere, and the correct degree of rigidity, which is instrumental in determining the bubble expansion. Fixing the image is accomplished by overall triggered release of the remaining gas and then leaving it to diffuse out without heating.

Gases used in such processes include nitrogen generated from a variety of compounds. Examples include diazo compounds such as *p*-diazodimethyl- and *p*-diazodiethylaniline zinc chloride, *p*-diazodiphenyl-amine sulfate, *p*-diazoethylhydroxyethylaniline zinc chloride, 4-benzoyl-amino-2,5-diethoxybenzene diazonium chloride, and 7-dimethylamino-8-methoxy-3-oxo-dihydro-1,4-thiazine-6-diazonium chloride.[1-3]

Azide derivatives, such as 1-carbazido-1-naphthol, 1,4-dicarbazido-2,3-dihydroxyfuran, and 2-amino-1-carbazidobenzene,[4] are also used as sources for nitrogen gas as are diazo oxides, diazo sulfonates, azides and quinone diazides,[5] and, in particular, sodium 1,2-naphthoquinone-2-diazide-5-sulfonate.[6]

Hydrogen halide and carbon dioxide gases are also cited in the literature as suitable agents for vesicular imaging. Usually the hydrogen halide triggered by the imaging energy undergoes an acid–base reaction with a carbonate salt, causing the release of carbon dioxide gas which is amplified to form the visible image.

Compounds suitable for the release of hydrogen halide include poly(vinyl chloride), poly(vinylidene chloride), chlorinated paraffin or rubber, *N*-halosuccinimide, and carbon tetrahalides and their derivatives.[7] The compounds which can release carbon dioxide include alkali or alkaline earth carbonates, bicarbonates, oxalates, and tartrates.

Commercialized examples of vesicular imaging include the process marketed by the Kalvar Corporation under the Kalvar[8] name. This process utilized the triggered release of nitrogen gas from a diazo compound to form images and found application in microfilming,[9] motion pictures,[10] CRT recording,[11] and laser recording.[12,13]

International Business Machines Corporation, on the other hand, published[7] the use of hydrogen halide in conjunction with carbon dioxide to form a vesicular image.

### 1.1.1.2. Stabilizing Images

The use of the triggered release of a gas as a tool to stabilize images has also been reported. Imaging systems based on silver halide salts as the light-sensitive medium suffer greatly from the printout of the background. This problem is most critical for systems which are dry-processed, such as the thermographic silver halide systems. The printout of the background is due to the photoreduction of the silver halide salt to silver metal.

A number of routes have been published which minimize such a reaction. Eastman Kodak in 1962 published[14] the use of microcapsules as a delivery system for a fixing solution which would stabilize the unexposed silver halide salt. The microcapsules containing the fixing solution are ruptured on demand by the thermally triggered release of nitrogen gas from incorporated azo compounds such as azobisisobutyronitrile.

NCR has published the use of photochromic compounds such as indolinospiropyrans as formers of colored images. These images are stabilized by the thermally triggered release of a variety of gases including nitrogen dioxide,[15] sulfur dioxide, iodine, or boron trichloride.[16] Alternatively, the triggered release of a gas can initiate the delivery of a fixing solution by means of microcapsule rupture.[17]

### 1.1.2. Applications within Polymer Science

The most common use of the triggered release of gases in polymer science is found in the plastics industry. The formation of synthetic foams requires the controlled release of a gas within the polymeric matrix to produce the cavities and vesicules. Compounds which release the gases are known as chemical blowing agents. Such compounds can be either organic or inorganic and are sensitive to heat. They are characterized by their triggering temperature, at which the gas is released. This temperature determines their practical use as blowing agents.

Useful blowing agents should also possess acceptable solubility in plasticizers and organic solvents. Compounds reported in the literature as useful blowing agents include carbonates, bicarbonates, nitrates, peroxides, azo compounds, urea derivatives, hydrazines, semicarbazides, azides, N-nitroso compounds, and triazoles.[18,19]

## 1.2. CHEMICAL REACTIONS FOR THE RELEASE OF GASES

A variety of known organic compounds can be used as reservoirs for certain gases. The release of a gas is controlled by a triggering agent, usually in the form of excitation energy.

In this section highlights of such compounds will be reviewed. Emphasis will be given to reactions which would find industrial applications because of their practical triggering conditions and triggering agents. As a result, thermally and photochemically triggered release of gases will occupy a major part of this section. Reference to other triggering agents, such as electric currents, will also be made. A recent review by Reid[192] discusses the release of gases from heterocyclic compounds.

### 1.2.1. Thermally Triggered Release

#### 1.2.1.1. Carbon Dioxide from Carboxylic Acids

The most common compounds capable of releasing carbon dioxide gas under the action of heat are carboxylic acids (**1-1**) in a reaction termed decarboxylation.[20]

$$RCOOH \rightarrow RH + CO_2$$

**1-1**

Decarboxylation reactions are well known. A carboxylic group adjacent to an active methylene group will be easily lost. A useful generalization is that decarboxylation is facilitated by a multi-bond linkage on the $\beta$-position such as that in **1-2**.

$$Y \frac{}{\parallel} XCH_2COOH$$
$$R$$

**1-2**

X and Y can be part of a ring, with substituents which may reinforce the decarboxylation reaction. For example, if X and Y complete an aromatic ring, electron-withdrawing groups on the ring will provide a driving force for the reaction.

Together with the release of carbon dioxide gas during the decarboxylation reaction, a new molecule is also released: its nature obviously depends on the parent acid and hereafter will be termed the secondary molecule. The following are examples of decarboxylation reactions.

$$HOOCCH_2COCH_3 \rightarrow CO_2 + CH_3COCH_3$$

**1-3**

$$HOOCCH_2COOH \rightarrow CO_2 + CH_3COOH$$

**1-4**

$$HOOCCH_2CN \rightarrow CO_2 + CH_3CN$$

**1-5**

The elimination of carbon dioxide from such acids may involve the formation of a quasi-six-membered ring in the transition state.

Malonic acid, being activated for a facile release of carbon dioxide as well as containing an active methylene group, is a versatile synthetic building unit which can be modified into another structure through the release of carbon dioxide gas. For example, malonic acid may may be modified to release $\beta$-methylvaleric acid[21] (**1-6**) and carbon dioxide or pelargonic acid[22] (**1-7**) and carbon dioxide.

$$CH_3CH_2CH(CH_3)CH(COOH)_2 \rightarrow CO_2 + CH_3CH_2CH(CH_3)CH_2COOH$$

**1-6**

$$CH_3(CH_2)_5CH_2CH(COOH)_2 \rightarrow CO_2 + CH_3(CH_2)_7COOH$$

**1-7**

Good leaving groups on the $\beta$-carbon of a carboxylic acid facilitate the thermally triggered release of carbon dioxide and the secondary molecule, an olefin. Halogens are commonly used as good leaving groups, as with the case of brominated *trans*-cinnamic acid[23] (**1-8**).

$$C_6H_5CHBrCHBrCOONa \rightarrow CO_2 + C_6H_5CH = CHBr + NaBr$$

**1-8**

Charged leaving groups are also used in facilitating the release of carbon dioxide from carboxylic acids, for example, the addition product (**1-9**) of $\alpha$-chlorocarboxylic acid and pyridine $N$-oxide. The driving force of such a reaction is the formation of stable molecules with no charge separation as a result of the release of carbon dioxide; this is termed oxidative decarboxylation.[24]

$$C_5H_4\overset{+}{N}-O-CR^1RCOOH \overset{T^\circ C}{\longrightarrow} C_5H_4N + R^1RCHO + CO_2$$
$$X^-$$

**1-9**

| X | $R^1$ | R | $T(^\circ C)$ | Time (h) | Percent $CO_2$ |
|---|-------|---|---------------|----------|----------------|
| Br | $C_6H_5$ | H | 80 | 6 | 46 |
| Br | $C_6H_5$ | H | 110 | 8 | 57 |
| Br | $C_2H_5$ | H | 80 | 24 | 50 |
| Br | $CH_3$ | $CH_3$ | 138–146 | 9 | 100 |
| Cl | H | H | 138–146 | 4 | 100 |

The effect of substituents on the heat-triggered release of carbon dioxide in decarboxylation reactions can be correlated with relative rates. For

*Table 1.1.* Relative Decarboxylation Rates of
Unsaturated Acids

| Acid | Relative rate |
|---|---|
| $CH_2=CHCH_2COOH$ | 1 |
| $CH_2=CHC(CH_3)_2COOH$ | 5.6 |
| $CH_3CH=CHC(CH_3)_2COOH$ | 0.84 |
| $CH_2=CCH_3C(CH_3)_2COOH$ | 168 |
| $CH_3CH=C(C_2H_5)C(CH_3)_2COOH$ | 54.1 |
| 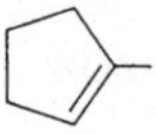 —$C(CH_3)_2COOH$ | 49.5 |
| 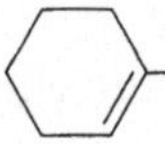 —$C(CH_3)_2COOH$ | 20.4 |
| $C(CH_3)_2COOH$  | 68.9 |

example, $\beta,\gamma$-unsaturated carboxylic acids may vary in their decarboxylation rates[25] depending on the steric environment about the double bond, as shown in Table 1.1.

The effect of substituents on the rates for aromatic carboxylic acids is a function of the electron affinity of the substituent and its position on the ring with respect to the carboxyl group.[26] Electron-donating groups in the position para to the carboxyl group seem to enhance the carbon dioxide release rate (Table 1.2).

Smith and Kelly[27] have published a compilation of gas-phase thermal decomposition data for a variety of acids, with reference to temperature and activation energies and entropies.

*Table 1.2.* Effect of Substituents on the Release of Carbon Dioxide
from Aromatic Carboxylic Acids

| Substituent | Relative rate | |
|---|---|---|
| | Benzoylacetic acid | 2,2-Dimethyl-3-phenyl-3-butenoic acid |
| $m$-$NO_2$ | 0.29 | 0.30 |
| $p$-Cl | 0.40 | 0.67 |
| H | 1.00 | 1.00 |
| $p$-$CH_3$ | 1.52 | 0.85 |
| $p$-$CH_3O$ | 2.25 | 7.75 |

Pyrolysis of metal salts[28] of carboxylic acids also provides a route to carbon dioxide and the corresponding metal oxide. The use of metal salts rather than the free acids to release carbon dioxide eliminates the presence of a free acid in the medium and thus allows the release of the carbon dioxide at an alkaline pH, as shown in the following examples.[29]

$$(CH_3COO)_2Ca \rightarrow CH_3COCH_3 + CaO + CO_2$$

$$(C_6H_5COO)_2Ca \rightarrow (C_6H_5)_2CO + CaO + CO_2$$

$$(HCOO)_2Zn \rightarrow CH_2O + ZnO + CO_2$$

$$(C_{17}H_{35}COO)_2Mg \rightarrow (C_{17}H_{35})_2CO + MgO + CO_2$$

### 1.2.1.2. Carbon Dioxide from Four-Membered Lactones

Oxetan-2-ones, under mild temperatures, are susceptible to decarboxylation with the release of carbon dioxide and the corresponding olefin.[30,31] 4-Methyl-4-phenyloxetan-2-one (**1-10**), for example, undergoes thermal ring opening[32] with the release of carbon dioxide and the corresponding olefin.

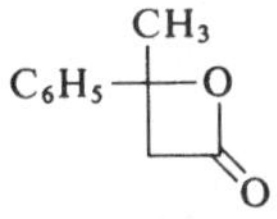

$$\rightarrow C_6H_5(CH_3)C{=}CH_2 + CO_2$$

**1-10**

The nature of substituents on the ring is directly related to the ease of the carbon dioxide release. The effect of halogens at the 3-position of oxetan-2-one (**1-11**) on the ring opening reaction and the carbon dioxide (olefin) release was studied by Brady and Patel.[32] The reaction was monitored by the disappearance of the carbonyl infrared band, with the carbonyl band of 2-heptanone as an internal reference. Substituents at the 4-position of the ring, as in **1-12**, may also show a stabilizing effect.

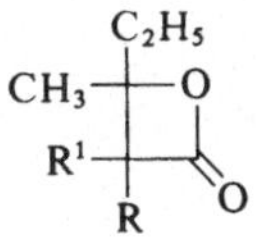

$$\xrightarrow{80°C} R^1RC{=}C(CH_3)C_2H_5 + CO_2$$

**1-11**

| $R^1$ | R | Relative rate |
|-------|-----|---------------|
| H | H | 100 |
| H | Cl | 6 |
| Cl | Cl | 1 |

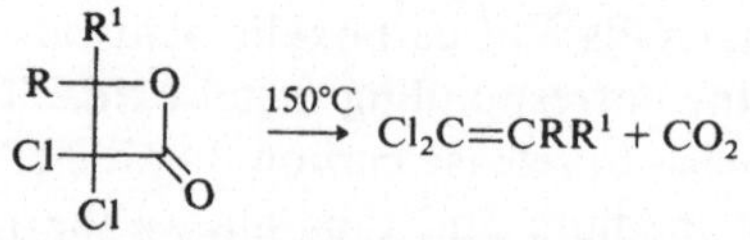

$$\underset{\textbf{1-12}}{\overset{R^1}{\underset{Cl}{\overset{R}{\underset{Cl}{\Big|}}}}}\overset{O}{\underset{O}{\Big|}} \quad \xrightarrow{150°C} \quad Cl_2C{=}CRR^1 + CO_2$$

| $R^1$ | R | Relative rate |
|-------|-----|---------------|
| $CH_3$ | $C_2H_5$ | 116 |
| $CH_3$ | $CH_2Cl$ | 17 |
| $CH_2Cl$ | $CH_2Cl$ | 1 |

Further comparison of the effect of substituents at the 3- and 4-positions can be found. For example, 4,4-dimethyloxetan-2-one (**1-13**) readily releases carbon dioxide at room temperature while 4-methyl-4-ethyloxetan-2-one does not release carbon dioxide until the temperature reaches 50°C.

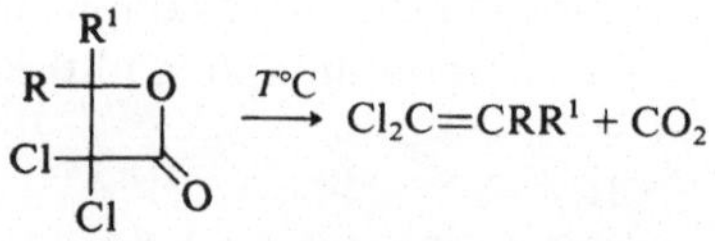

$$\underset{\textbf{1-13}}{\overset{R^1}{\underset{Cl}{\overset{R}{\underset{Cl}{\Big|}}}}}\overset{O}{\underset{O}{\Big|}} \quad \xrightarrow{T°C} \quad Cl_2C{=}CRR^1 + CO_2$$

| $R^1$ | R | $T$ (°C) |
|-------|-----|----------|
| $CH_3$ | $CH_3$ | 25 |
| $CH_3$ | $C_2H_5$ | 50 |
| $CH_3$ | $CH_2Cl$ | 170 |
| $CH_3$ | $COCH_3$ | 170 |

The variation of substituents at the 3- and 4-positions allows further control of the temperature at which the carbon dioxide and olefin may be released, as shown for the example of oxetan-2-one (**1-14**).

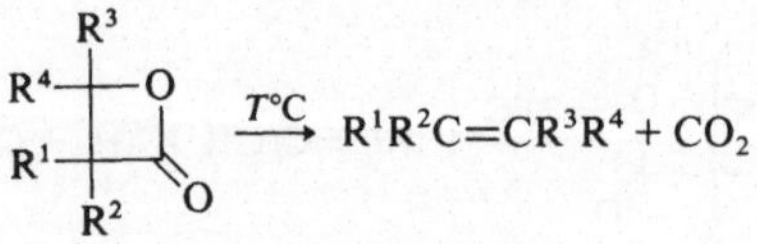

$$\underset{\textbf{1-14}}{\overset{R^3}{\underset{R^2}{\overset{R^4}{\underset{R^1}{\Big|}}}}}\overset{O}{\underset{O}{\Big|}} \quad \xrightarrow{T°C} \quad R^1R^2C{=}CR^3R^4 + CO_2$$

| $R^1$ | $R^2$ | $R^3$ | $R^4$ | $T$(°C) |
|-------|-------|-------|-------|---------|
| Cl | Br | H | $C_6H_5$ | 100 |
| Cl | Br | H | $CH(CH_3)_2$ | 160 |
| Cl | $CH_3$ | H | $CCl_3$ | No reaction |

A general conclusion from the previous examples is that electronegative groups on the oxetan-2-one ring decrease the rate of the carbon dioxide-olefin release during the decarboxylation reaction. However, if an extension of conjugation results, it serves as a driving force for the reaction. This is perhaps an indication of a reaction mechanism whereby the 4-position assumes some positive charge.

### 1.2.1.3. Carbon Dioxide from Bicyclic Lactones

Unsaturated bicyclic lactones undergo a thermally triggered ring opening reaction with the release of carbon dioxide and the corresponding diene.

Bicyclic heptenones and octenones such as 2-oxabicyclo[2.2.2]oct-5-en-3-one (**1-15**) release carbon dioxide and cyclohexadiene on heating.[33,34] Such bicyclooctenones are produced by the Diels-Alder reaction of $\alpha$-pyrones with strong dienophiles such as maleic anhydride, benzoquinones, and cyclopentadiene.

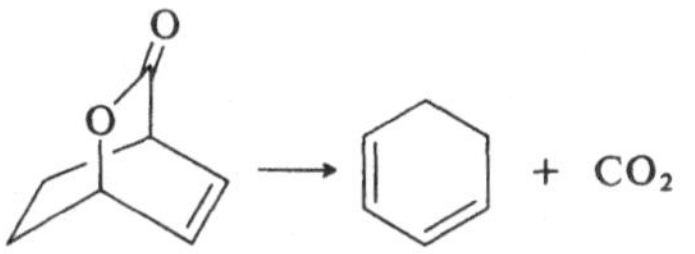

**1-15**

Fine control of the triggering temperature can be accomplished with a variety of modifications to the bicyclic structure, as in **1-16**,[33,34] **1-17**,[35] and **1-18**.[36]

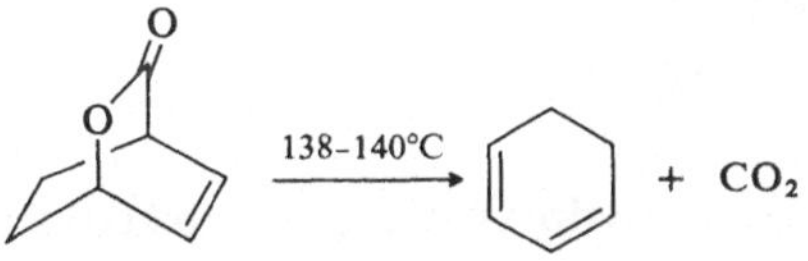

**1-16**

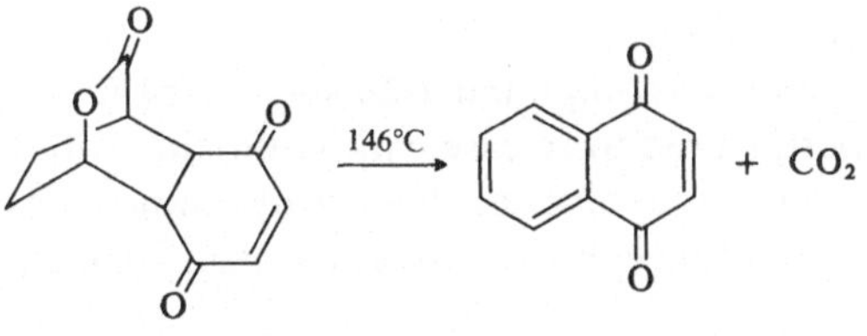

**1-17**

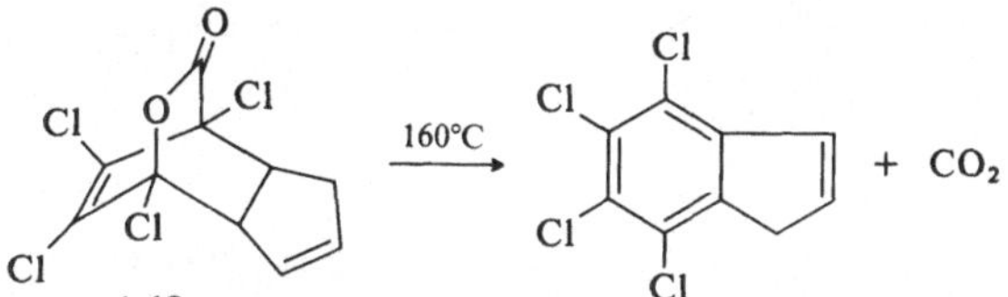

**1-18**

Lactones **1-17** and **1-18** can be prepared by the thermal reaction of $\alpha$-pyrone with 1,4-benzoquinone[35] at 100°C and the reaction of 3,4,5,6-tetrachloro-$\alpha$-pyrone with cyclopentadiene at 49°C, respectively.

Cyclic lactones with a peroxide bond follow similar routes. Adam and Rucktaschel have reported that 4,4-di-$n$-butyl-1,2-dioxalan-3,5-dione (**1-19**) smoothly releases carbon dioxide under temperatures between 70 and 80°C with the formation of a secondary polymeric molecule.[37]

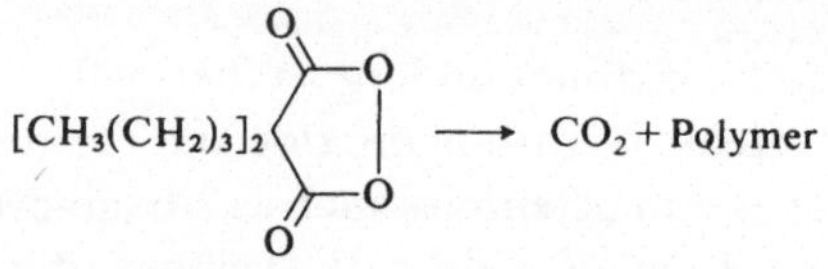

$$[CH_3(CH_2)_3]_2 \quad\longrightarrow\quad CO_2 + Polymer$$

**1-19**

## 1.2.1.4. Carbon Dioxide from Heterocycles

Heterocyclic rings with a lactone bond configuration undergo a thermally triggered loss of carbon dioxide accompanied by ring contraction. Cyclic carbonates (**1-20**) release carbon dioxide with the formation of cyclic ethers.[38]

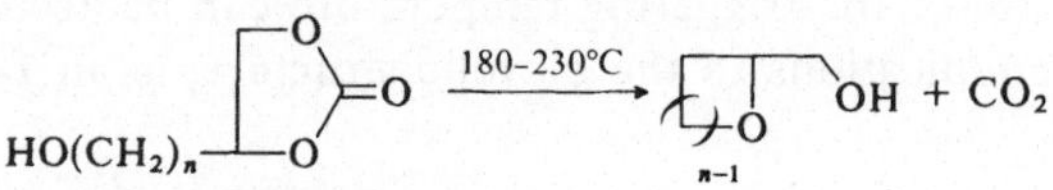

$$HO(CH_2)_n \xrightarrow{180-230°C} \quad OH + CO_2$$

**1-20**

The presence of a free hydroxymethyl group activates the cyclic carbonate. The ease of the above thermal release of carbon dioxide and the corresponding cyclic ether together with high yields (34–78%) make this route synthetically useful.

Sulfur analogues of compound **1-20** such as oxathiolan-5-ones (**1-21**) can be thermally triggered to release carbon dioxide with the formation of thiiranes.[39,40] In the presence of bis(diethylamino)phosphine, thiirane releases elemental sulfur with the formation of the corresponding olefin.

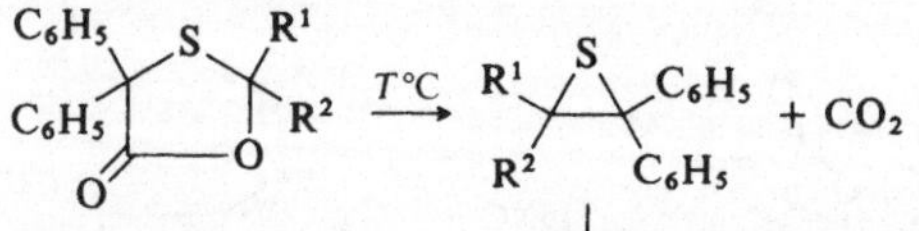

$$\begin{array}{c} C_6H_5 \quad S \quad R^1 \\ C_6H_5 \quad\quad R^2 \\ O \end{array} \xrightarrow{T°C} \begin{array}{c} R^1 \quad S \quad C_6H_5 \\ R^2 \quad C_6H_5 \end{array} + CO_2$$

**1-21**

$$S + (C_6H_5)_2C = CR_1R_2$$

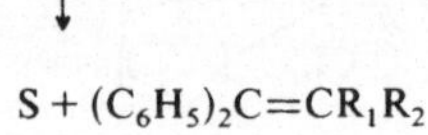

| $R^1$ | $R^2$ | $T(°C)$ | Time (h) |
|-------|-------|---------|----------|
| H | $C_6H_5$ | 150–160 | 2 |
| 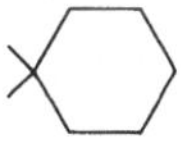 | | 160–200 | 5 |
| 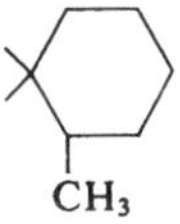 | | 220–240 | 6 |
| 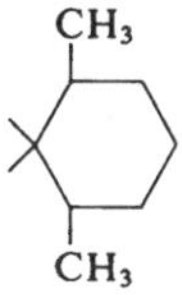 | | 180–230 | 4 |

Useful oxathiolan-5-ones (**1-21**) may be prepared through the condensation of thiobenzylic acid with ketones.[41]

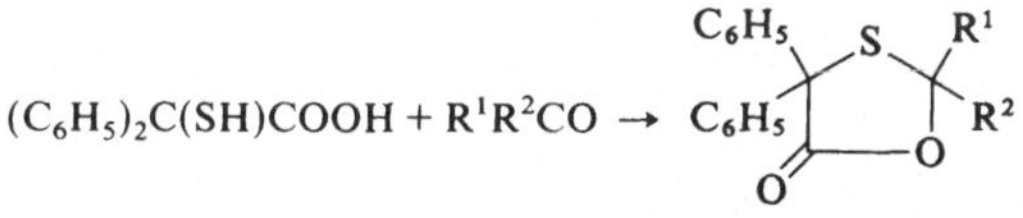

**1-21**

Analogues of oxathiolan-5-one (**1-21**) with the sulfur atom at a different position within the ring, such as ethylene monothiocarbonate (**1-21a**), undergo a similar release of carbon dioxide with the formation of thiirane[42–44] in high yields, between 75 and 80%.

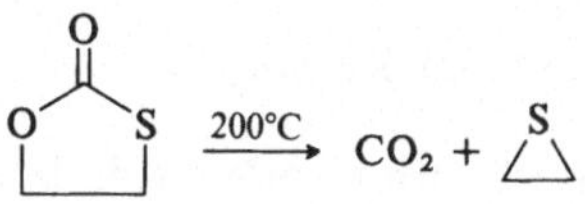

**1-21a**                          88% yield

Reynolds *et al.*[43,45,46] compared the behavior of cyclic thiocarbonates to that of their open-chain analogues, such as **1-22**, under alkaline catalysis. Their results are given in Table 1.3.

$$R^1COXCH_2CH(YH)R \rightarrow CO_2 + \underset{R}{\overset{S}{\triangle}} + R^1H$$

| $R^1$ | X | Y | Compound |
|-------|---|---|----------|
| $C_2H_5O$ | S | O | Ethyl 2-hydroxyethyl thiocarbonate (**1-22**) |
| $C_2H_5NH$ | O | S | Ethyl 2-mercaptoethyl carbamate (**1-23**) |
| $C_6H_5NH$ | O | S | 2-Mercaptoethyl phenylcarbamate (**1-24**) |

*Table 1.3.* Triggering Temperature for the Release of Carbon Dioxide from Open Chain Thiocarbonates

| Compound | mg/mol $CH_3ONa$ | Triggering temperature (°C) |
|----------|------------------|----------------------------|
| **1-21** | 0.0 | No reaction |
| | 1.0 | 189 |
| | 20.0 | 140 |
| **1-22** | 0.0 | 224 |
| | 1.0 | 214 |
| | 20.0 | 115 |
| **1-23** | 0.0 | No reaction |
| | 20.0 | 153 |
| **1-24** | 20.0 | 153 |
| | 50.0 | 202 |

Oxazolinones (**1-26**) are also sensitive to thermally and photochemically triggered release of carbon dioxide.[47]

$$\underset{\textbf{1-26}}{\overset{\displaystyle R^2 \diagdown \quad N \!=\!\! = \!\! = \!\! -R^1}{\underset{R^3}{\diagup \; O \diagdown \; O}}} \xrightarrow{115°C} CO_2 + R^1C \equiv \overset{+}{N} - \overset{-}{C}R^2R^3$$

                                                                        **1-25**

### 1.2.1.5. Carbon Monoxide from Ketones

The triggered release of carbon monoxide known as decarbonylation can be accomplished using a parent molecule with a carbonyl group.

Substituents at the carbonyl group may be separate groups or part of a cyclic configuration. Electron-withdrawing groups or those imposing steric strain facilitate the extrusion of carbon monoxide.

The triggered release of carbon monoxide can be accomplished using keto acids or esters, e.g., **1-27**.[48] These are prepared by the condensation of diethyl oxalate with active methylene groups.

$$CH_3CH(COOC_2H_5)COCOOC_2H_5 \rightarrow CO + CH_3CH(COOC_2H_5)_2$$

**1-27**

Further examples that utilize keto esters are the reactions of **1-28**, **1-29**, and **1-30**. The condensation product of diethyl oxalate with ethyl stearate (**1-28**)[49] and with phenylacetic acid ester (**1-29**)[50] can thermally release carbon monoxide.

$$C_{16}H_{33}CH(COOC_2H_5)COCOOC_2H_5 \rightarrow CO + C_{16}H_{33}CH(COOC_2H_5)_2$$

**1-28**

$$C_6H_5CH(COOC_2H_5)COCOOC_2H_5 \rightarrow CO + C_6H_5CH(COOC_2H_5)_2$$

**1-29**

Diethyl diketosuccinate (**1-30**) thermally releases carbon monoxide and a monoketo ester (**1-31**) which can decarbonylate further to form the oxalate ester (**1-32**).

$$C_2H_5OCOCOCOCOOC_2H_5 \xrightarrow{125°C} CO + C_2H_5OCOCOCOOC_2H_5$$

**1-30**                                         **1-31**

$$(COOC_2H_5)_2 + CO \xleftarrow{180°C}$$

**1-32**

In certain cases decarbonylation can be accelerated by the addition of powdered soft glass.[51]

### 1.2.1.6. Carbon Monoxide from Bicyclic Adducts

A carbonyl group at a bridgehead in a bicyclic structure is thermally sensitive and can be lost as carbon monoxide. These bicyclic adducts are generally obtained by the Diels-Alder reactions of cyclopentadienone or its derivatives with dienophiles such as olefins, acetylenes, and aziridines.[52-55] The presence of a C,C-double bond at the C-2 and C-3 positions is essential for the release of carbon monoxide. A further driving force is the presence of another double bond at the C-5 and C-6 positions which allows aromatized products to be formed along with carbon monoxide.

Ogliaruso et al.[55] have published a detailed review of the chemistry of cyclopentadienes with examples of bicycloheptenones and their ability to release carbon monoxide.

Bicyclo[2.2.1]hept-2-en-7-ones (**1-33**) and bicyclo[2.2.1]hept-2,5-diene-7-ones (**1-34**) are well known to thermally release carbon monoxide gas. Their chemistry was reviewed by Allen[56] in 1945.

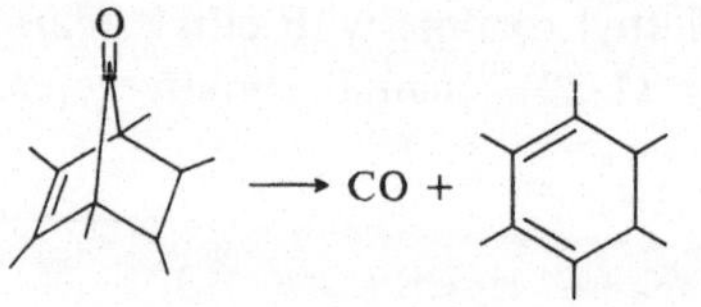

**1-33** 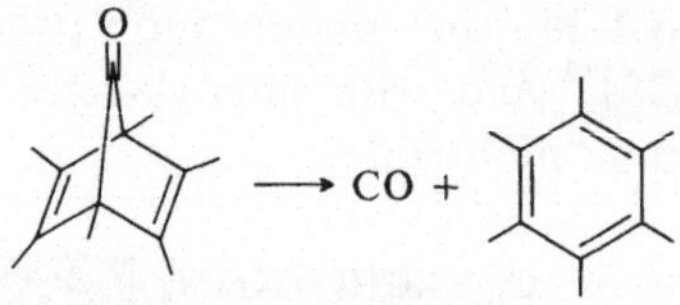                                                      **1-34**

Furthermore, the cycloaddition of substituted cyclopentadienone (**1-35**) to activated azirine is accompanied by the release of carbon monoxide and the formation of 2*H*-azepines.[57] The reaction proceeds smoothly in refluxing toluene over 2–4 days.

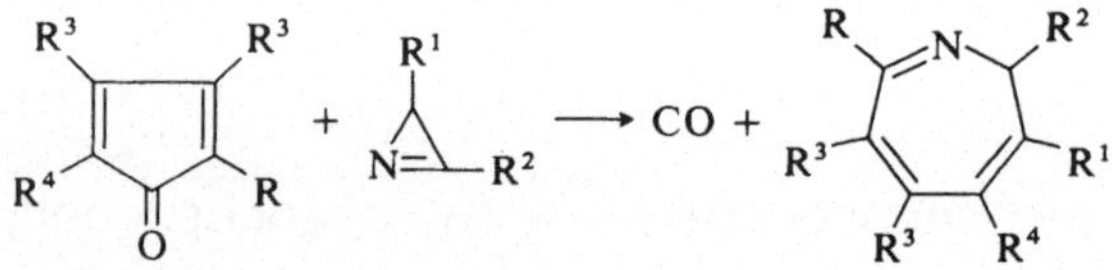

**1-35**

$R = C_6H_5, CH_3, C_2H_5$
$R^1 = C_6H_5, C_6H_5CH_2$
$R^2 = H, CH_3, CH_2OH, C_6H_5$
$R^3 = C_6H_5$

An analogous reaction between cyclopropene and cyclopentadienones (**1-35**) affords the corresponding bicycloadducts.[58-63]

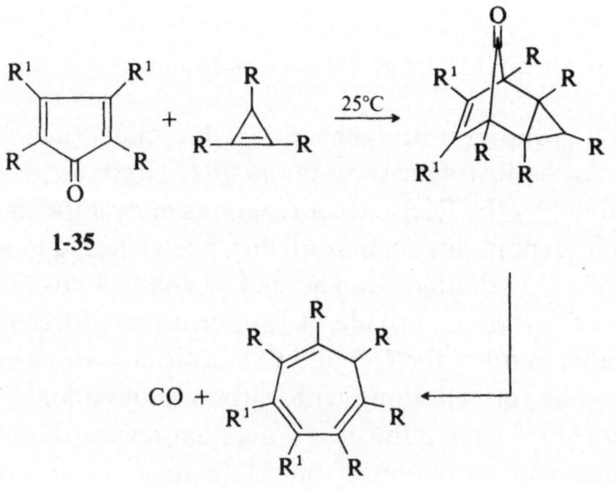

**1-35**

$R = CH_3, R^1 = C_6H_5$

Such a cycloadduct with a bridgehead carbonyl group smoothly releases carbon monoxide in refluxing benzene over four hours or in refluxing toluene or xylene instantaneously.

### 1.2.1.7. Sulfur Monoxide from Heterocycles

The thermally controlled decomposition of thiiran-1-oxide affords a route to sulfur monoxide. In recent years the properties of sulfur monoxide gas have been investigated extensively,[61-63] but because of its high reactivity, its chemical properties are still obscure. Usually, sulfur monoxide disproportionates to elemental sulfur and sulfur dioxide.

Hartzel and Page[64,65] have reported a number of ethylene episulfoxides (1-36), prepared by the oxidation of ethylene episulfides using sodium metaperiodate. These release sulfur monoxide, detected as sulfur and sulfur dioxide.[69]

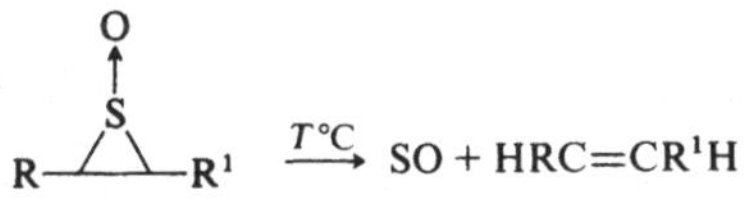

| R | $R^1$ | $T(°C)$ |
|---|---|---|
| $C_6H_5$ | $C_6H_5$ | 85 |
| $CH_3$ | $CH_3$ | 150 |
| H | H | 100 |
| $C_6H_5$ | H | 100 |
| $CH_3$ | H | 100 |
| $C_6H_{12}$ | $C_6H_{12}$ | 100 |

Due to the thermal instability of sulfur monoxide, researchers have used trapping with dienes as a method to prove its release. Dodson and Sauers[66] thermally trapped sulfur monoxide with 2,3-dimethylbutadiene.

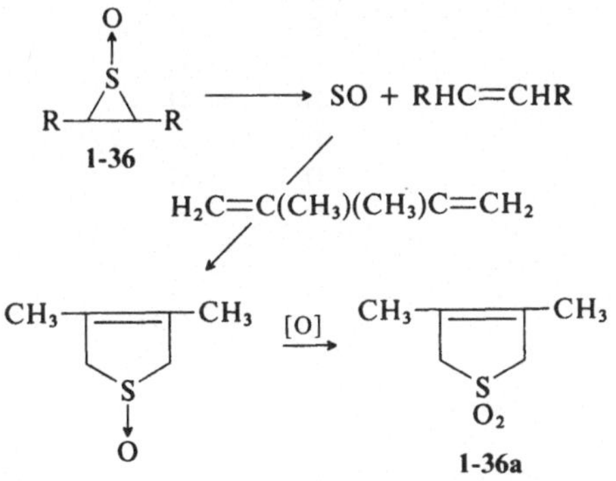

## 1.2.1.8. Sulfur Dioxide from Three-Membered Heterocycles

Unlike sulfur monoxide, sulfur dioxide is a more stable gas which has been well studied. Organic heterocycles with sulfur dioxide in their skeletal structure may be induced to liberate this gas.

Thiirane- and thiirene-1,1-dioxides provide heat-sensitive latent sources for sulfur dioxide. Thiirene-1,1-dioxide (**1-37**), reported by Carpino *et al.*,[67] undergoes a thermally triggered release of sulfur dioxide at mild temperatures. Substituents about the C,C-double bond affect the release of sulfur dioxide and acetylene derivatives.

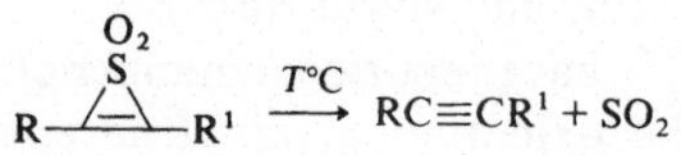

$$R \overset{\overset{\displaystyle O_2}{\underset{\displaystyle\triangle}{S}}}{=} R^1 \xrightarrow{T^\circ C} RC{\equiv}CR^1 + SO_2$$

**1-37**

| $R^1$ | R | $T$ (°C) |
|-------|-----|----------|
| $C_6H_5$ | $C_6H_5$ | 116–126 |
| $CH_3$ | $CH_3$ | 101–102 |
| $CH_3$ | $C_6H_5$ | 117–118 |

Thiirene-1,1-dioxides may be synthesized[68] by a modified Ramberg-Bäcklund reaction from the $\alpha,\alpha'$-dihalo sulfone parent using triethylamine as a base.

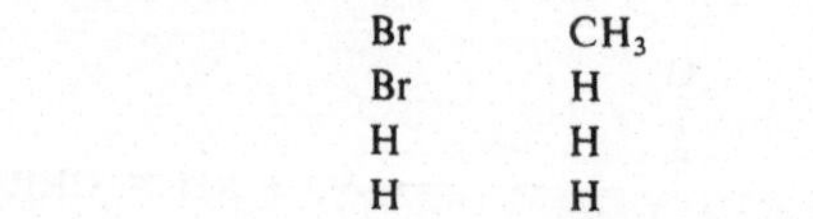

$$R^1 \overset{\overset{\displaystyle O_2}{\underset{\displaystyle\underset{H \quad R^2}{\triangle}}{S}}}{=} R^3 \xrightarrow{T^\circ C} SO_2 + R^1HC{=}CR^2R^3$$

**1-38**

| $R^1$ | $R^2$ | $R^3$ | $T$(°C/mm Hg) |
|-------|-------|-------|---------------|
| $CH_3$ | Br | $CH_3$ | 62–63 |
| $CH_3$ | Br | H | 57–58 |
| Br | H | H | 73 |
| I | H | H | 78–81 |
| H | H | H | 19 |
| $C_2H_5$ | H | H | (80/0.2) |
| $C_6H_5$ | H | H | 39 |
| $C_6H_5CH_2$ | H | H | 49–50 |
| $C_6H_5CH_2$ | H | $CH_3$ | (62/10) |
| $C_6H_5CH_2$ | $CH_3$ | $C_2H_5$ | 77–78 |
| 7,7-Dimethyl-2-oxo-bicyclo[2.2.1]hept-1-yl | H | H | 83–85 |
|  | H | $CH_3$ | 77–79 |

Thermochemical stability of thiirene-1,1-dioxides has been examined by Mackle *et al.*[69] They have reported heats of combustion and formation of $\alpha,\beta$-unsaturated sulfones, including divinyl, phenylvinyl, *p*-tolylvinyl, *cis*- and *trans-p*-tolyl-$\beta$-styryl, *trans*-phenyl-$\beta$-styryl, *trans*-propenyl, *p*-tolyl, *p*-tolylisopropenyl, but-1-enyl-*p*-toyl, and *p*-tolylisobutenyl sulfones.

Thiirane-1,1-dioxides (**1-38**) at mild temperatures release sulfur dioxide and olefins in a smoothly controlled reaction. Most reactions proceed at the melting or boiling points of the dioxide.[67]

A general route to these thiirane-1,1-dioxides is by the action of sulfonyl chloride with diazoalkanes at 0°C.[70] A review of the mechanism of the decomposition has been published by Paquette.[68]

## 1.2.1.9. Sulfur Dioxide from Five-Membered Heterocycles

Sulfur dioxide undergoes reversible addition to dienes, forming 2,5-dihydrothiophene-1,1-dioxides[71] (**1-39**).

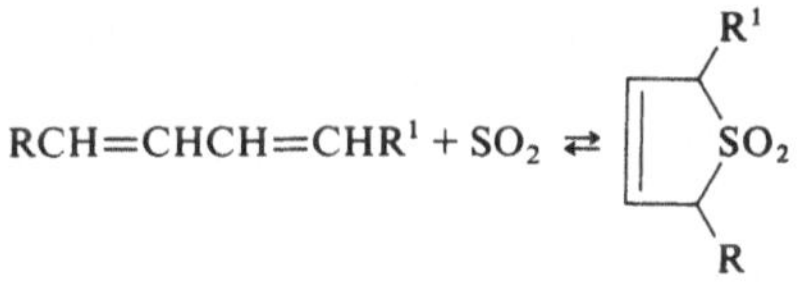

$$RCH{=}CHCH{=}CHR^1 + SO_2 \rightleftarrows$$

1-39

These reversible reactions were examined by Cava *et al.*[72,73] with the aim of extending the scope of such a reaction. Their study revolved about dioxides (**1-40**) and (**1-41**) which on thermal triggering release sulfur dioxide and benzocyclobutenes.

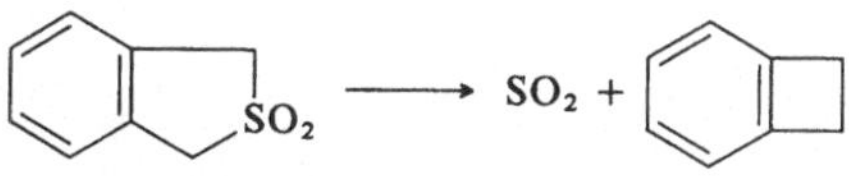

1-40

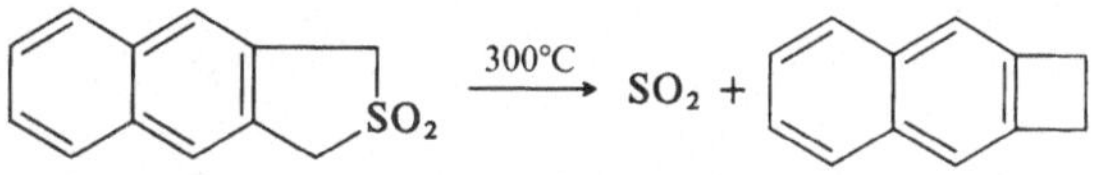

1-41

The synthesis of 1,3-dihydroisothianaphthene-2,2-dioxide was accomplished by oxidation of 1,3-dihydroisothianaphthene with peracetic acid. The intermediate was in turn prepared from sodium sulfide and $\alpha,\alpha'$-dibromo-*o*-xylene.

Cava *et al.*[74] also prepared a disulfone (**1-42**) which thermally released sulfur dioxide and benzo[1.2:4.5]dicyclobutane.

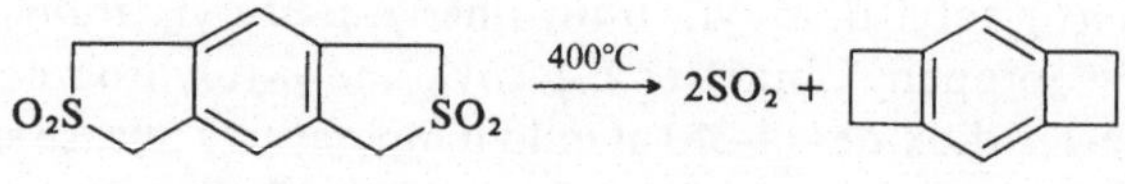

Further examples have been cited in the literature[75] which follow similar reaction patterns with the release of sulfur dioxide and a cyclobutane derivative. Bicyclic adducts with the sulfur dioxide at a bridgehead thermally release sulfur dioxide gas more readily than structures with the sulfur dioxide group within a monocyclic system. This is due to the relief of the steric strain upon the loss of the sulfur group.

The following (**1-43**-**1-47**) are some further examples:

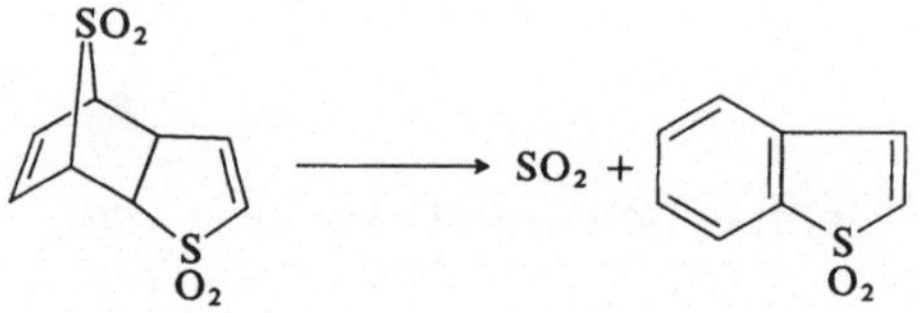

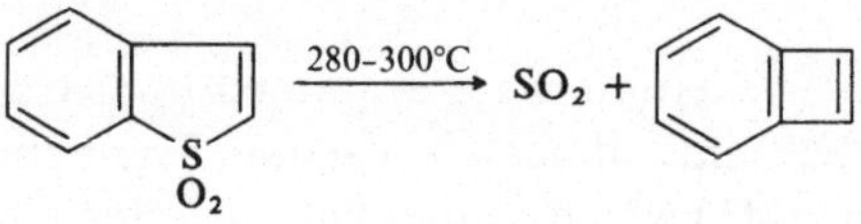

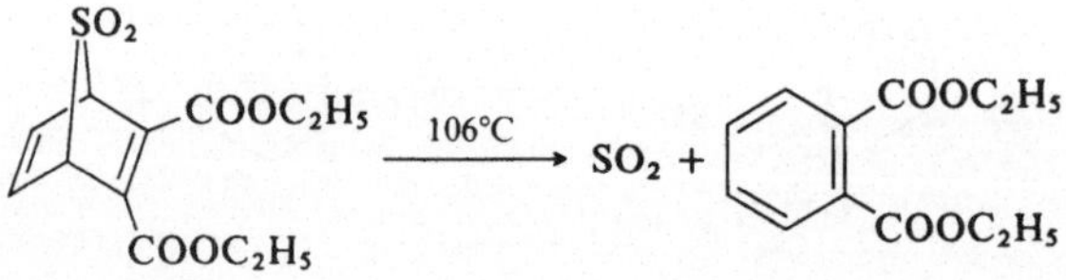

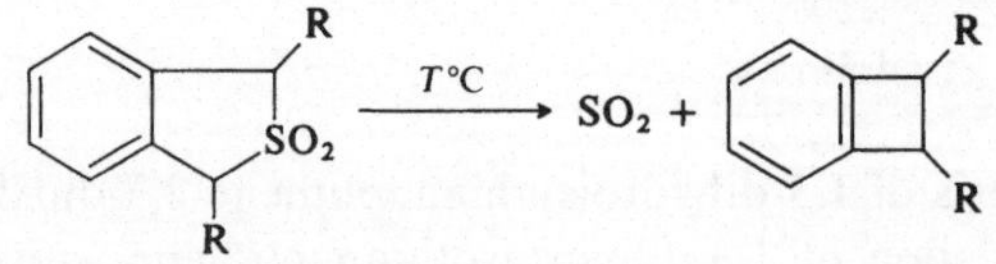

**1-46**    R = H, $T$ = 280
**1-47**    R = $C_6H_5$, $T$ = 250

Backer and Blass have reported[76] that 2,5-dihydrothiophene dioxide (1-48) decomposes thermally with the release of sulfur dioxide and butadiene. The diene was trapped with activated olefins in a Diels-Alder reaction. A variety of substituents were introduced onto the ring which appears in the diene.

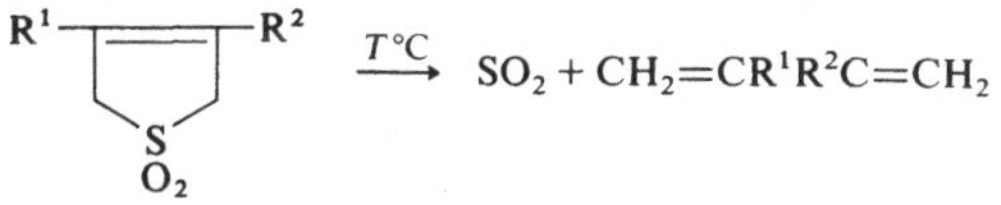

$$R^1 \text{---} R^2 \quad \xrightarrow{T\,^\circ C} \quad SO_2 + CH_2{=}CR^1R^2C{=}CH_2$$

**1-48**

| $R^1$ | $R^2$ | $T(^\circ C)$ |
|-------|-------|---------------|
| H | Cl | 125–145 |
| H | S(CH$_2$)$_3$CH$_3$ | 155 |
| CH$_3$ | Cl | 140–145 |
| CH$_3$ | SCH$_3$ | 150–160 |
| CH$_3$ | SC$_2$H$_5$ | 140 |
| CH$_3$ | SO$_2$CH$_2$CH$_3$ | 160 |
| CH$_3$ | SCH(CH$_3$)$_2$ | 160 |
| CH$_3$ | S(CH$_2$)$_3$CH$_3$ | 145–150 |
| CH$_3$ | SC$_6$H$_5$ | 140–150 |
| CH$_3$ | NC$_4$H$_4$ | 140–142 |

Burke and Carlos[77] reported that 1,3,2,4-dioxathiazole S-oxide (1-49) undergoes thermally triggered release of sulfur dioxide and isocyanates in quantitative yields, making this reaction a synthetic route to isocyanates. These compounds, unlike those previously mentioned, contain sulfur in its monoxide state and on thermal activation a ring oxygen is removed along with the sulfur atom. The 1,3,2,4-dioxathiazole S-oxides can be prepared from hydroxamic acids and thionyl chloride or by a dipolar cycloaddition reaction of nitrile oxides with sulfur dioxide.

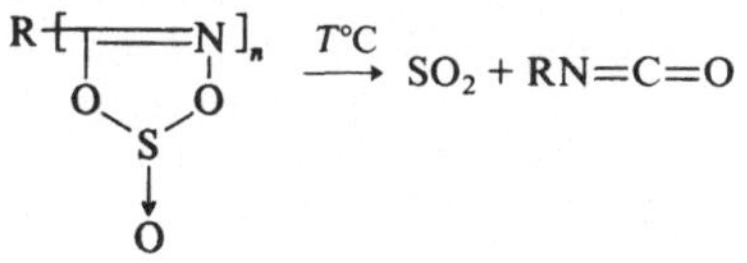

$$R{\left[ \text{---} N \right]}_n \quad \xrightarrow{T\,^\circ C} \quad SO_2 + RN{=}C{=}O$$

**1-49**

$n = 1, R = H, T = 80, \%\,yield = 90$
$n = 2, R = 1,4{-}C_6H_4, T = 130, \%\,yield = 99$

## 1.2.1.10. Sulfur Dioxide from Large Heterocycles

Large heterocycles with a sulfur dioxide unit can be induced to expel sulfur dioxide in a manner similar to smaller members of the family. Few compounds possessing the thiapine ring are known. However, the *S*-dioxide derivative (**1-50**) of this seven-membered heterocycle is a stable compound, which slowly releases sulfur dioxide at its melting point.[78] The benzo derivatives are also known to undergo a similar reaction.

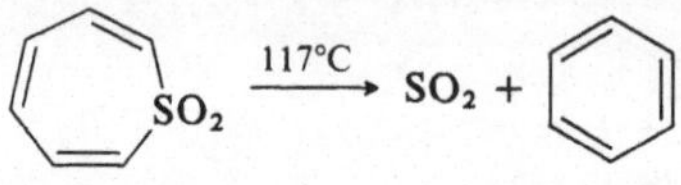

$$\text{(structure)} \xrightarrow{117°C} SO_2 + \text{(structure)}$$

**1-50**

Thiadiazepine ring structures with a sulfur dioxide group also follow the general pattern of thermally releasing sulfur dioxide and stable pyridazine structures. Loudon and Young[79] have examined the thermal release of sulfur dioxide from 2,7-dihydro-3,6-diphenyl-1,4,5-thiadiazepine (**1-51**) yielding pyridazine. The reaction proceeds smoothly in refluxing ethanol. A variety of derivatives have been reported with alkyl and halo substituents. Carbocyclic[80] derivatives (**1-52**) give rise to anthracenes and naphthalenes.

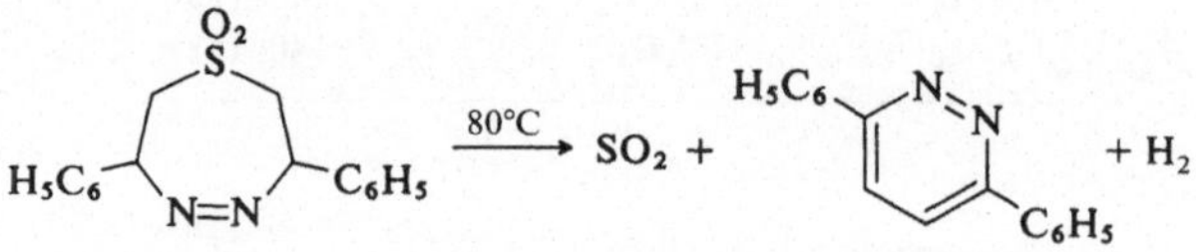

$$\text{(structure)} \xrightarrow{80°C} SO_2 + \text{(structure)} + H_2$$

**1-51**

$$\text{(structure)} \xrightarrow{100°C} SO_2 + \text{(structure)}$$

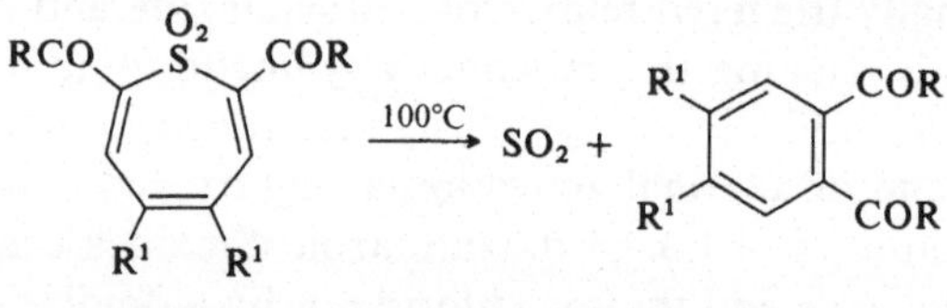

**1-52**

$$R^1 = C_6H_5; \; C_{12}H_8$$

Six-membered heterocycles with the sulfur dioxide group also thermally release sulfur dioxide. Sultones (**1-53**), for example, afford a controlled release of sulfur dioxide together with furan derivatives.[81]

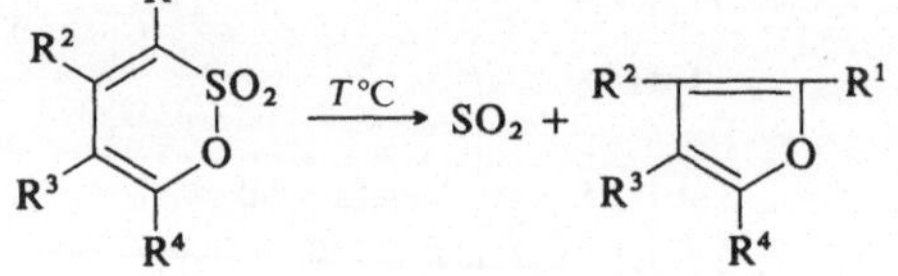

$$\text{(structure)} \xrightarrow{T°C} SO_2 + \text{(structure)}$$

**1-53**

| $R^1$ | $R^2$ | $R^3$ | $R^4$ | $T(°C)$ |
|---|---|---|---|---|
| $CH_3$ | H | $CH_3$ | H | 230 |
| $CH_3(CH_2)_2$ | H | $CH_3$ | H | 150–160 |
| $CH_3(CH_2)_3$ | H | $CH_3$ | H | 150–160 |
| $C_6H_5$ | H | $CH_3$ | H | 230 |
| $CH_3$ | H | $C_6H_5$ | H | 230 |
| $CH_3$ | $CH_3$ | $CH_3$ | $CH_3$ | 200–230 |
| $CH_3$ | $CH_3$ | H | $CH_3$ | 230 |
| $CH_3$ | $CH_3CH_2$ | H | $CH_3$ | 230 |
| $CH_3$ | H | $-CH_2CH_2CH_2-$ | | 190 |
| $CH_3$ | H | $-CH_2(CH_2)_2CH_2-$ | | 100 |
| $CH_3CH_2$ | H | $-CH_2(CH_2)_2CH_2-$ | | 190 |
| $C_6H_5$ | H | $-CH_2(CH_2)_2CH-$ | | 200–230 |
| $CH_3$ | $CH_3CH_2$ | $CH_3CH_3$ | $CH_2$ | 185–190 |
| $CH_3$ | $-CH_2(CH_2)_2CH_2-$ | | $CH_3$ | 210 |
| $CH_3$ | $CH_3$ | $-CH_2(CH_2)_2CH_2-$ | | 200 |
| $-CH_2(CH_2)_2CH_2-$ | | $-CH_2(CH_2)_2CH_2-$ | | 230 |

Sultams (**1-54**), on the other hand, release sulfur dioxide[82,83] together with pyrroles at 100–200°C.

**1-54**

| R | $R^1$ | $R^2$ | $R^3$ | $R^4$ | $T(°C)$ |
|---|---|---|---|---|---|
| $C_6H_5$ | H | $CH_3$ | H | $CH_3$ | 200 |
| $C_6H_4CH_2$ | H | $CH_3$ | H | $CH_3$ | 200 |
| 2-Pyridylmethyl | H | $CH_3$ | H | $CH_3$ | 220 |
| $p\text{-}CH_3C_6H_4$ | H | $CH_3$ | $CH_3$ | $CH_3$ | 110–120 (0.06 mm Hg) |
| $m\text{-}(CH_3)_2NC_6H_4$ | H | $CH_3$ | $CH_3$ | $CH_3$ | 100–140 (0.07 mm Hg) |
| $C_6H_5$ | H | $CH_3$ | $+CH_2)_2CHCH_3CH_2-$ | | 250 |
| $C_6H_4CH_2$ | $CH_3$ | H | $CH_3$ | $CH_3$ | 250 |
| $C_6H_5$ | H | $CH_3$ | H | $+CH_2)_6CH_3$ | 190 |
| $C_6H_5$ | H | $CH_3$ | H | $+CH_2)_{10}CH_3$ | 300 |
| $p\text{-}CH_3OC_6H_4$ | H | $CH_3$ | H | $CH_3$ | 260 |
| $p,p'\text{-}(C_6H_4)_2$ | H | $CH_3$ | H | $CH_3$ | 280 |

## 1.2.1.11. Sulfur Dioxide from Open-Chain Sulfones

Allylic structures with a sulfur dioxide group within the bond arrangement (**1-55**) undergo a thermally triggered rearrangement with the release of sulfur dioxide and olefins.[84] The reaction provides a novel way of lengthening the carbon chain of an olefin and is general for a variety of allyl sulfones (**1-56**), where reaction temperatures can be varied though the choice of substituents.

$$C_6H_5SO_2CH_2CH=CH_2 \longrightarrow SO_2 + C_6H_5CH_2CH=CH_2$$

**1-55**

$$CH_2=CR^1CHR^2SO_2R^3 \xrightarrow{T°C} SO_2 + CH_2=CR^1CHR^2R^3$$

**1-56**

| $R^1$ | $R^2$ | $R^3$ | $T(°C)(mm\ Hg)$ |
|-------|-------|-------|-----------------|
| Cl | H | $CH_2CH_2OCOCH_3$ | 170–190 (100) |
| Cl | H | $n\text{-}C_6H_{13}$ | 175–200 (100) |
| H | H | $CH_2CH=CH_2$ | 190–145 (760) |
| H | H | $CH_2C_6H_5$ | 200–215 (50) |
| H | $CH_3$ | $CH_2C_6H_5$ | 210 (50) |
| H | H | $CH_3$ | 200 (760) |
| $CH_3$ | H | $CH_2C(CH_3)=CH_2$ | 374–383 (760) |
| H | H | $CH_2COOC_2H_5$ | 380–387 (760) |
| H | H | $t\text{-}C_7H_{15}$ | 190–245 (760) |
| CN | H | $CH_3$ | 190–240 (90) |
| H | H | $CH_2COCH_3$ | 220–240 (220) |
| H | F | $CH_2C_6H_5$ | 210–240 (200) |

Furthermore, methyl alkyl sulfites (**1-57**) undergo thermally triggered release of sulfur dioxide and olefins.[85]

$$R^1CH_2R^2CHOSOOR^3 \xrightarrow{T°C} SO_2 + R^3OH + R^1CH=CHR^2$$

**1-57**

| $R^1$ | $R^2$ | $R^3$ | $T(°C)$ | Percent yield(olefin) |
|-------|-------|-------|---------|-----------------------|
| $C_6H_5$ | $CH_3$ | $CH_3$ | 245 | 90 |
| $C_6H_5$ | $CH_3$ | $CH_3$ | 160 | 50 |
| $C_6H_4CH_2$ | H | $CH_3$ | 260 | 45 |
| 2-Phenylcyclohexyl | | | | |
|    *cis* | | $CH_3$ | 170 | 93 |
|    *trans* | | $CH_3$ | 200 | 88 |

## 1.2.1.12.  Nitrogen from Azoalkanes

Azoalkanes (**1-58**), known[86,87] since 1909, have gained increased interest during the past ten years.[88-91] These compounds tend to lose nitrogen under a variety of conditions in a clean reaction which also produces radicals.

$$R^1 \underline{\!\!^a\!\!} N{=}N \underline{\!\!^b\!\!} R^2 \rightarrow R^{1.} + R^{2.} + N_2$$

**1-58**

The azoalkanes discussed in this section will be limited to those where the azo group is directly bonded to saturated carbon atoms.

The ease of removal of nitrogen from azoalkanes (**1-58**) is a function of the nature of the $R^1$ and $R^2$ groups[92,89] and their ability to stabilize the radicals produced from the homolytic fission of bonds $a$ and $b$.

The major effect of these substituents occurs when they are present on the carbons carrying the radical electrons, rather than on a remote site (even in conjugation). Engel,[88] in his review, has compiled activation and solvent parameters for symmetrical and unsymmetrical acyclic *trans*-azoalkanes.

The ability of $R^1$ and $R^2$ to stabilize the radical[93,94] by resonance, as well as the size of these groups, influences the rate of release of nitrogen. Larger rings, in general, increase the release rate, as do unsaturated substituents at the carbon carrying the radical.

Steric strain about the azo group can be used to accelerate the thermal release of nitrogen and the radicals. For example,[95,96] azoalkane (**1-58**) with $R^1$ and $R^2$ as $CH_3[(CH_3)_3CCH_2]_2C-$ releases nitrogen 57,000 times faster than its methyl analogue. There is a fine balance between the increase of steric strain to accelerate the reaction rate and its opposite effect through steric inhibition of resonance stabilization of the radicals. When steric considerations are equal, the resonance is the determining factor.[97] The following are examples of azoalkanes from the literature:

$$R^1R^2R^3CN{=}NCR^4R^5R^6 \rightarrow R^1R^2R^3C{\cdot} + R^4R^5R^6C{\cdot} + N_2$$

**1-59**

| $R^1$ | $R^2$ | $R^3$ | $R^4$ | $R^5$ | $R^6$ | Reference |
|-------|-------|-------|-------|-------|-------|-----------|
| F | F | F | F | F | F | 98 |
| $CH_3$ | $CH_3$ | CN | $CH_3$ | $CH_3$ | CN | 99 |
| $C_6H_5CO$ | $CH_3$ | $CH_3$ | $C_6H_5CO$ | $CH_3$ | $CH_3$ | 100 |
| $HC{\equiv}C$ | $CH_3$ | $CH_3$ | $HC{\equiv}C$ | $CH_3$ | $CH_3$ | 94 |
| H | H | H | $CH_3$ | $CH_3$ | H | 102 |
| $C_2H_5$ | $C_2H_5$ | H | $C_6H_5$ | $C_2H_5$ | $C_2H_5$ | 101 |
| $C_6H_5$ | H | H | $C_6H_5$ | $CH_3$ | $CH_3$ | 103 |

$$p\text{-}XC_6H_4CH_2N{=}NCH_2(p\text{-}XC_6H_4) \xrightarrow{T\,^{\circ}C} (p\text{-}XC_6H_4CH_2)_2 + N_2$$

**1-60**

| X | $k \times 10^4 \ (s^{-1})^a$ | |
|---|---|---|
| | $T = 150°C$ | $T = 165°C$ |
| H | 3.00 | 11.2 |
| CH$_3$ | 3.16 | 12.2 |
| CH$_3$O | 3.86 | 16.1 |
| Cl | 3.88 | 14.9 |
| C$_6$H$_5$ | 5.49 | 20.9 |

$^a$ Refs. 92 and 96.

$$R(CH_3)_2CN{=}NC(CH_3)_2R \rightarrow [R(CH_3)_2C]_2 + N_2$$

**1-61**

| R | $k \times 10^2 \ (s^{-1}) \ (100°C)$ |
|---|---|
| H | 1 |
| CH$_3$COO | 3.9 |
| CH$_3$ | 5.6 |
| CH$_3$COOCH$_2$ | 6.7 |
| CH$_3$(CH$_2$)$_2$ | 14.0 |
| CH$_3$O | 58.0 |
| (CH$_3$)$_3$C | 70.0 |
| C$_6$H$_5$CH$_2$ | 89.0 |
| C$_6$H$_5$O | 280.0 |
| Cl | 700.0 |
| CH$_3$COS | 3600.0 |
| (CH$_3$)$_2$CCH$_2$ | 7200.0 |
| C$_6$H$_5$S | 3700.0 |
| CH$_3$S | $1 \times 10^5$ |
| C$_2$H$_5$COO | $1.4 \times 10^6$ |
| CN | $1.7 \times 10^6$ |
| C$_6$H$_5$ | $2.3 \times 10^7$ |
| H$_2$C=CH | $5.0 \times 10^7$ |
| HC≡C | $5.1 \times 10^7$ |

In general, azoalkanes are excellent nitrogen and radical precursors. In particular, azopropanes are much more sensitive than azocumenes, diphenylazomethanes, or diphenylazoethanes. The rate differences are mostly due to resonance and inductive contributions to the radical stability and partly predictable steric contributions.

Steric and resonance effects play equivalent roles in azoalkanes within a cyclic structure. Engel, in his review,[88] has tabulated activation parameters for the thermal release of nitrogen from monocyclic, ring-fused, bridged, bicyclic, and polycyclic azoalkanes.

Pyrazolines, which fall within the class of cyclic azoalkanes, release nitrogen as a function of the pH of the medium. The reaction involves ring contraction and the formation of cyclopropanes, following the isomerization of the 2-pyrazoline (1-62) to the 1-pyrazoline (1-63). Most reactions of pyrazolines, described in a number of reviews,[104–107] proceed at temperatures close to 200°C.

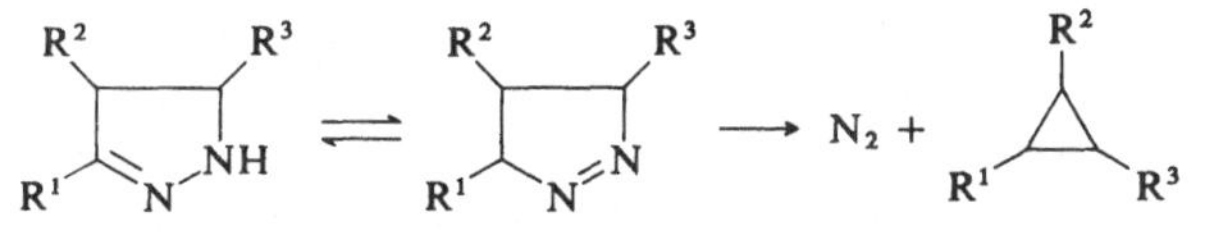

| 2-Pyrazoline | 1-Pyrazoline |
|---|---|
| 1-62 | 1-63 |

The thermal triggering temperature can be lowered to values below 200°C by activation of the pyrazoline rings 1-64 and 1-65 with electron-withdrawing groups.

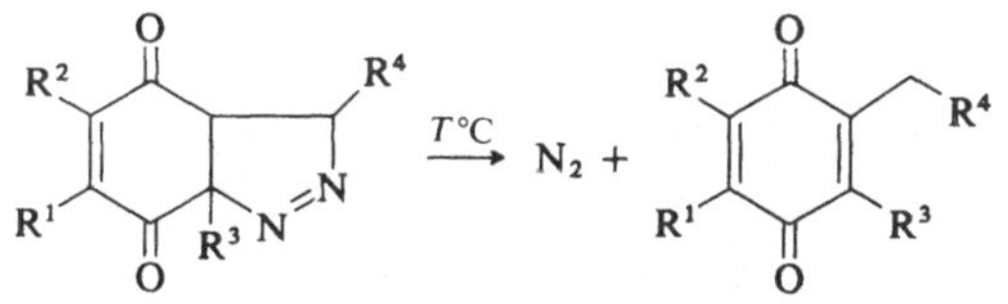

1-64

| $R^1$ | $R^2$ | $R^3$ | $R^4$ | $T(°C)^a$ |
|---|---|---|---|---|
| $-C_4H_4-$ | | H | H | 25 |
| $-C_4H_4-$ | | $CH_3$ | H | 89 |
| $-C_4H_4-$ | | $(C_6H_5)_2CH$ | H | 190 |
| $CH_3O$ | H | $CH_3CO$ | H | 80 |
| $CH_3O$ | H | $CH_3CO$ | $CH_3$ | 80 |
| $CH_3O$ | H | $CH_3CO$ | $CH_3$ | 80 |
| $CH_3O$ | H | $CH_3CO$ | $C_2H_5$ | 80 |

[a] Refs. 108 and 109.

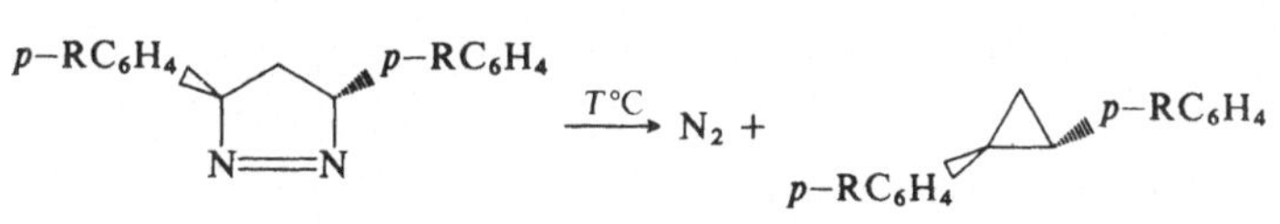

3-5-Diphenyl-1-pyrazoline

1-65

| R | $k \times 10^4\,(\mathrm{s}^{-1})$ | |
|---|---|---|
| | $T = 70°C$ | $T = 85°C$ |
| H | 2.73 | 15.2 |
| $CH_3$ | 3.43 | 15.2 |
| $CH_3O$ | 3.97 | 18.1 |
| Cl | 5.43 | 24.0 |

## 1.2.1.13. Nitrogen from Diazo Heterocycles

1,2,3-Benzothiazoles (**1-66**) have been reported[104] to lose nitrogen via thioketo dipoles which can be trapped by dipolarophiles.

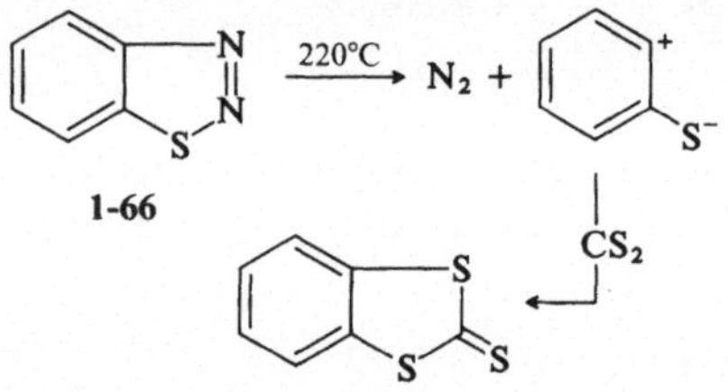

88% yield

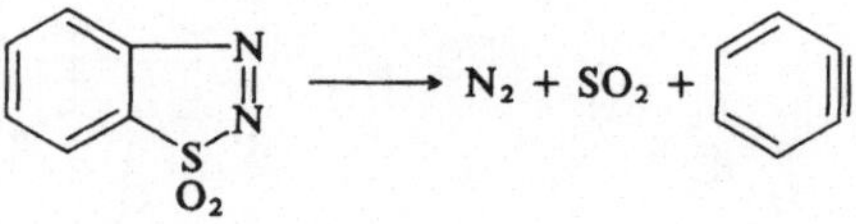

Analogously, 1,2,3-benzothiadiazole-1,1-dioxide (**1-67**), described by Witting and Hoffmann,[110,111] forms nitrogen together with sulfur dioxide and benzyne.

Other heterocycles[112] with both lactone and azo groups (**1-68**) undergo thermally triggered release of nitrogen, carbon dioxide, and olefins. Such reactions proceed at the melting points of these compounds, approximately 190–192°C.

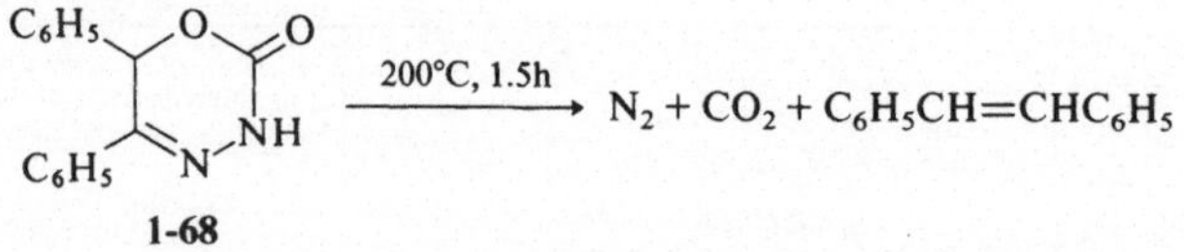

## 1.2.1.14. Nitrogen from Triazo Heterocycles

N-Aryl- (**1-69**) and N-acyltriazolines are known to undergo thermally triggered release of nitrogen and aziridines via 1,3-dipolar intermediates

(1-70). Huisgen[104] has reported the trapping of the 1,3-dipolar intermediate
with dipolarophiles.

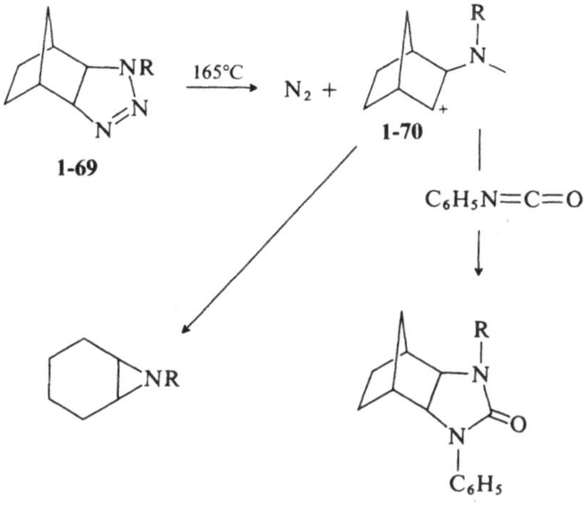

Awad et al.,[113] in their investigation, have reported a number of N-
aryltriazolines (1-72) which thermally release nitrogen and aziridines (1-73)
when heated to their melting points. These triazolines were prepared through
a 1,3-dipolar addition of phenyl azide to unsaturated dipolarophiles as
described by Alder and Stein.[114]

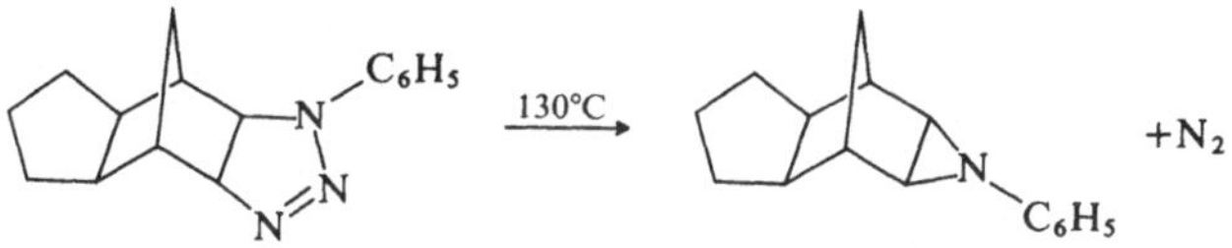

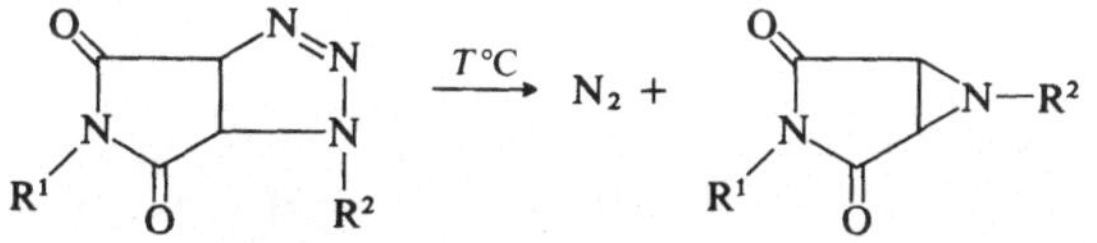

| $R^1$ | $R^2$ | $T$ (°C) |
|---|---|---|
| $C_6H_5$ | $C_6H_5$ | 156 |
| $p\text{-}ClC_6H_4$ | $C_6H_5$ | 175 |
| $p\text{-}CH_3OC_6H_4$ | $C_6H_5$ | 150 |
| $p\text{-}CH_3C_6H_5$ | $C_6H_5$ | 159 |
| $C_6H_5$ | $p\text{-}NO_2C_6H_4$ | 191 |
| $p\text{-}CH_3C_6H_4$ | $p\text{-}NO_2C_6H_4$ | 185 |
| $p\text{-}CH_3OC_6H_4$ | $p\text{-}NO_2C_6H_4$ | 182 |
| $p\text{-}ClC_6H_4$ | $p\text{-}NO_2C_6H_4$ | 184 |

Benzo derivatives of triazoles (**1-74**), reported by Crow and Wentrup,[115] undergo thermally triggered release of nitrogen and nitriles. The high yields (80–100%) make the route synthetically valuable, although the temperatures utilized are relatively high (500–800°C).

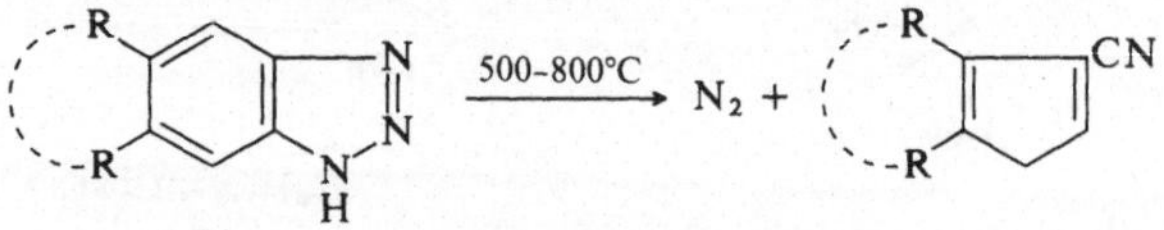

**1-74**

R···R=

Benzo

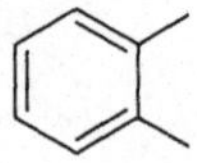

Naphtho

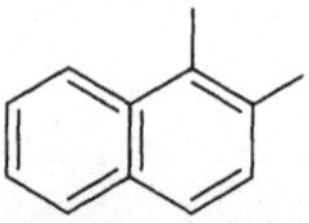

Phenanthro

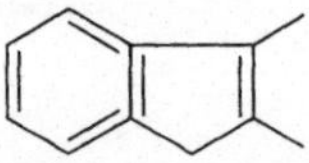

Fluoreno

### 1.2.1.15. Nitrogen from Tetrazo Heterocycles

1,2,4,5-Tetrazines (**1-75**) easily add to olefins to form bicyclic adducts with the azo group at the bridgehead. These compounds, because of bridgehead strain, release nitrogen and pyrazines easily.[116-118]

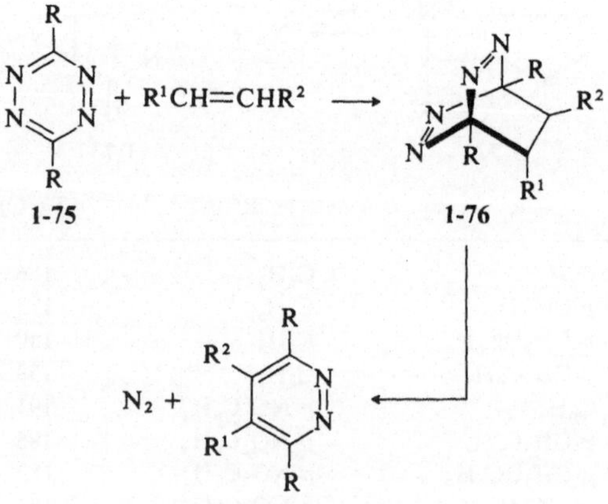

**1-75**                                                        **1-76**

$N_2$ +

$R = C_6H_5$

The route to bicyclic tetrazine derivatives **1-76** is most facile when R is an electron-withdrawing and $R^1(R^2)$ an electron-donating group. For example, a polyfluoro group activates the tetrazine ring more than a phenyl group. Carboni and Lindsey[116] have described the reaction of two tetrazines with a variety of dienophiles. The reaction is accompanied by a color change from the bluish red tetrazine color to a colorless medium with the evolution of nitrogen.

Synthetic routes to tetrazines (**1-75**) are well known in the literature.[119] On heating, these compounds undergo a reverse electrocyclization reaction to yield the corresponding organic nitriles and nitrogen.

$$\text{1-75} \xrightarrow{T^{\circ}C} 2RCN + N_2$$

**1-75**

| R | $R^1$ | $R^2$ | $T$ (°C) |
|------|-------|-------|-------------|
| $CHFCF_3$ | $C_6H_5$ | H | 20 (3 min.) |
| $C_6H_5$ | $C_6H_5$ | H | — |
| $C_6H_5$ | $C_6H_5$ | $C_6H_5$ | 86 (3 days) |
| $C_6H_5$ | H | H | — |
| $CHFCF_3$ | $CH=CH_2$ | H | — |
| $CHFCF_3$ | $CH_3$ | H | — |
| $C_6H_5$ | CN | H | 100 (6 days) |

## 1.2.1.16. Nitrogen from Azides and Diazomethanes

Boyer and Canter[120] compiled in 1954 a review on alkyl and aryl azides, which includes a section on their pyrolyses.

Azides thermally release nitrogen to give nitrenes. In general, alkyl azides[120] are less stable than their aryl analogues and may be detonated. Controlled release of nitrogen is more likely from aryl azides. When the azido group is attached to an aryl group, substituents at the position ortho to the group decrease its stability.

Heating at moderate temperatures releases nitrogen; however, rapid heating may cause an explosion. Most aromatic azides undergo decomposition at temperatures between 100 and 200°C. The loss of nitrogen from arylazides **1-77** and **1-78** may also involve the co-release of secondary molecules such as heterocyclic rings, amines, or azo compounds.[121] For example, azo groups in the position ortho to the azido group **1-79** undergo

a cyclization reaction with the release of nitrogen and benzotriazole derivative (1-80).[122,123]

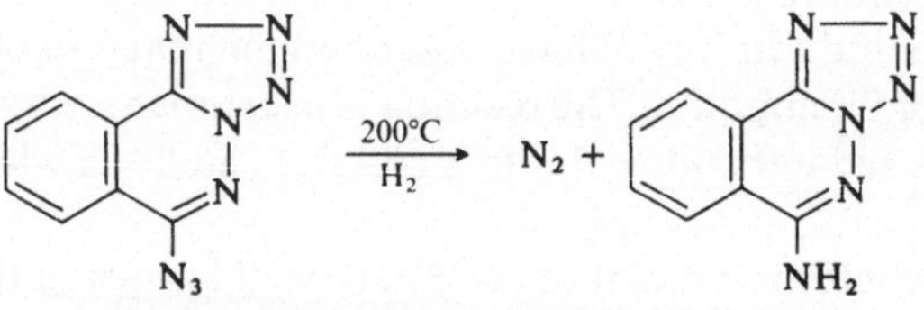

1-77

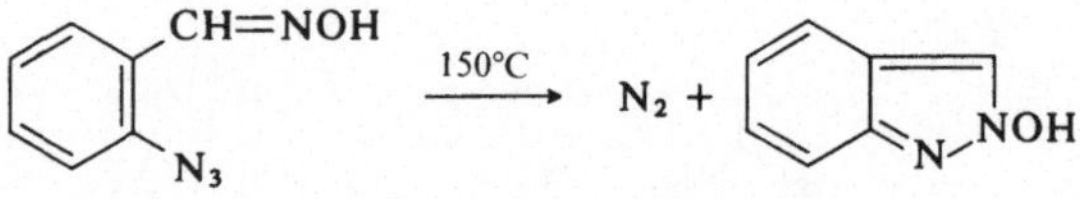

1-78

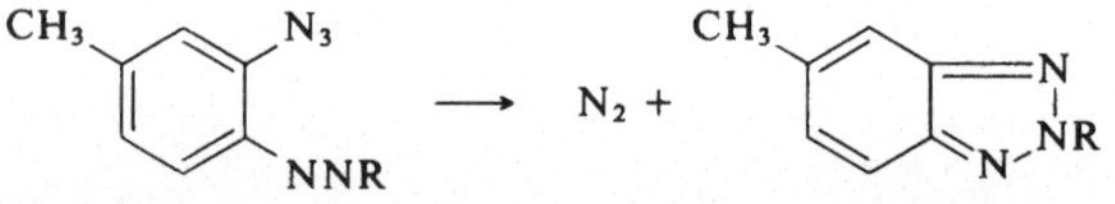

1-79                        1-80

$o$-Nitrophenyl azide (1-81) at temperatures between 85 and 95°C undergoes thermally triggered release of nitrogen with the formation of ben-
zofuroxan.[124]

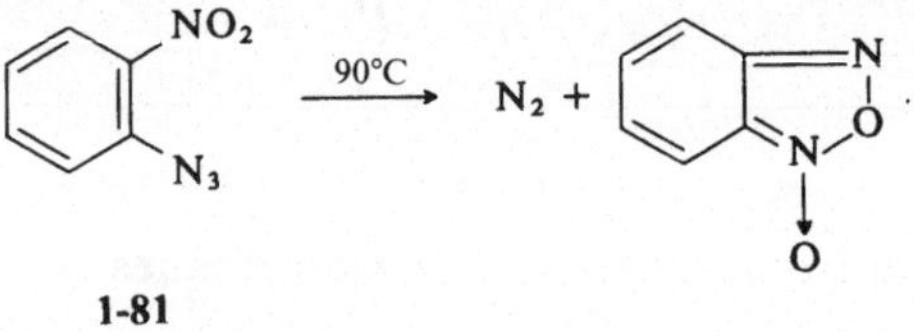

1-81

The Curtius rearrangement[125,126] of acyl azides (1-82) gives isocyanates
and nitrogen.

$$RCON_3 \rightarrow N_2 + RN=C=O$$

1-82

The reaction is a useful route to nitrogen and isocyanates since it does not
require water, which may hydrolyze the latter. However, if the isocyanates
need to be hydrolyzed, the reaction may be carried out in water or alcohol.
The usefulness of the Curtius reaction arises from its generality, and it may
be applied to almost any carboxylic acid containing aliphatic, aromatic, or
heterocyclic groups.

Diazomethane derivatives (**1-83**) can thermally undergo release of nitrogen and olefins. The reaction is dependent on the nature of the R-groups, which stabilize the carbene formed during the loss of nitrogen.

$$RRC=\overset{+}{N}=\overset{-}{N} \rightarrow N_2 + RRC=CRR$$

**1-83**

Such a reaction can be catalyzed by sulfur compounds, as reported by Benati et al.[127] Examples of these catalysts include dibenzyl sulfide (**1-84**), benzylphenyl sulfide (**1-85**), thiolan (**1-86**), and 1,2,3-benzothiadiazole (**1-87**).

$(C_6H_5CH_2)_2S$      $C_6H_5SCH_2C_6H_5$

**1-84**          **1-85**

**1-86** 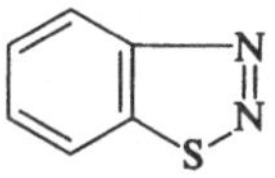          **1-87**

This thermally triggered reaction proceeds at 80°C in chlorobenzene, where diazodiphenylmethane releases nitrogen and carbene. The latter can undergo further reactions with the solvent, oxygen, and the heterocyclic catalysts.

### 1.2.1.17. Nitrogen from Diazonium Salts and *N*-Nitroso Compounds

Aryl diazonium salts, particularly fluoro-, chloro-, and bromoborates, as well as fluorophosphates, undergo thermally triggered release of nitrogen in the solid state with formation of the corresponding aryl halides.

$$R\overset{+}{N}_2BF_4^- \rightarrow RF + N_2 + BF_3$$

**1-88**

The 1927 reaction, known as the Schiemann reaction,[128,129] which uses the fluoroborate salts **1-88**, is a route to aryl fluorides. The mechanism is of the $S_N1$ type, through the formation of an aryl cation. Roe in his review[129] has compiled a list of aryl derivatives of diazonium salts with reference to substituents, yields, and decomposition temperatures.

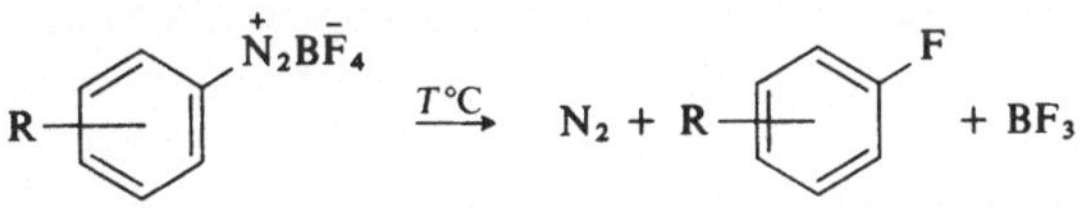

**1-89**

| R | $T(°C)$ | Percent yield (aryl fluorides) |
|---|---------|-------------------------------|
| o-COOH | 125 | 19 |
| p-COOH | No reaction | 0 |
| o-Br | 156 | 81 |
| p-Br | 133 | 75 |
| p-OH | No reaction | 0 |
| p-CH₃O | 139 | 67 |
| o-CH₃O | 125 | 54–67 |
| m-NO₂ | 170–180 | 43–54 |
| p-NO₂ | 156 | 40–58 |
| o-CH₃ | 106 | 90 |
| m-CH₃ | 108 | 87 |
| p-CH₃ | 110 | 70 |

Rutherford *et al.*[130] have reported a range of reactions of diazonium hexafluorophosphates (**1–90**) which upon thermal triggering release aryl fluorides together with nitrogen. A comparison with the analogous fluoroborate salts have shown that the fluorophosphates provide a more facile reaction.

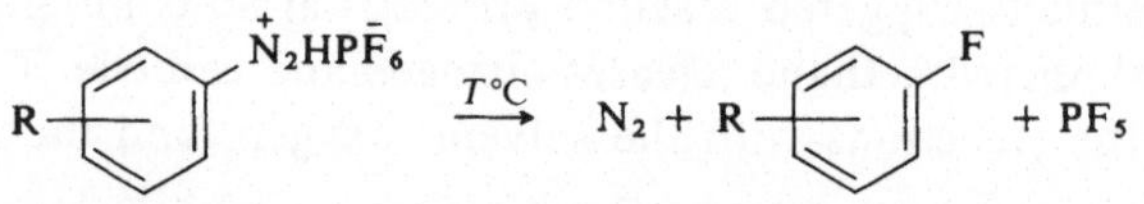

| R | $T(°C)$ | Percent yield (aryl fluorides) |
|---|---------|-------------------------------|
| o-COOH | 129 | 78 |
| p-COOH | 150 | 64 |
| o-Br | 156 | 77 |
| p-Br | 144 | 79 |
| p-OH | 120–130 | 10–20 |
| p-Cl | 137 | 60 |
| o-CH₃O | 120 | 60 |
| p-CH₃O | 149 | 70 |
| o-NO₂ | 161 | 10–20 |
| o-CH₃ | 110 | 60 |
| m-CH₃ | 110 | 57 |
| p-CH₃ | 112 | 71 |

Olah and Tolgyeshi[131] have extended the scope of the Schiemann reaction to include aryl diazonium tetracholoro- and tetrabromoborates (**1-91**), which release the corresponding aryl halide together with nitrogen. The reaction can be carried out smoothly in the presence of inert diluents such as high-boiling aliphatic hydrocarbons.

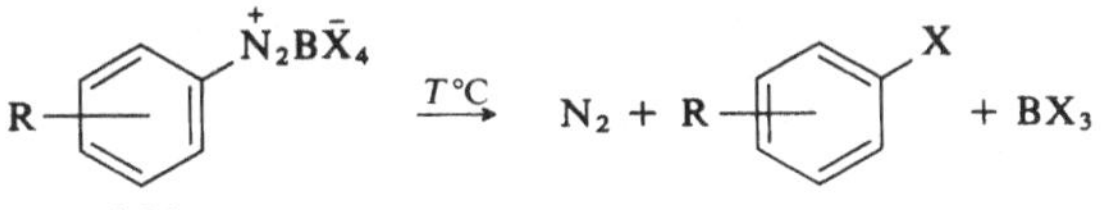

$$R \overset{+}{\text{—}} \langle \text{—} \rangle \text{—} \overset{+}{N_2} B \bar{X}_4 \quad \xrightarrow{T°C} \quad N_2 + R \text{—} \langle \text{—} \rangle \text{—} X + BX_3$$

**1-91**

| R | X | T(°C) | Percent yield (aryl halide) |
|---|---|-------|-----------------------------|
| H | Cl | 86 | 79 |
| o-CH$_3$ | Cl | 36 | 77 |
| p-CH$_3$ | Cl | 87 | 48 |
| p-NO$_2$ | Cl | 101 | 66 |
| o-NO$_2$ | Cl | 90$^a$ | 59 |
| m-Br | Cl | 34 | 75 |
| p-F | Cl | 106 | 81 |
| H | Br | 82 | 71 |
| p-CH$_3$ | Br | 90 | 80 |
| o-NO$_2$ | Br | 100 | 51 |
| p-F | Br | 130 | 96 |

$^a$ Should be handled with care.

N-Nitrosoaryl compounds (**1-92**) undergo clean, facile, and quantitative thermally triggered release of nitrogen together with aryl radicals.[132] The rate of such release is independent of the solvent.

$$2RN(NO)COCH_3 \xrightarrow{2H^+} R{-}R + N_2 + 2CH_3COOH$$

**1-92**

These compounds are obtained by the nitrosation of amides derived from a variety of acids. The following (**1-93**) are some examples.

$$R^1N(NO)COCH_3 + R^2H \rightarrow R^1R^2 + N_2 + CH_3COOH$$

**1-93**

| R$^1$ | R$^2$ | Percent yield (R$^1$R$^2$) |
|-------|-------|---------------------------|
| C$_6$H$_5$ | C$_6$H$_{15}$ | 5 |
| p-CH$_3$C$_6$H$_4$ | C$_6$H$_5$ | 16–37 |
| m-CF$_3$C$_6$H$_4$ | C$_6$H$_5$ | 35 |
| m-NO$_2$C$_6$H$_4$ | C$_6$H$_5$ | 63 |
| p-BrC$_6$H$_4$ | o,p-(NO$_2$)$_2$C$_6$H$_3$ | 34.4 |

## 1.2.1.18. Nitrogen from Miscellaneous Compounds

Nitrogen can be released from the compounds listed in Tables 1.4 and 1.5, which have been selected as representatives of their respective families.

*Table 1.4.* Triggering Temperature for the Release of Nitrogen Gas

| Compound[a,b] | Triggering temperature (°C) |
|---|---|
| *Azo compounds* | |
| Azodicarbonamide | |
| $H_2NCONNCONH_2$ | 190–230 |
| Azobisisobutyronitrile | |
| $NCC(CH_3)_2NNC(CH_3)_2CN$ | 85–120 |
| Diazoaminobenzene | |
| $C_6H_5NHNNC_6H_5$ | 103 |
| *N-Nitroso compounds* | |
| $N,N'$-Dimethyl-$N,N'$-dinitrosoterephthalamide | |
| $p,p'$-$[CH_3N(NO)OC]_2C_6H_4$ | 195 |
| $N,N'$-Dinitrosopentamethylenetetramine | |

| 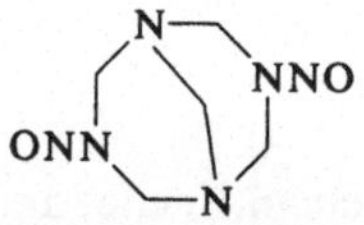  | 160–200 |
|---|---|
| *Sulfonyl hydrazides* | |
| Benzenesulfonylhydrazide | |
| $C_6H_5SO_2NHNH_2$ | 90–100 |
| $p$-Toluenesulfonylhydrazide | |
| $p$-$CH_3C_6H_4SO_2NHNH_2$ | 103–110 |
| Benzene-1,3-disulfonylhydrazide | |
| $m$-$(H_2NNHSO_2)_2C_6H_4$ | 146 |
| Diphenylsulfone-3,3'-disulfonylhydrazide | |
| $[m$-$(H_2NNHSO_2)_2C_6H_3]_2$ | 148 |
| 4,4'-Oxybis(benzenesulfonylhydrazide) | |
| $(C_6H_4SO_2NHNH_2)_2O$ | 150 |
| $p$-Toluenesulfonylsemicarbazide | |
| $p$-$CH_3C_6H_4SO_2NHNHCONH_2$ | 210–230 |
| 5-Morpholyl-1,2,3,4-thiatriazole 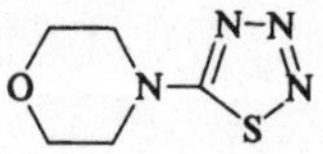 | 115 |

| 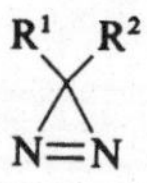  | 100–145 |
|---|---|

$R^1$ = F, Cl, Br; $R^2$ = F, $C_6H_5$

| 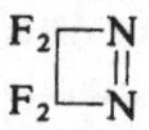  | 120 |
|---|---|

*Table 1.4.* (cont.)

| Compound[a,b] | Triggering temperature (°C) |
|---|---|
| 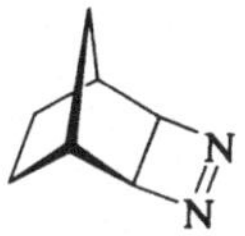 | 119 |
| 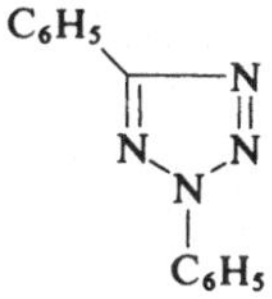 | 25 |
| 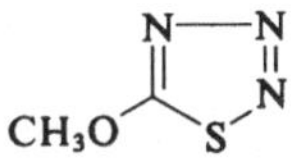 | 25 |
|  <br><br>$R^1, R^2, R^3, R^4 = H, CH_3, C_6H_5$ | 90–237 |
| 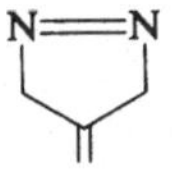 | 188 |
| 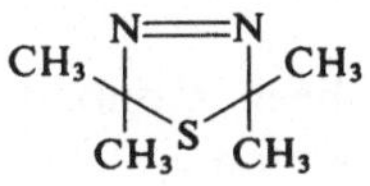 | 25 |
| 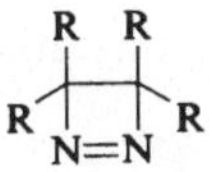 $R = H, C_6H_5$ | 62–79 |
| 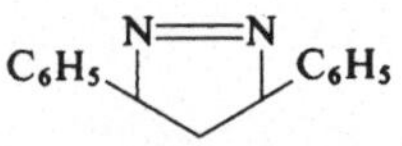 | 80 |
| 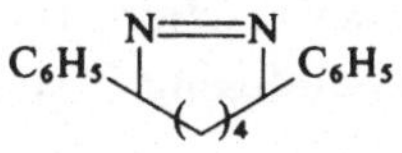 | 80 |
|  | 3.5 |

*continued*

Table 1.4. (cont.)

| Compound[a,b] | Triggering temperature (°C) |
|---|---|
| 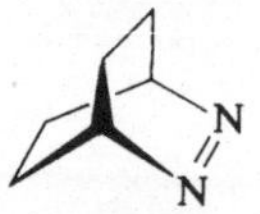 | 240 |
| 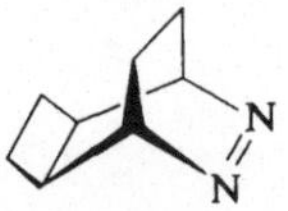 | 150 |
| 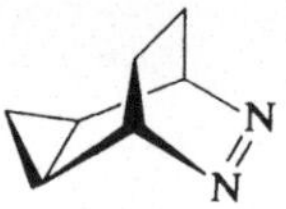 | 25 |
| 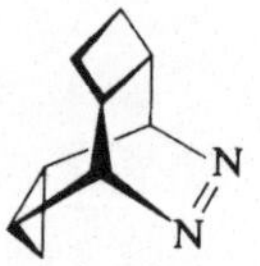 | 199 |

[a] Ref. 133.
[b] Ref. 142.

Other classes of compounds which have been reported to release nitrogen include substituted isocyanates,[134] ammonium carbonylsulfonate,[135] substituted ureas,[136] 2,4-dioxo-1,2-dihydro-4-benzoxazine derivatives,[137] substituted triazines,[138] ammonium 5-azidotetrazole,[139] cyanamide,[140] and azodicarboamide derivatives.[141]

### 1.2.1.19. Carbonyl Sulfide from Xanthates

The thermal triggering of xanthates (**1-94**) to release olefins, mercaptans, and carbonyl sulfide is known as the Chugaev reaction. The chemistry of the Chugaev reaction was reviewed by DePuy and King[143a] and Nace[143b] in the 1960s.

Alkyl xanthates can be prepared from the reaction of the corresponding alcohol and carbon disulfide in an alkaline medium followed by an alkylation reaction. The thermal reactions of xanthates are analogous to those of carboxylic esters. However, the elimination reaction is easier with xanthates and the required temperatures are usually lower. Studies by Bader and Bourns[144] have substantiated the proposed involvement of the sulfur atom of the C=S group in the cyclic intramolecular transition state. The thermal stability of xanthates (**1-94**) was found to be related to the electronic character of the R-group. The more electron withdrawing the R-group is, the higher the induced positive charge on the carbon atom and hence the higher the driving force for bond cleavage.

$$RSCSOCH_2R \xrightarrow{100-225°C} RSH + COS + CH_2{=}R$$

**1-94**

A number of xanthates and their thermal properties have been reported in the literature. Benkeser and Hazdra[145] have reported the factors affecting the Chugaev reaction for 1-alkylcyclohexanols (**1-95**) and alkylcyclohexyl-carbinols (**1-96**).

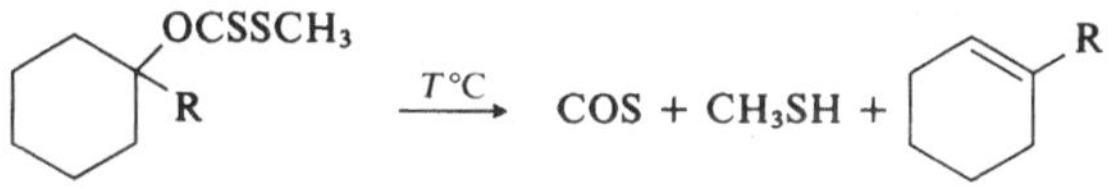

**1-95**

| R | $T(°C)$ |
|---|---|
| $CH_3$ | 200 |
| $C_2H_5$ | 200 |
| $(CH_3)_2CH$ | 100 |

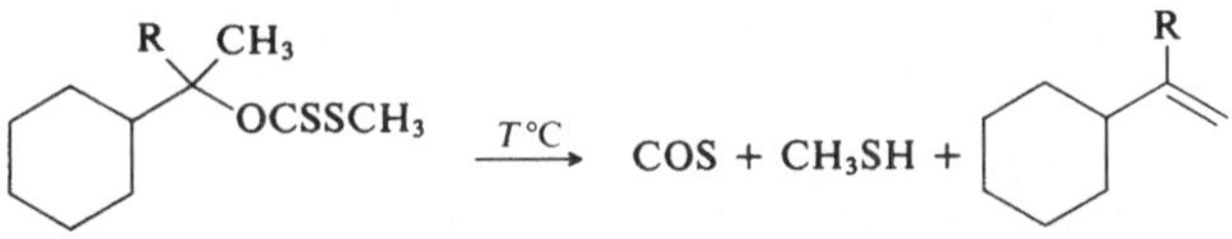

**1-96**

| R | $T(°C)$ |
|---|---|
| H | 250 |
| $CH_3$ | 150 |

## 1.2.1.20. Oxygen from Endoperoxides

Endoperoxides formed by the cycloaddition of molecular oxygen to the $\pi$-system of polynuclear aromatic compounds are sensitive to thermal triggering which releases oxygen. A review of such endoperoxides and their properties has been published by Bergmann and McLean.[146]

Anthracene is a common molecule suitable for the trapping of molecular oxygen and the formation of endoperoxides. 9,10-Endoperoxides of substituted anthracenes lose molecular oxygen quantitatively under thermal triggering. The substituents at the 9- and 10-positions play an important role; for example, phenyl groups at both the 9- and 10-positions enhance the quantitative release reaction, which is more facile than with the monosubstituted analogue, while nonsubstituted anthracene shows no release reaction. The common driving force for the release of the oxygen molecule is the re-aromatization of the ring system. A number of other derivatives of anthracene endoperoxides (**1-97**) and naphthacene endoperoxides (**1-98**) have been reported[146] to undergo similar thermal release of oxygen.

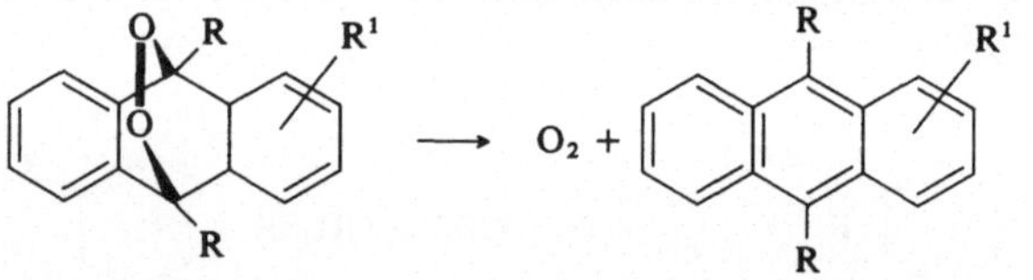

**1-97**

| R | Percent yield (oxygen)[a] |
| --- | --- |
| H | 0 |
| 9-Methyl | 0 |
| 9-Phenyl | 12 |
| 9-Phenyl-10-methyl | 20 |
| 9-Phenyl-10-ethyl | 35 |
| 9,10-Diphenyl | 96 |
| 9,10-Di(o-tolyl) | 83 |
| 9,10-Di(m-tolyl) | 93 |
| 9,10-Di(p-tolyl) | 94 |
| 9,10-Di-α-naphthyl | 90 |
| 9,10-Di-β-naphthyl | 95 |
| 9,10-Diphenyl-1,4-dimethoxy | 98 |
| 9-Phenyl-10-carbomethoxy | 60 |
| 9,10-Diphenyl-2-carbomethoxy | 92 |

[a] Ref. 146.

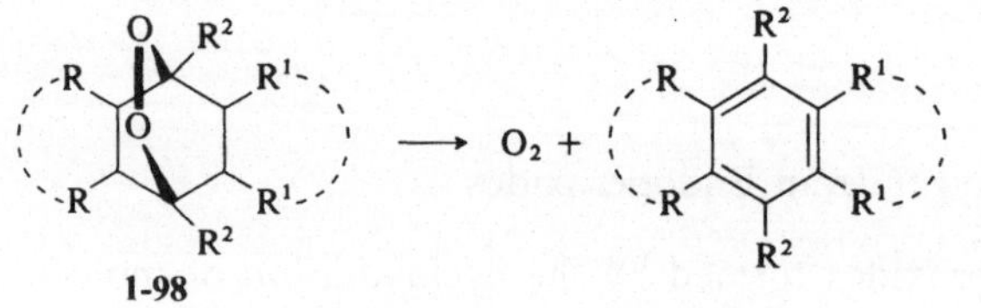

**1-98**

R = phenyl; R¹ = naphthyl

| Substituents | Percent yield (oxygen)$^a$ |
|---|---|
| 5,6,11-Triphenyl | 15 |
| 5,6,11,12-Tetraphenyl | 80 |
| 5,11-Di(p-tolyl)-2,12-diphenyl | 77 |
| 5,6,11,12-Tetraphenyl-2,8-dimethyl | 66 |
| 11-(p-Tolyl)-5,6,12-triphenyl-2-methyl | 64 |
| 5,11-Di(p-tolyl)-6,12-diphenyl-2,8-dimethyl | 74 |
| 5,11-Di(β-naphthyl)-6,12-diphenyl | 80 |
| 1,2,3,4-Tetrahydro-6,11-diphenyl | 80 |
| 2,6,8,12-Tetraphenyl-5,11-dibiphenyl | 70 |
| 5,11-Di(p-carboxyphenyl)-6,12-diphenyl | 59 |
| Hydrogens | 0 |
| 5,11-Diphenyl | 0 |
| 6,11-Diphenyl | 0 |

$^a$ Ref. 146.

## 1.2.2. Photochemically Triggered Release

### 1.2.2.1. Carbon Dioxide from Heterocycles and Cyclic Peroxides

Heterocycles with a lactone group undergo a photochemically triggered release of carbon dioxide. For example, oxazolones (**1-99**) at a temperature of $-190°C$ release carbon dioxide and 1,3-dipoles.[147] Similar reactions are discussed in Section 1.2.1.4.

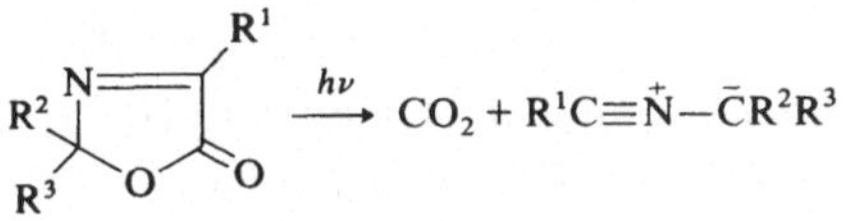

**1-99**   $R^1, R^2, R^3 = C_6H_5$

Azalactones (**1-100**), described by Barton and Willis,[148] undergo photo-chemically triggered release of carbon dioxide together with an azine molecule, under irradiation with a medium-pressure mercury arc lamp in cyclohexane.

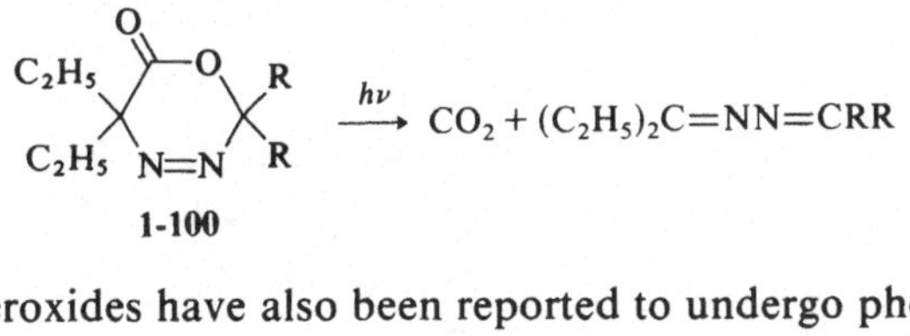

**1-100**

Cyclic peroxides have also been reported to undergo photochemically triggered release of carbon dioxide with the formation of the corresponding carboxylic acids in high yields.

1,2-Dioxalan-3,5-dione was found by Adam and Rucktaschel[149] to thermally release carbon dioxide. The high formation yields of the carboxylic acid made this a viable synthetic route. This photochemically triggered release of carbon dioxide. For example, oxazolones **1-99** at a temperature medium.

4,4-Di-*n*-butyl-1,2-dioxalan-3,5-dione (**1-101**), prepared from di-*n*-butylmalonic ester, photochemically gave carbon dioxide and 2-*n*-butyl-2-

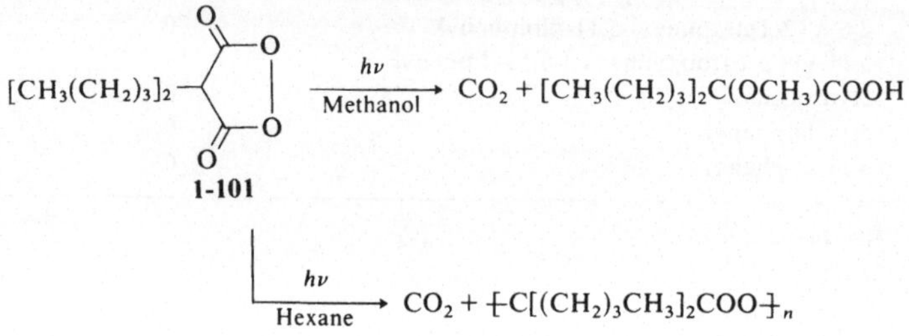

$$[CH_3(CH_2)_3]_2 \xrightarrow[\text{Methanol}]{h\nu} CO_2 + [CH_3(CH_2)_3]_2C(OCH_3)COOH$$

**1-101**

$$\xrightarrow[\text{Hexane}]{h\nu} CO_2 + \{C[(CH_2)_3CH_3]_2COO\}_n$$

methoxycaproic acid in methanol and a polyester resin in non-nucleophilic solvents. The photolysis reactions were carried out for 1–2 hours at 35–40°C in a quartz vessel using 254- and 310-nm irradiation from a 450-W Hanovia mercury arc.

## 1.2.2.2. Carbon Monoxide from Bicyclic Adducts and Heterocycles

The photochemically triggered release of carbon monoxide from bicyclic structures is usually accompanied by relief of steric strain or by an increase in the aromatic character of the products.

These bicyclic adducts (**1-102**) are products of the reaction of cyclopentadienone and cyclopropene as described by Hassner and Anderson.[150]

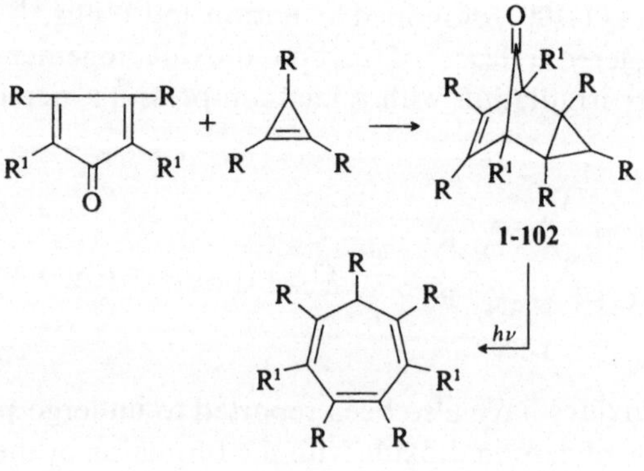

**1-102**

$$R = C_2H_5, C_6H_5; R^1 = CH_3, C_6H_5, H$$

Four-membered heterocycles with a ketonic group photochemically release carbon monoxide. For example, tetramethyloxetan-3-one (**1-103**) undergoes photochemical elimination of carbon monoxide with the formation of oxiranes[151] when irradiated with an unfiltered Hanovia 450-W Type L lamp.

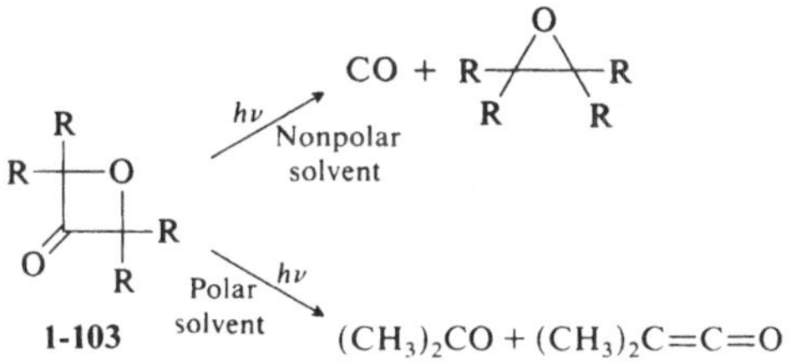

In polar solvents the yields of acetone were quantitative with a quantum yield equal to unity. As the solvent becomes less polar, the acetone yield drops to less than 50%.

Oxetan-3-one **1-103** can be obtained from the oxidative cyclization of tetramethylacetate with lead tetraacetate.

### 1.2.2.3. Carbon Monoxide from Cyclic Ketones

Cyclic ketones (**1-104**), under photochemical triggering, give carbon monoxide and olefins. Blacet and Miller[152] have reported a comparative study of cyclic ketones with variation in ring size. Quantum yields as a function of temperature and triggering-light wavelength have also been reported (Table 1.5) and show that the smaller the ring, the lower the reaction efficiency towards carbon monoxide release.

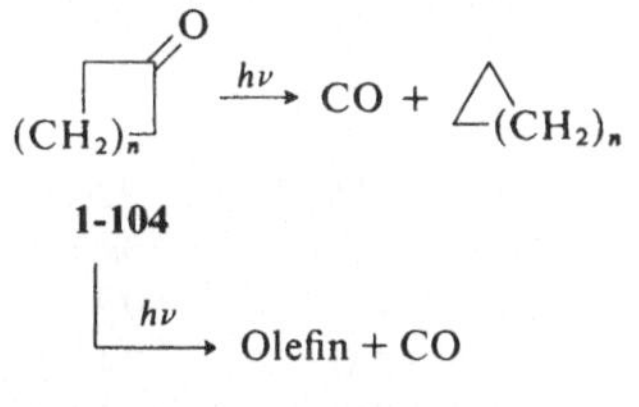

1,3-Cyclobutanediones (**1-105**), which undergo facile but more complex photochemically induced release of carbon monoxide, have been studied in detail by Heller and Srinivasan.[153]

Turro *et al.*[154] have examined several tetrasubstituted 1,3-cyclobutanediones (**1-106**) in inert solvents in order to establish a simplified reaction route. Irradiation was carried out for 2–8 hours using a 450-W Hanovia lamp.

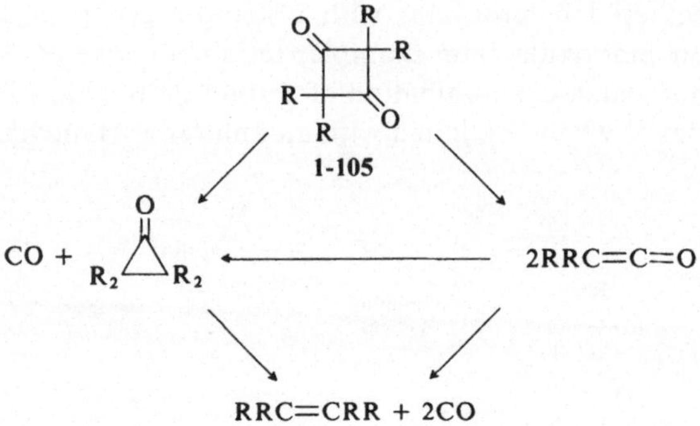

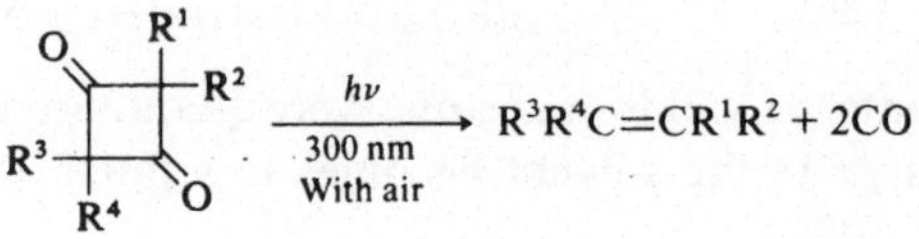

$$1\text{-}105$$

$$CO + R_2\!\!\bigtriangleup\!\!R_2 \longleftarrow 2RRC{=}C{=}O$$

$$RRC{=}CRR + 2CO$$

$$\xrightarrow[\text{With air}]{\substack{h\nu \\ 300\,nm}} R^3R^4C{=}CR^1R^2 + 2CO$$

**1-106**     $R^1 = R^2 = R^3 = R^4 = CH_3;\ \phi = 0.38$
         $R^1R^2 = R^3R^4 = \text{cyclohexane};\ \phi = 0.31$

*Table 1.5.* Quantum Yields for the Release of Carbon Monoxide from Cyclic Ketones (**1-104**)

| | | | | |
|---|---|---|---|---|
| **$n = 3$** | | | | |
| Wavelength (nm) | 313 | 313 | 265 | 254 |
| Temperature (°C) | 125 | 300 | 100–300 | 100–300 |
| $\phi$ (CO) | 0.22 | 0.91 | 0.78 | 0.81 |
| $\phi$ (ethane) | 0.02 | 0.02 | 0.02 | 0.03 |
| $\phi$ (propylene) | 0.03 | 0.03 | 0.02 | 0.02 |
| $\phi$ (cyclopentane) | 0.06 | 0.31 | 0.28 | 0.21 |
| $\phi$ (1-pentene) | 0.12 | 0.59 | 0.45 | 0.32 |
| $\phi$ (polymerization) | 0.01 | 0.02 | 0.03 | 0.25 |
| | | | | |
| **$n = 2$** | | | | |
| Wavelength (nm) | 313 | 313 | 265 | 254 |
| Temperature (°C) | 125 | 225 | 100–300 | 100–300 |
| $\phi$ (CO) | 0.33 | 0.42 | 0.61 | 0.74 |
| $\phi$ (ethylene) | 0.15 | 0.34 | 0.25 | 0.21 |
| $\phi$ (cyclobutane) | 0.26 | 0.17 | 0.27 | 0.21 |
| $\phi$ (polymerization) | 0.01 | 0.10 | 0.21 | 0.43 |
| | | | | |
| **$n = 1$** | | | | |
| Wavelength (nm) | 313 | 313 | 265 | |
| Temperature (°C) | 100–300 | 100 | 100–300 | |
| $\phi$ (CO) | 0.35 | 0.40 | 0.53 | |
| $\phi$ (ethylene) | 0.51 | 0.54 | 0.53 | |
| $\phi$ (cyclopropane) | 0.13 | 0.14 | 0.17 | |
| $\phi$ (propylene) | 0.01 | 0.01 | 0.12 | |
| $\phi$ (polymerization) | 0.21 | 0.25 | 0.24 | |

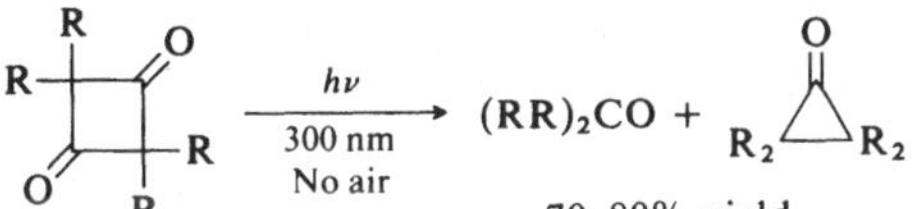

**1-106**        R = CH$_3$

## 1.2.2.4. Sulfur Dioxide from Sulfonyloxy Compounds

Feigenbaum *et al.*[155] have reported that 2-arenesulfonyloxy-2-cyclo-hexenone (**1-107**) undergoes a photochemically triggered release of sulfur dioxide gas and an enol. The quantum yield was directly related to the wavelength. Longer wavelengths gave higher quantum yields; for example, at 313 nm $\phi$ = 0.125, while at 254 nm $\phi$ = 0.05.

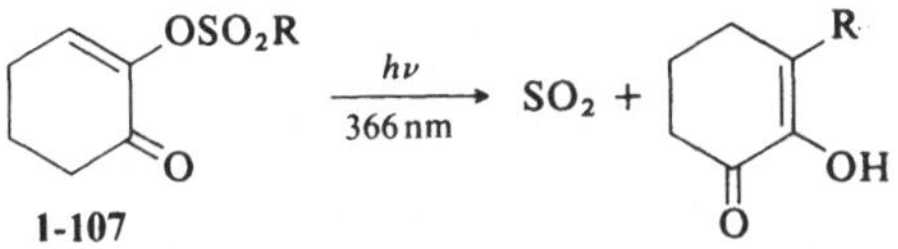

**1-107**

| R | Percent yield (dione) |
|---|---|
| C$_6$H$_5$ | 40 |
| p-CH$_3$C$_6$H$_4$ | 50 |
| p-CH$_3$OC$_6$H$_4$ | 50 |

2-Arenesulfonyloxy-2-cyclohexenone can be easily prepared from the corresponding cyclohexan-1,2-dione by the reaction with arylsulfonyl chloride.

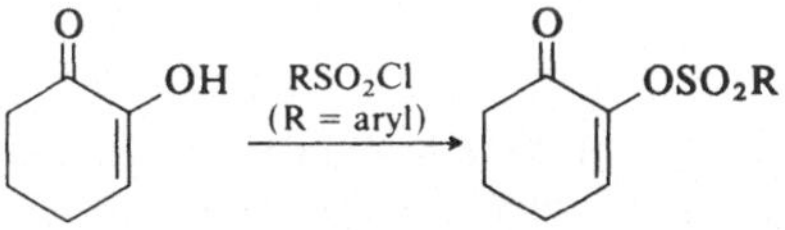

**1-107**

## 1.2.2.5. Nitrogen from Azoalkanes

Excellent reviews[88,91] on the photochemistry of azoalkanes are available, covering the literature through 1980. An extensive discussion of the photochemically triggered release of nitrogen is found in the review by Meier and Zeller.[142]

Photochemical reactions of acyclic azo compounds have been known for over 40 years. The irradiation of an azo compound, which can exist as either the *cis* or the *trans* isomer, produces a vibrationally excited singlet state. These hot states may either decompose with the release of nitrogen or be deactivated by collision with inert molecules. It has been found that high pressure causes more deactivation, favoring isomerization between the *cis* and *trans* forms. Shorter-wavelength irradiation produces hot excited states with greater vibrational energy so they are likely to decompose and release nitrogen. The quantum yields of such excitations reflect the energy content in the reaction and thus the rate of decomposition and nitrogen release.

The photochemistry of cyclic azoalkanes has also been investigated,[156,157] especially that of the three-membered rings of diazirines (1-108). These are considered to be useful carbene and nitrogen precursors. The photochemically triggered release of nitrogen and carbene from substituted diazirines proceeds through a diazomethane and subsequent rearrangement.

$$
\underset{\substack{\displaystyle N=N \\ \textbf{1-108}}}{\overset{\displaystyle R}{\triangle}} \xrightarrow{h\nu} RC{=}\overset{+}{N}{=}\overset{-}{N} \xrightarrow{h\nu} RC{:} + N_2
$$

R = $C_6H_5$, 4-$CH_3C_6H_4$, 4-$CH_3OC_6H_4$, 4-$CH_3COOC_6H_4$, 3-$ClC_6H_4$, 4-pyridyl, and 3-pyridyl

Smith and Knowles[157] have reported synthetic routes utilizing the photochemical reactions of a number of 3-aryl-3$H$-diazirines (1-108). The photolysis reactions were carried out in *n*-hexane using a medium-pressure Hanovia mercury vapor UV lamp in a Pyrex cooling jacket.

Frey and Penny[158] have reported an extensive study of the photochemical properties of 3-aryl-3-chlorodiazirine using a cadmium-ion laser at 325 nm and a nitrogen laser at 337 nm. The quantum yield of the decomposition was determined, at room temperature with the 325-nm irradiation, to be $0.945 \pm 0.02$.

The photochemically triggered release of nitrogen from larger cyclic azoalkanes is also known.[88,142] The reaction is dependent on the size and the steric strain in the ring as reflected by the quantum yields (Table 1.6).

### 1.2.2.6. Nitrogen from Diazonium Salts

Aryl diazonium salts are well known[159] to undergo photochemical decomposition to release nitrogen and the corresponding aryl radicals. These diazonium salts are generally prepared by treatment of aromatic amine salts with sodium nitrate solution in presence of an acid at 0°C.

Several salts have been reported, with such anions as halides, oxalate, citrate, fluoroborate, and zinc chloride. The simple salts with single anions

*Table 1.6.* Quantum Yields for the Release of Nitrogen from Cyclic Azoalkanes

| Azo compounds[a] | | | Quantum yield ($\phi$) |
|---|---|---|---|
| 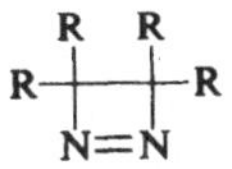 | **(1-109)** | | 0.52 |
|  | **(1-110)** | R = Aryl | 0.98 |
| 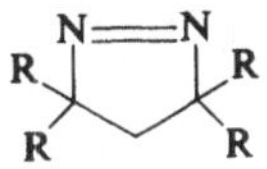 | **(1-111)** | R = Alkyl | 0.12 |
| 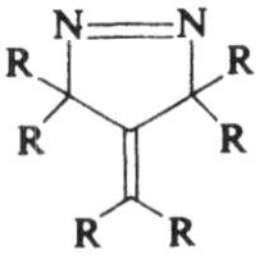 | **(1-112)** | | 0.45 |
| 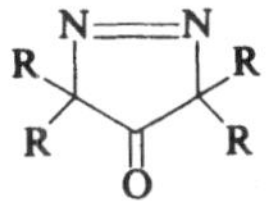 | **(1-113)** | | 0.01 |
| 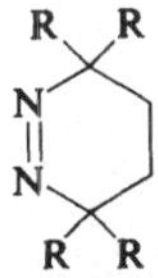 | **(1-114)** | | 0.002 |
|  | | | |
| $n = 1$ | **(1-115)** | | 1.00 |
| $n = 2$ | **(1-116)** | | 0.02 |

[a] Refs. 80 and 91.

are claimed to exhibit the most sensitivity to light and thus are very unstable. Comparatively more stable salts are the fluoroborates, fluorosulfonates, and zinc chlorides.

Like other light-sensitive compounds, diazonium salts are affected by light in the blue and near-ultraviolet regions. The basic reaction affords the release of nitrogen and aryl radicals, which combine with nucleophiles in the solid or solution phase.[160]

$$R\overset{+}{N}_2\overset{-}{X} \xrightarrow[\text{300-400 nm}]{h\nu} R\cdot + X\cdot + N_2$$

**1-117**

$$R\overset{+}{N}_2\overset{-}{X} \xrightarrow[\text{Solution}]{h\nu} ROH + N_2 + HX$$

**1-117**

$$R\overset{+}{N}_2\overset{-}{X} \xrightarrow[\text{Solid}]{h\nu} RX + N_2$$

**1-117**

The effect of substituents[160-163] on the aromatic ring (**1-118**) on the light sensitivity of the diazonium salt has been extensively studied. In

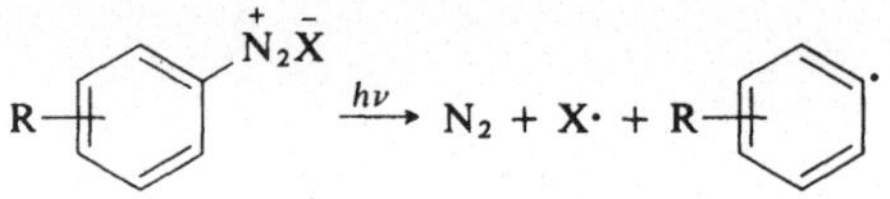

**1-118**

| R | UV light sensitivity |
|---|---|
| $(CH_3CH_2)_2N$ | $p > m > o$ |
| $CH_3O$ | $o > m > p$ |
| $CH_3$ | $o > m > p$ |
| $Cl$ | $o > m > p$ |
| $Br$ | $o > m > p$ |
| $COOH$ | $o > m > p$ |
| $SO_3H$ | $o > p > m$ |

general, ortho substituents on the aromatic ring enhance the photochemical sensitivity as compared to the meta and para isomers.

In the crystalline state, the photolysis products of diazonium salts are simlar to the thermal products. Diazonium hexafluorophosphates and fluoroborates (**1-119**) gave the corresponding aryl fluorides. The irradiation light is usually in the blue region of the spectrum, at about 350 nm. Ando[159] has reported the effects of substituents and the counterion on the reaction time.

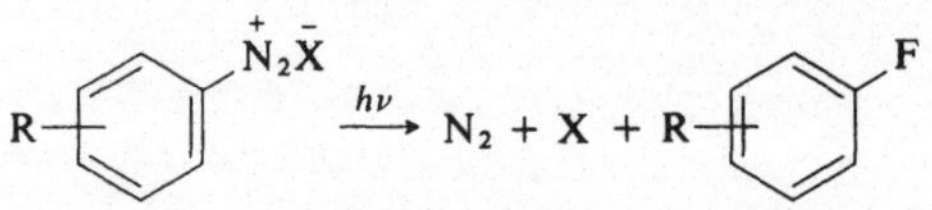

**1-119**    $X^- = BF_4^-, PF_6^-; X = BF_3, PF_5$

Furthermore, Tsunoda and Yamaoka[164] have investigated the effects of substituents on the quantum yields of diazonium salts; some of their results are given in Table 1.7. The sensitivity of these diazonium salts can

*Table 1.7.* Quantum Yields for the Release of Nitrogen Gas
from Diazonium Salts (**1-119**)[a]

| Ring substituents | Absorption wavelength (nm) | Quantum yield ($\phi$) |
|---|---|---|
| H | 264 | 0.38 |
| 4-Cl | 283 | 0.29 |
| 3-Cl | 266 | 0.41 |
| 2-Cl | 269 | 0.18 |
| 4-Br | 293 | 0.35 |
| 4-F | 269 | 0.50 |
| 4-CH$_3$ | 279 | 0.23 |
| 3-CH$_3$ | 279 | 0.35 |
| 2-CH$_3$ | 269 | 0.30 |
| 3-CH$_3$O | 275 | 0.24 |
| 2-CH$_3$O | 355 | 0.18 |
| 4-OH | 350 | 0.54 |
| 4-(CH$_3$)$_2$N | 385 | 0.58 |
| 3-NO$_2$ | 232 | 0.17 |
| 2-NO$_2$ | 280 | 0.10 |

[a] Ref. 164.

be extended into the visible region[165] through spectral sensitization with
dyes and activators (e.g., amines, thiourea, and benzenesulfinate).

Diazonium salts when prepared from aminophenols are known as
diazophenols or quinone azides (**1-120**), where the negative charge of the
phenolic oxygen is delocalized on the diazo nitrogens. The photolytically
triggered release of nitrogen from such compounds is accompanied by a
ring contraction reaction, forming cyclopentadiene carboxylic acid.

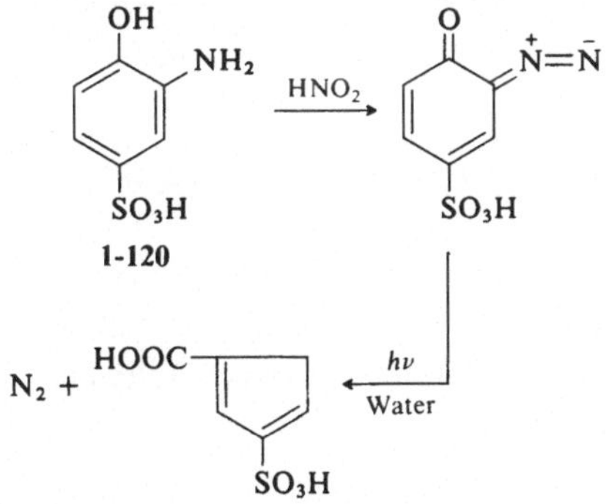

## 1.2.2.7. Nitrogen from Azides and Heterocycles

The photochemically triggered release of nitrogen and nitrenes from
light-sensitive azides is known from only a few examples. Recently,

photolysis of a series of *o*-azidobiphenyls (**1-121**) produced high yields of carbazoles.[167]

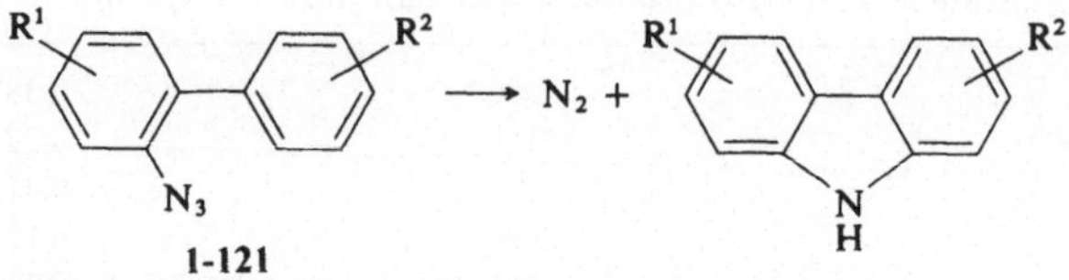

1-121

| R¹ | R² | Percent yield (carbazole) |
|---|---|---|
| H | H | 77 |
| H | 5-Br | 23 |
| H | 3,5-Br₂ | 57 |
| 4-NO₂ | 4-NO₂ | 66 |
| H | 3-NO₂ | 52 |
| H | 3,5-(NO₂)₂ | 85 |

Heterocycles with an azo group within their molecular structure can be photochemically induced to release nitrogen gas with the formation of ring contraction products.

Pyrazolenines (**1-22**), for example, can be readily converted into cyclopropenes and nitrogen.[168,169]

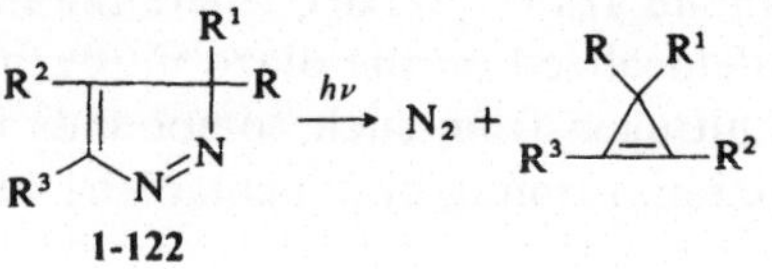

1-122

$R^1 = R = CH_3$, $R^2 = H$, $R^3 = CH_3$
$R^1 = R = CH_3$, $R^2$-$R^3 = +CH_2+_5$
$R^1 = $ 9-fluorenyl, $R = H$, $R^2 = R^3 = COOCH_3$
$R^1 = R = C_6H_5$, $R^2 = R^3 = COOCH_3$

Pyrazolines (**1-124**) also photochemically release nitrogen and cyclopropanes in high yields. Such a reaction is general for the pyrazoline ring independent of the ring substituents, as exemplified by (**1-123**),[170] (**1-125**),[171] (**1-126**),[172,173] and (**1-127**).[174]

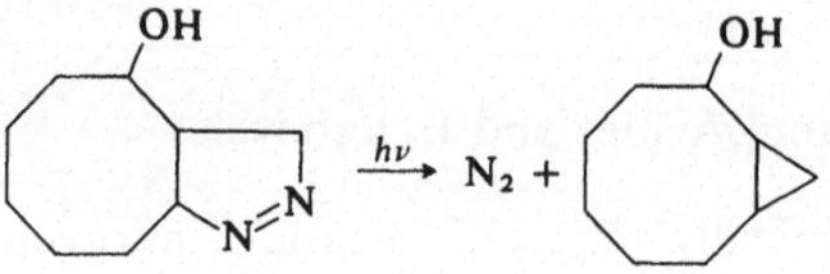

1-123

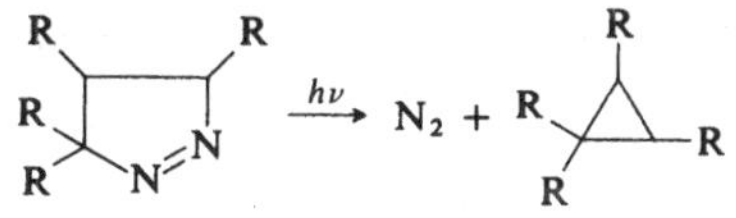

1-124    R = CH$_3$

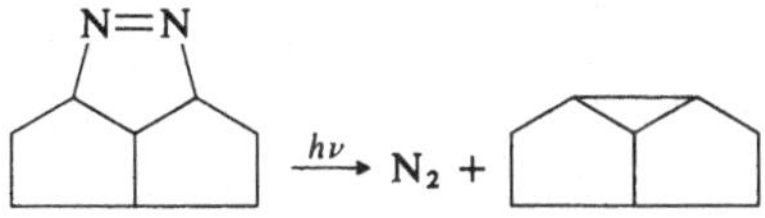

1-125

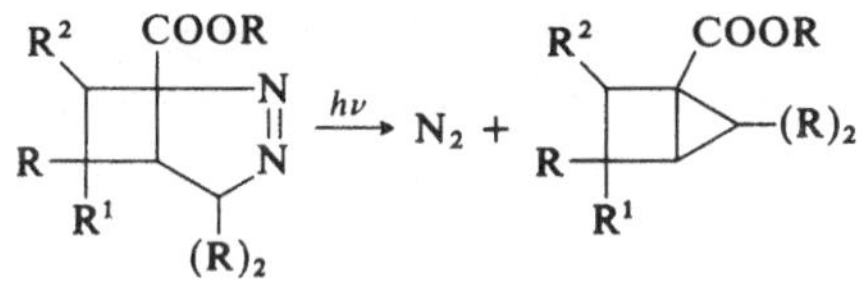

1-126    R = R$^1$ = CH$_3$, R$^2$ = H
1-127    R = CH$_3$, R$^1$ = R$^2$ = Cl

The triazoline ring structures (1-128 to 1-130) can, as shown by Scheiner,[175] be converted efficiently into aziridines by photolysis, with a quantitative loss of nitrogen gas. The photolysis was studied in inert solvents (acetone, toluene, p-dioxane, etc.) with a GE sunlamp and at a reaction temperature which did not exceed 35°C.

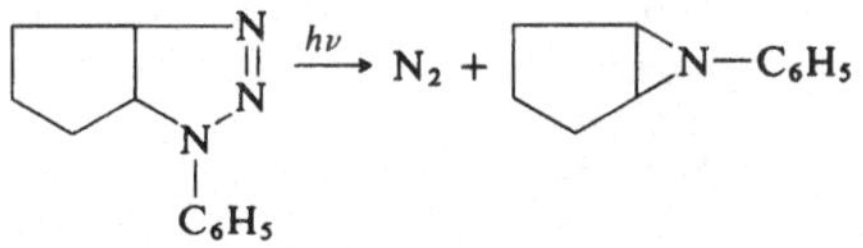

1-128

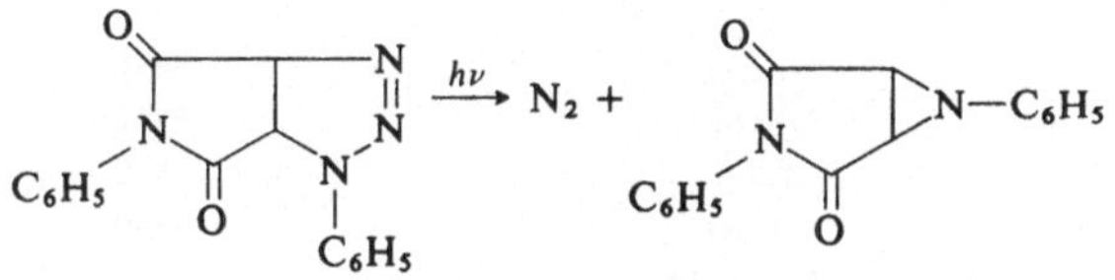

1-129

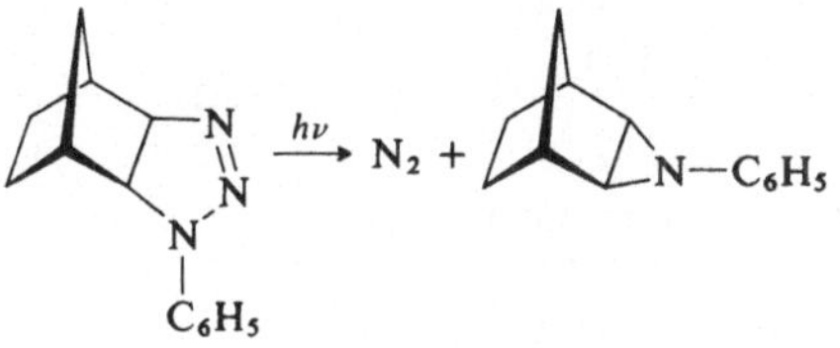

1-130

### 1.2.3. Electrochemically Triggered Release of Carbon Dioxide

Electrolysis of carboxylic acids is a well-known chemical process. Such a process utilizes the electric current as the triggering agent for the release of carbon dioxide gas.

The Kolbe reaction[176,177] (1-131), first reported in 1849 uses a variety of carboxylic acid salts in an electrolytic bath, for example, alkyl-, aryloxy-,[178,179] phenyl-,[179,180] diphenyl-,[180,181] and trifluoroacetic acids,[182] as well as $\alpha$-acylamino-,[183] $\alpha$-alkoxy-,[184,185] and $\alpha$-cyanocarboxylic acids.[186] No coupling reaction was reported with the use of $\alpha$-haloacids.[187,188] Substituents further removed from the carboxylic group than the $\alpha$-position seem, in general, to exert little if any influence on the Kolbe reaction.

The general experimental details are described in Chapter 2.

$$2RCOO^- \rightarrow R{-}R + CO_2$$

1-131

Table 1.8. Triggering Time for the Release of Gases Ultrasonically

| Compound | Reaction medium | Triggering time (min) |
|---|---|---|
| Benzene | Air/Ag$^+$ | 20.0 |
| | Nitrogen/Ag$^+$ | 10.0 |
| | Argon/Ag$^+$ | 3.9 |
| | Nitrogen | 6.0 |
| | Argon | 5.0 |
| Bromobenzene | Air/Ag$^+$ | 6.0 |
| | Nitrogen/Ag$^+$ | 10.0 |
| | Argon/Ag$^+$ | 5.0 |
| | Nitrogen | 10.0 |
| | Argon | 5.0 |
| Phenol | Argon/Ag$^+$ | 40.0 |
| | Nitrogen | 40.0 |
| Tropolone | Nitrogen | 20.0 |
| | Argon | 20.0 |
| Pyridine | Air/Ag$^+$ | 40.0 |
| | Nitrogen/Ag$^+$ | 40.0 |
| | Argon/Ag$^+$ | 10.0 |
| | Nitrogen | 54.0 |
| | Argon | 20.0 |
| Pyrazine | Air/Ag$^+$ | 45.0 |
| Pyrrole | Air/Ag$^+$ | 21–56 |
| | Nitrogen/Ag$^+$ | 56.0 |
| | Argon/Ag$^+$ | 30.0 |
| | Nitrogen | 40.0 |
| | Argon | 20.0 |

### 1.2.4. Ultrasonic-Wave Triggered Release of Gases

It has been reported[189] that ultrasonic waves trigger halogenated compounds, for example, bromo- and iodobenzene, $\alpha$-iodothiophene, and $\alpha$-bromofuran, in the presence of silver nitrate to release halogens and acetylene radicals, the latter trapped as the silver salts. Further experiments have also revealed[190] that the halo group is not a necessity in the case of pyridine and pyrrole.

Currell and Zechmeister[191] have reported that the silver ions in the reaction served as a trapping agent and not as a required reactant. Ultrasonic generators used for the triggering had a nominal power output of 200 watts at 1000 kilocycles/s.

Table 1.8 lists some compounds that release gases under ultrasonic wave triggering.

## REFERENCES

1. A. Baril and E. Klein, U.S. Pat. 2,976,145 (1961).
2. R. J. Bruni and C. R. Morgan, U.S. Pat. 2,923,703 (1960).
3. U. Vahtra and R. M. Lindquist, French Pat. 1,308,936 (1962).
4. Kodak-Pathe, U.S. Pat. 3,143,418 (1964).
5. Kodak Ltd., British Pat. 962,557 (1964).
6. C. E. Herrick and B. I. Halperin, British Pat. 850,954 (1960).
7. IBM Corp., German Pat. 1,178,298 (1964).
8. R. T. Nieset, *J. Photogr. Sci. 10*, 118 (1962).
9. M. G. Anderson and E. E. Sandlin, *Vesicular Systems for High Resolution Generation Printing*, NMA Proceedings of the 1964 Convention, Vol. 13 (1964).
10. R. T. Nieset and N. T. Notley, *S. SMPTE 74*, 786 (1965).
11. R. G. O'Lone, *Cathode Ray Display Speeds—SRAM Data*, Aviation Week and Space Technology, March, 1967.
12. H. Inaba, T. Kobayashi, K. Yamawaki and A. Serigiyamo, *Infrared Physics 7*, 145 (1967).
13. J. F. Forkner and D. D. Lowenthal, *Appl. Optics 6*, 1419 (1967).
14. Eastman Kodak, British Pat., 1,016,823 (1962).
15. (a) NCR Corp., British Pat. 1,109,554 (1968); (b) NCR Corp., U.S. Pat. 3,364,023 (1968).
16. (a) NCR Corp., British Pat. 1,073,999 (1967); (b) NCR Corp., U.S. Pat. 3,359,103 (1967).
17. NCR Corp., U.S. Pat. 3,072,481 (1963).
18. H. R. Lasman, *Encyclopedia of Polymer Science and Technology*, Vol. 2, pp. 532–565, Interscience, New York (1965).
19. C. J. Benning, *Plastics Foams*, Vol. 2, J. Wiley and Sons, New York (1969).
20. R. C. Fuson, *Reactions of Organic Compounds*, p. 665, J. Wiley and Sons, New York (1962).
21. E. B. Vliet, C. S. Marvel, and C. M. Hsueh, *Org. Synthesis 2*, 416 (1943).
22. E. E. Reid and J. R. Ruhoff, *Org. Synthesis 2*, 474 (1943).
23. S. J. Cristol and W. P. Norric, *J. Am. Chem. Soc. 72*, 2645 (1953).
24. T. Cohen and I. H. Song, *J. Org. Chem. 31*, 3058 (1966).
25. D. B. Bigley and R. W. May, *J. Chem. Soc., Sect. B*, 557 (1967).
26. D. B. Bigley and J. C. Thurman, *Tetrahedron Lett.*, 2377 (1967).

27. G. G. Smith and F. W. Kelly, in: *Progress in Physical Organic Chemistry* (A. Streitwieser, Jr. and R. W. Taft, eds.), Vol. 2, p. 153, Wiley-Interscience, New York (1971).

28. R. C. Fuson, *Reactions of Organic Compounds*, p. 669, J. Wiley and Sons, New York (1962).

29. A. G. Dobson and H. H. Hatt, *Org. Synthesis 33*, 84 (1953).

30. H. E. Zauger, *Org. Reactions 8*, 365 (1954).

31. Y. Etiene and N. Flecher, in: *Heterocyclic Compounds* (A. Weiseberger, ed.), Vol. XIX, Part 2, pp. 793–880, Interscience, New York, (1964).

32. W. T. Brady and A. D. Patel, *J. Org. Chem. 37*, 3536 (1972).

33. G. Marel, *Chem. Ber. 96*, 1441 (1963).

34. J. Schreiber, W. Leimgruber, M. Pesaro, P. Schudel, T. Threlfall, and A. Eschenmoser, *Helv. Chim. Acta 44*, 540 (1961).

35. P. Bosshard, S. Fumagalli, R. Good, N. Trueb, W. V. Philipsborn, and C. H. Eugster, *Helv. Chim. Acta 47*, 769 (1964).

36. P. Bosshard, S. Fumagalli, R. Good, W. Trueb, W. V. Philipsborn, and C. H. Eugster, *Chem. Ber. 96*, 1441 (1963).

37. W. Adam and R. Rucktaschel, *J. Am. Chem. Soc. 93*, 557 (1971).

38. D. B. Pattison, *J. Am. Chem. Soc. 79*, 3455 (1957).

39. D. H. Barton and B. J. Willis, *J. Chem. Soc., Chem. Commun.*, 1225 (1970).

40. C. T. Peterson, *Acta Chim. Scand. 22*, 247 (1968).

41. H. Becker and A. Bistrzycki, *Ber. 49*, 3149 (1914).

42. D. D. Reynolds, D. L. Fields, and D. L. Johnson, *J. Org. Chem. 26*, 5130 (1960).

43. D. D. Reynolds, M. K. Massad, D. L. Fields, and D. L. Johnson, *J. Org. Chem. 26*, 5109 (1961).

44. D. D. Reynolds, *J. Am. Chem. Soc. 79*, 4951 (1957).

45. D. D. Reynolds, D. L. Johnson, and D. L. Fields, *J. Org. Chem. 26*, 5125 (1961).

46. D. D. Reynolds, D. L. Fields, and D. L. Johnson, *J. Org. Chem. 26*, 5116 (1961).

47. W. Sieber, P. Gilgen, S. Chaloupka, H-J. Hansen, and H. Schmid, *Helv. Chim. Acta 56*, 1679 (1973).

48. W. W. Hartman and M. R. Brethen, *Org. Synthesis, Coll.*, Vol II, 297 (1943).

49. D. E. Floyd and S. E. Miller, *Org. Synthesis 34*, 13 (1954).

50. P. A. Levene and G. M. Meyer, *Org. Synthesis, Coll.*, Vol. II, 288 (1943).

51. W. E. Bachman, J. W. Cole, and A. L. Wilds, *J. Am. Chem. Soc. 62*, 824 (1940).

52. M. L. Kloetzel, *Org. Reactions 4*, 1 (1940).

53. H. L. Holmes, *Org. Reactions 4*, 60 (1948).

54. J. A. Norton, *Chem. Rev. 31*, 319 (1952).

55. M. A. Ogliaruso, M. G. Romanelli, and E. I. Becker, *Chem. Rev. 65*, 261 (1956).

56. C. F. H. Allen, *Chem. Rev. 37*, 209 (1945).

57. A. Hassner and D. J. Anderson, *J. Org. Chem. 39*, 3070 (1974).

58. M. A. Battiste, *J. Am. Chem. Soc. 85*, 2175 (1963).

59. M. A. Battiste, *Tetrahedron Lett.*, 3795 (1964).

60. M. A. Battiste and T. J. Barton, *Tetrahedron Lett.*, 1227 (1967).

61. P. W. Schenik and R. Steudel, *Angew. Chem. Intl. Ed. Engl. 4*, 402 (1965).

62. A. Carrington and D. H. Levy, *J. Phys. Chem. 71*, 5 (1966).

63. A. Carrington, D. H. Levy, and T. A. Miller, *Proc. Roy. Soc. 293A*, 108 (1966).

64. G. E. Hartzel and J. N. Page, *J. Am. Chem. Soc. 88*, 2616 (1966).

65. G. E. Hartzel and J. N. Page, *J. Org. Chem. 32*, 459 (1967).

66. R. M. Dodson and R. F. Sauers, *J. Chem. Soc., Chem. Commun.*, 1189 (1967).

67. L. A. Carpino, L. V. McAdams III, R. H. Rynbrandt, and T. W. Spiewak, *J. Am. Chem. Soc. 93*, 476 (1971).

68. L. A. Paquette, *Acc. Chem. Res. 1*, 209 (1968).

69. H. Mackle, D. V. McNally, and W. V. Steele, *Trans. Faraday Soc. 65*, 2060 (1969).

70. G. Optiz and K. Fischer, *Angew. Chem. Intl. Ed. Engl. 4*, 70 (1965).
71. R. Livingstone, in: *Chemistry of Carbon Compounds* (E. H. Rodd, ed.), Vol. IV, Part A, pp. 119–327, Elsevier, New York (1967).
72. M. P. Cava and A. A. Deana, *J. Am. Chem. Soc. 81*, 4266 (1959).
73. M. P. Cava and R. L. Shirley, *J. Am. Chem. Soc. 82*, 654 (1960).
74. M. P. Cava, A. A. Deana, and K. Muth, *J. Am. Chem. Soc. 82*, 2524 (1960).
75. B. P. Stark and A. J. Duke, *Extrusion Reactions*, pp. 72–74, Pergamon Press, Oxford (1967).
76. H. J. Backer and A. H. Blass, *Recl. Trav. Chim. Pays-Bas 61*, 785 (1942).
77. E. H. Burke and D. D. Carlos, *J. Heterocycl. Chem. 7*, 177 (1970).
78. W. L. Mock, *J. Am. Chem. Soc. 89*, 1281 (1967).
79. J. D. Loudon and L. B. Young, *J. Chem. Soc.*, 5496 (1963).
80. J. D. Loudon and L. B. Young, *J. Chem. Soc.*, 3262 (1962).
81. T. Mortel and P. E. Verkade, *Recl. Trav. Chim. Pays-Bas 70*, 35 (1950).
82. B. Helferich, R. Dhein, K. Geist, H. Junker, and D. Wiehle, *Ann. 646*, 45 (1961).
83. B. Helferich and W. Klebert, *Ann. 657*, 79 (1962).
84. E. M. LaCombe and B. Stewart, *J. Am. Chem. Soc. 83*, 3457 (1961).
85. C. C. Price and G. Berti, *J. Am. Chem. Soc. 76*, 1213 (1954).
86. J. Thiele, *Chem. Ber. 42*, 2575 (1909).
87. E. L. Allard, J. E. Oberlander, and P. F. Rankin, *J. Am. Chem. Soc. 100*, 4910 (1978).
88. P. S. Engel, *Chem. Rev. 80*, 99 (1980).
89. T. Koeing, *Free Radicals*, 1 (1973).
90. P. S. Engel and C. Steel, *Acc. Chem. Res. 6*, 275 (1973).
91. K. McKenzie, in: *The Chemistry of the Hydrazo, Azo and Azoxy Groups* (S. Patai, ed.), J. Wiley and Sons, New York (1975).
92. B. K. Bandlish, A. W. Garner, M. L. Hodges, and J. W. Timberlake, *J. Am. Chem. Soc. 97*, 5856 (1975).
93. S. F. Nelsen and P. D. Bartlett, *J. Am. Chem. Soc. 88*, 137 (1966).
94. P. E. Engel and D. J. Bishop, *J. Am. Chem. Soc. 97*, 6754 (1975).
95. A. W. Garner, J. W. Timberlake, P. S. Engel, and R. A. McLaugh, *J. Am. Chem. Soc. 97*, 7377 (1975).
96. W. J. Timberlake and A. W. Garner, *J. Org. Chem. 41*, 1666 (1976).
97. W. Duismann and C. Ruchardt, *Chem. Ber. 106*, 1083 (1973).
98. E. Leventhal, C. R. Simonds, and C. Steel, *Can. J. Chem. 40*, 930 (1962).
99. F. M. Lewis and M. S. Matheson, *J. Am. Chem. Soc. 71*, 747 (1949).
100. G. C. Overberger and H. Biletch, *J. Am. Chem. Soc. 73*, 4880 (1951).
101. G. C. Overberger and A. V. Giulio, *J. Am. Chem. Soc. 81*, 2154 (1959).
102. H. C. Ramsperger, *J. Am. Chem. Soc. 51*, 2134 (1929).
103. K. R. Kopecky and T. Gillian, *Can. J. Chem. 47*, 2371 (1969).
104. R. Huisgen, *Angew. Chem. Intl. Ed. Engl. 2*, 565 (1963).
105. T. L. Jacobs, in: *Heterocyclic Compounds* (R. C. Ederfield, ed.), Vol. 5, pp. 76–83, 108–109, J. Wiley and Sons, New York (1975).
106. S. G. Beech, J. H. Turnbull, and W. Wilson, *J. Chem. Soc.*, 4686 (1952).
107. R. Huisgen, *Angew. Chem. 67*, 439 (1955).
108. F. M. Dean, P. G. Jones, R. B. Morton and P. Sinisunthorn, *J. Chem. Soc.*, 5336 (1963).
109. L. F. Fieser and J. L. Hartwell, *J. Am. Chem. Soc. 57*, 1479 (1935).
110. G. Wittig and R. W. Hoffmann, *Angew. Chem. 73*, 435 (1961).
111. G. Wittig and R. W. Hoffmann, *Chem. Ber. 95*, 2729 (1962).
112. M. Rosenblum and H. Moltzan, *Chem. Ind.*, 1480 (1956).
113. W. I. Awad, S. M. A. R. Omran, and F. Nageib, *Tetrahedron 19*, 1591 (1963).
114. K. Adler and G. Stein, *Ann. 105*, 1 (1933).
115. W. D. Crow and C. Wentrup, *J. Chem. Soc., Chem. Commun.*, 1026 (1968).

116. R. A. Carboni and R. V. Lindsey, *J. Am. Chem. Soc. 81*, 4342 (1959).

117. M. Avram, I. G. Dinvlescu, E. Marica, and C. D. Nenitzeschi, *Chem. Ber. 95*, 2248 (1962).

118. J. Saver, A. Mielert, D. Lang, and D. Peter, *Chem. Ber. 98*, 1435 (1965).

119. J. G. Erickson, P. E. Wiley, and V. P. Wystrach, *The Chemistry of Heterocyclic Compounds*, Vol. X, Chapter IV, Interscience, New York (1956).

120. (a) J. H. Boyer and F. C. Canter, *Chem. Rev. 54*, 1 (1954); (b) T. S. Stevens and W. E. Watts, *Selected Molecular Rearrangements*, pp. 45–52, Van Nostrand Reinhold Co., London (1973).

121. L. Wolff, *Ann. 394*, 23 (1912).

122. T. Zincke and A. T. Lawson, *Ber. 20*, 1176 (1887).

123. T. Zincke and H. Jaenke, *Ber.* 540 (1888).

124. E. Noelting and K. Kohn, *Chem.-Ztg. 18*, 1095 (1894).

125. D. V. Banthorpe, in: *The Chemistry of the Azido Group* (S. Patai, ed.), pp. 397–405, Interscience, New York (1971).

126. P. A. S. Smith, *Org. Reactions 3*, 337 (1946).

127. L. Benati, P. C. Montevecchi, and P. Spagnolo, *J. Chem. Soc. Perkin Trans. 2*, 1437 (1981).

128. G. Balz and G. Schiemann, *Ber. 60B*, 1186 (1927).

129. A. Roe, *Org. Reactions 5*, 193 (1949).

130. K. G. Rutherford, W. Redmond, and J. Rigamonti, *J. Org. Chem. 26*, 5149 (1961).

131. G. A. Olah and W. S. Tolgyeshi, *J. Org. Chem. 26*, 2053 (1961).

132. O. C. Dermer and M. T. Edmison, *Chem. Rev. 57*, 77 (1957).

133. J. Stepek and H. Daoust, *Advances for Plastics*, pp. 114–119, Springer-Verlag, New York (1983).

134. H. Ulrich, U.S. Pat. 3,562,269 (1971).

135. D. H. Heinert, U.S. Pat. 3,553,113 (1971).

136. F. Muller and F. J. Maurin, British Pat. 1,297,973, 1,301,587 (1972).

137. W. D. Wirth, E. Muller, and H. Rohr, U.S. Pat. 3,725,321 (1973).

138. J. E. Herweh and A. G. Poshkus, U.S. Pat. 3,374,188, 3,374,189, 3,374,190 (1968); 3,492,301 (1970).

139. F. D. March and D. H. Thatcher, U.S. Pat. 3,374,188 (1968).

140. F. D. March, U.S. Pat. 3,410,658 (1968).

141. S. S. Adcock British Pat. 1,351,463 (1974).

142. H. Meier and K. P. Zeller, *Angew. Chem. Intl. Ed. Engl. 16*, 835 (1977).

143. (a) C. H. Depuy and R. W. King, *Chem. Rev. 60*, 431 (1960); (b) H. R. Nace, *Organic Reactions*, Vol. 12, pp. 75–100, J. Wiley and Sons, New York (1962).

144. R. F. W. Bader and A. N. Bourns, *Can. J. Chem. 39*, 348 (1961).

145. R. A. Benkeser and J. J. Hazdra, *J. Am. Chem. Soc. 81*, 228 (1959).

146. W. Bergmann and M. J. McLean, *Chem. Rev. 28*, 367 (1941).

147. W. Sieber, P. Gilgen, S. Chaloupka, H. J. Hansen, and H. Scmid, *Helv. Chim. Acta 56*, 1679 (1973).

148. D. H. R. Barton and B. J. Willis, *J. Chem. Soc., Chem. Commun.*, 1225 (1970).

149. W. Adam and R. Rucktaschel, *J. Am. Chem. Soc. 93*, 557 (1971).

150. A. Hassner and D. J. Anderson, *J. Am. Chem. Soc. 94*, 8255 (1972).

151. D. J. Wagner, C. A. Stout, S. Searles, and G. S. Hammond, *J. Am. Chem. Soc. 88*, 1242 (1966).

152. F. E. Blacet and A. Miller, *J. Am. Chem. Soc. 79*, 4327 (1957).

153. I. Heller and R. Srinivasan, *J. Am. Chem. Soc. 87*, 1144 (1965).

154. N. J. Turro, P. L. Leermakers, H. R. Wilson, D. C. Weckers, and G. F. Vesley, *J. Am. Chem. Soc. 87*, 2613 (1965).

155. A. Feigenbaum, J. P. Pete, and D. Scholler, *Tetrahedron Lett.*, 537 (1979).

156. (a) H. M. Frey, *Adv. Phot. Chem. 4*, 225 (1966); (b) S. Braslavsky and J. Heicklen, *Chem. Rev. 77*, 473 (1977).

157. R. A. Smith and J. R. Knowles, *J. Chem. Soc., Perkin Trans. 2*, 686 (1975).
158. H. M. Frey and D. E. Penny, *J. Chem. Soc., Faraday Trans. 1*, 73, 2010 (1977).
159. W. Ando, in: *The Chemistry of Diazonium and Diazo Groups* (S. Patai, ed.), Part I, Chapter 4, p. 341, J. Wiley and Sons, New York (1978).
160. M. Andressen, *Chem. Zentr.* 66, 530 (1895).
161. M. S. Dinaburg, *Photosensitive Diazo Compounds*, Focal Press, New York (1964).
162. H. Zollinger, *Azo & Diazo Chemistry*, Interscience, London (1961).
163. D. J. Brown, *Chem. Ind.*, 22, 146 (1944).
164. T. Tsunoda and T. Yamaoka, *J. Photogr. Sci., Japan* 29, 197 (1966).
165. T. Yamase, T. Ikawa, H. Kokado, and E. Inove, *Photogr. Sci. Eng.* 17, 28 (1973).
166. O. Sus, *Ann.* 556, 65, 85 (1944).
167. P. A. Smith and B. B. Brown, *J. Am. Chem. Soc.* 73, 2435 (1951).
168. G. L. Closs and W. A. Boll, *Angew. Chem. Intl. Ed. Engl.* 2, 399 (1963).
169. G. Ege, *Tetrahedron Lett.*, 1667 (1963).
170. K. Wiberg and A. de Meijere, *Tetrahedron Lett.*, 59 (1969).
171. M. Scharz, A. Besold, and E. R. Nelson, *J. Org. Chem.* 30, 2425 (1965).
172. T. H. Kinstle, R. L. Welch, and R. W. Exley, *J. Am. Chem. Soc.* 89, 3660 (1961).
173. P. G. Gassman and K. T. Mansfield, *J. Org. Chem.* 32, 915 (1967).
174. M. Franck-Neumann, *Tetrahedron Lett.*, 2979 (1968).
175. P. Scheiner, *J. Org. Chem.* 30, 7 (1965).
176. H. Kolbe, *Ann.* 69, 257 (1849).
177. B. C. L. Weedon, in: *Advances in Organic Chemistry: Methods and Results* (R. A. Raphael, E. C. Taylor, and H. Wynberg, eds.), Vol. 1, p. 1, Interscience, New York (1960).
178. F. Fichter and K. Kestenholz, *Helv. Chim. Acta* 25, 785 (1942).
179. F. Fichter and H. Stenzl, *Helv. Chim. Acta* 22, 971 (1939).
180. R. P. Linstead, B. R. Shepard, and B. C. L. Weedon, *J. Chem. Soc.*, 3624 (1951).
181. (a) L. Riecoboni, *Gazz. Chim. Ital.* 70, 748 (1940); (b) A. J. Van Der Hoek and W. T. Nauta, *Recl. Trav. Chim. Pays-Bas* 61, 845 (1942).
182. F. Swartz, *Bull. Soc. Chim. Belg.* 42, 102 (1933).
183. R. P. Linstead, B. R. Shepard, and B. C. L. Weedon, *J. Chem. Soc.*, 2854 (1951).
184. D. A. Fairweather, *Proc. Roy. Soc. Edinburgh* 45, 23 (1925).
185. W. Von Miller and H. Hofer, *Ber.* 27, 461 (1894).
186. F. Fichter and A. Schnider, *Helv. Chim. Acta* 13, 103 (1930).
187. K. Elbs and K. Kaatz, *J. Prakt. Chem.* 55, 502 (1897).
188. P. Kaufler and C. Hertzog, *Ber.* 42, 3870 (1909).
189. L. Zechmeister and L. Wallcave, *J. Am. Chem. Soc.* 77, 2853 (1955).
190. L. Zechmeister and E. F. Magoon, *J. Am. Chem. Soc.* 78, 2149 (1956).
191. D. L. Currell and L. Zechmeister, *J. Am. Chem. Soc.* 80, 205 (1958).
192. S. T. Reid, in: *Advances in Heterocyclic Chemistry* (A. R. Katritzky, ed.), Vol. 33, p. 88, Academic Press, London (1983).

# 2

# Triggered Release of Acids, Bases, Radicals, Nitrenes, and Carbenes

In this chapter the thermal, photochemical, and electrochemical release of acids, bases, and radicals will be surveyed. The reactions which have been most frequently cited in the literature as the most promising sources for any of these species will be discussed.

A number of compounds will be described which can be precursors capable of releasing both acids and radicals. These will be discussed with reference to one aspect and cross-referenced to the other(s).

## 2.1. PROCESSES UTILIZING THE TRIGGERED RELEASE OF ACIDS, BASES, AND RADICALS

The triggered release of acids, bases, and radicals has found a number of industrial applications in the areas of polymer and imaging science.

Iodonium salts have been used as precursors for the release of strong acids to catalyze the cationic polymerization of epoxy resins. Schlessinger[1] described a photoresist system for electronic applications, comprising a multifunctional novolac-epoxy resin (2-1).

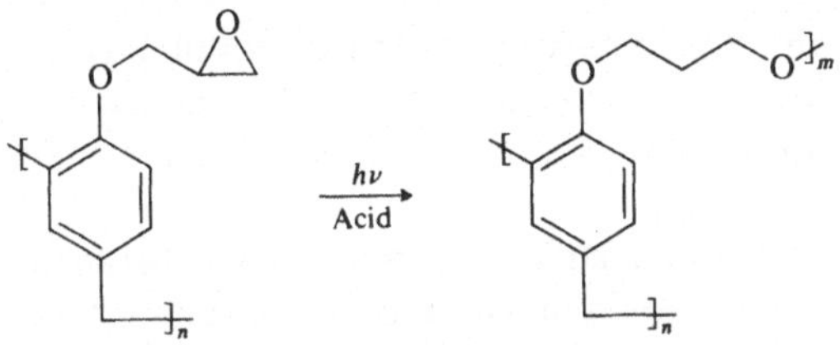

2-1

Vinyl ether difunctional monomers[2] (**2-2**) have found similar utility.

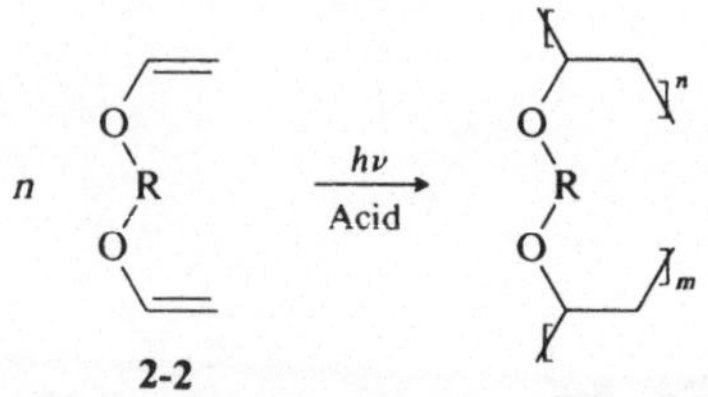

**2-2**

The acid-catalyzed rearrangement of *t*-butyl esters induced by the photodecomposition of diaryliodonium or triarylsulfonium salts was reported by Ito and Wilson[2] of IBM as a basis for a dry-developed photoresist (**2-3**), mainly for electronic circuit manufacturing.

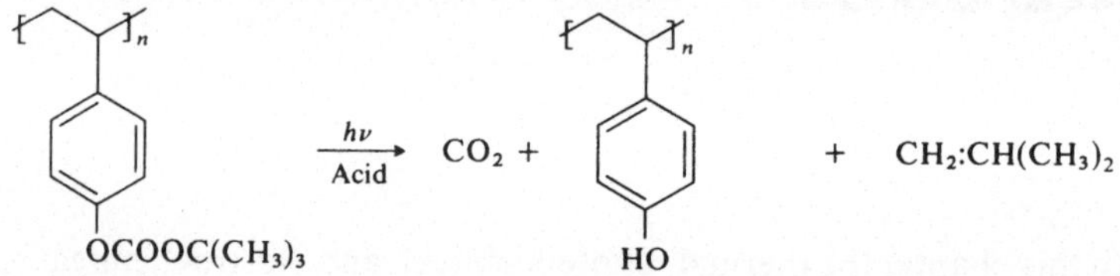

In the imaged areas of the light-sensitive film, the generation of the acid from the decomposition of the onium salt catalyzes the cleavage of the ester, which upon heating at 100°C for a few seconds decomposes into isobutene, carbon dioxide, and poly(4-hydroxystyrene). Ito and Wilson[2] also described the acid-catalyzed depolymerization of poly(phthalaldehyde) (**2-4**) as the basis for a positive-acting dry-developed photoresist system for microlithography.

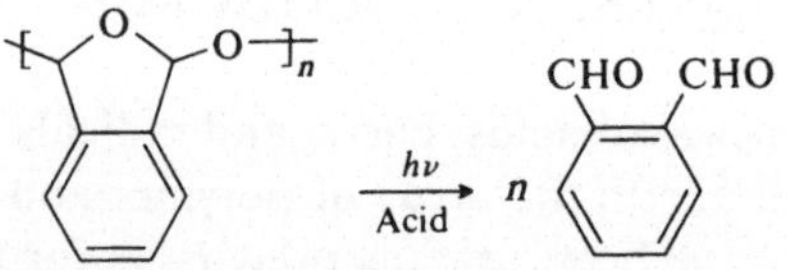

Diazonium salts have been reported by Winslow and Gatzke[3] as sources for the generation of acids in a photothermographic imaging system. Upon light exposure, the decomposition of the diazonium salts releases an acid which displaces nitric acid from a metal nitrate salt in an imagewise fashion. The nitric acid formed, on heating, oxidizes a leuco dye to the parent dye, thus forming a dye image.

The use of diazonium salts as coupling agents for the formation of dye images (**2-6**) is a well-known diazo process.[4] In such a process the coupling reaction between the diazonium salt (2-5) and the phenolic couplers is base catalyzed. Recently, derivatives of urea[5] and guanidine[6] have been used as precursors which thermally trigger the release of ammonia, the base catalyst for the coupling reaction.

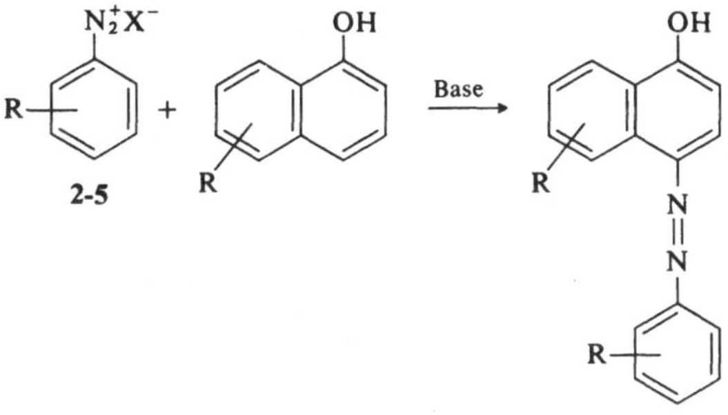

The Wolff rearrangement of diazoquinones (2-7) to release carboxylic acids has found a number of applications in positive-acting photoresists for printing plate technology. The presence of a diazoquinone, such as diazonaphthoquinone, with natural polymers, such as gelatin,[6] or synthetic polymers, such as vinyl–maleic acid copolymer and phenol–aldehyde condensate,[7] reduces the solubility of these polymers. Upon light exposure, the azido quinone rearranges with the release of the carboxylic acid functionality, which provides a solubility difference between the imaged and non-imaged areas in an alkaline solution. Neugebauer[7] describes hydroxy-(1,2,1′,2′)-pyridobenzimidazole esters of o-quinone diazide sulfonic acid as suitable sensitizers for photoresists. Once imaged, these can be developed with either organic solvents to produce a negative or with trisodium phosphate solution to give a positive-acting system.

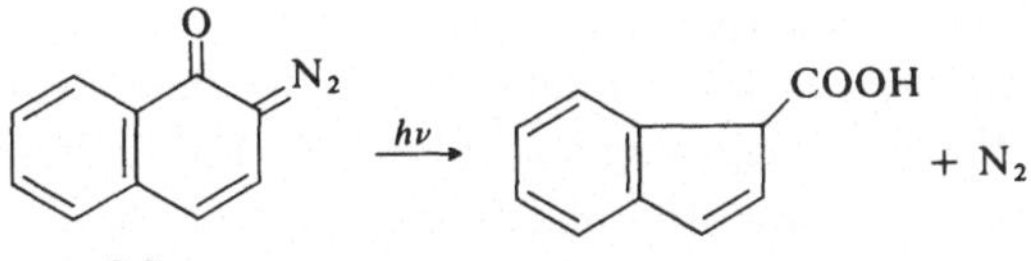

The generation of nitrenes via the decomposition of aryl azides has led to the application of aryl azides as photochemically triggered polymer cross-linkers. Examples include the use of p-azidodiphenylamine carboxylic acid, diazophenylamine carboxylic acid,[8a] and azidostyrylketones and azidostyrylazides[8b] as photo-hardeners with gelatin, glue, gum arabic,[10a] polyacrylamides and their derivatives,[10b] and acrylates.[10c] Coatings using such chemistry were recommended for the preparation of screen printing stencils[10d] and for sensitized offset printing plates.

Azides have also found applications in direct color forming materials discovered by Sagura and van Allan[9a] and Schoen.[9b] Upon exposure to the imaging radiation, the aryl azide (**2-8**) decomposes into the corresponding nitrene, which in the presence of suitable couplers, such as phenolics, undergoes an insertion reaction with the formation of a dye (**2-9**) in an imagewise fashion.

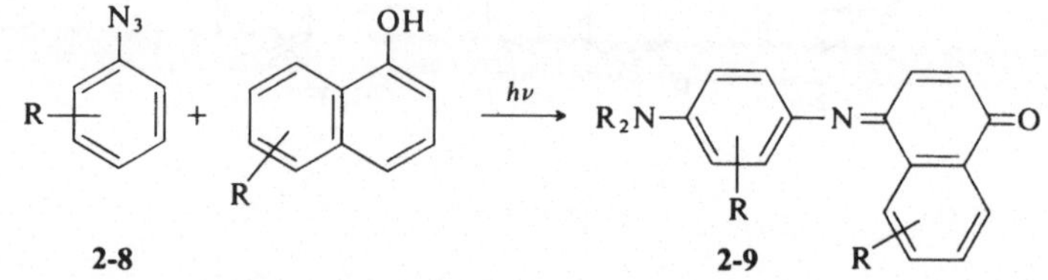

## 2.2. CHEMICAL REACTIONS FOR THE RELEASE OF ACIDS, BASES, RADICALS, NITRENES, AND CARBENES

### 2.2.1. Thermally Triggered Release

#### 2.2.1.1. Carboxylic Acids from Esters and Oxime Derivatives

Routes toward the thermal release of acids have not been as common as the analogous photochemical ones, partly due to the fact that the thermal triggering conditions under which the acids are released are also suitable for their decomposition.

A large number of reactions suitable for thermal release of acids are found in the early German chemical literature. In general, these reactions were published with few examples but a survey of them may suggest directions for further research.

2-Hydroxymethylene-1[$H$]-indanone (**2-10**) and its derivatives are unstable under thermal triggering conditions. At 130°C these undergo self-condensation, involving the loss of one molecule of formic acid.[11a]

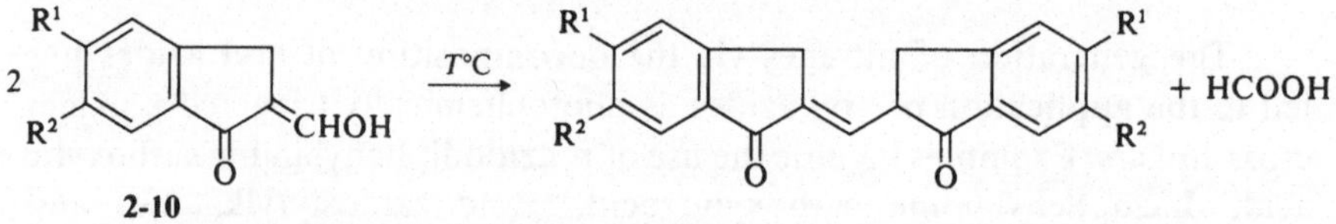

$$R^1 = R^2 = H, \ T = 130°C; \qquad R^1R^2 = -OCH_2-, \ T = 160–170°C$$

Another elegant route toward the release of carboxylic acids is by the thermal fragmentation of acyl derivatives of aldoximes (**2-11**). The reaction was first reported by Neber *et al.*,[11b] using *O*-benzoyl-4-nitrobenzaldoxime heated at 140°C for 30 minutes to release benzoic acid and the corresponding 4-nitrobenzonitrile (or benzoisoxazole depending on the nature of $R^6$) in quantitative yields. Similar reactions were also reported by Lindermann *et al.*[12] and Von Auwers and Krese.[13]

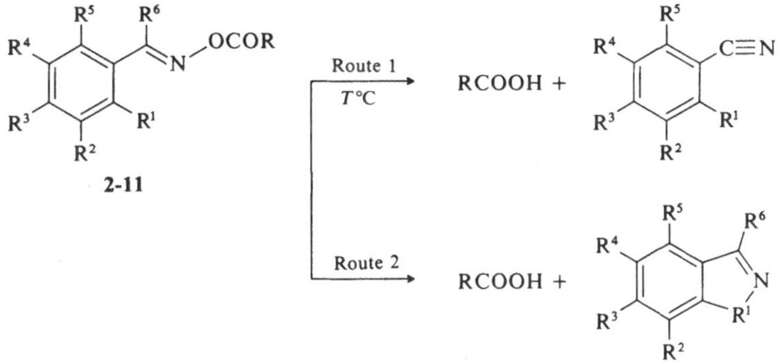

| R | $R^1$ | $R^2$ | $R^3$ | $R^4$ | $R^5$ | $R^6$ | $T(°C)$ | Reference | Route |
|---|---|---|---|---|---|---|---|---|---|
| $OCH_2CH_3$ | $NHCO_2CH_3$ | H | H | H | H | H | 120–130 | 13 | 1 |
| $CH_3$ | OH | Br | H | Br | H | H | 146 | 12a | 1 |
| $CH_3$ | OH | H | H | H | H | H | 130 | 12a | 1 |
| $CH_3$ | OH | H | H | $-C_4H_4-$ | | H | 120–125 | 12b | 1 |
| $CH_3$ | OH | H | H | OH | H | $CH_3$ | 160–165 | 12b | 2 |
| $C_6H_5$ | H | H | $NO_2$ | H | H | H | 140 | 11 | 1 |
| $CH_3$ | OH | H | OH | H | H | $CH_3$ | 140–145 | 12b | 2 |

Ambrose and Brady[14] in the 1950s extended the scope of the reaction and examined its kinetics and the effect of substituents on the release rate. Pyrolysis of acyl derivatives of aldoximes (**2-12**) to release acids and nitriles showed first-order kinetics as monitored by titration of the released acid, and the rate constants (Table 2.1) depended on the substituents on the aromatic ring. Such acyl oximes were prepared by literature methods of condensation followed by acylation.[15,16]

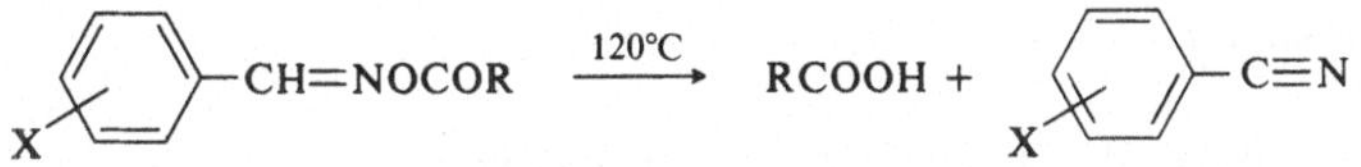

Reactivity vs. substituents

X = 2-$CH_3O$ > 4-$(CH_3)_2N$ > 4-$CH_3O$ > 3,4-$CH_2O_2$ > H > 3-$CH_3O$ > 2-$NO_2$ > 3-$NO_2$ ≃ 4-$NO_2$

R = $CCl_3$ ≫ $(C_2H_5)_2N$ > $C_2H_5O$ > 4-$NO_2C_6H_4$ > $C_6H_5$ > 4-$CH_3C_6H_4$ ≃ $ClCH_2$ > $CH_3$

*Table 2.1.* Rate of Acid Release from Acyl Oximes **2-12**

| R | X | Solid | Solution | | | | |
|---|---|---|---|---|---|---|---|
|  |  |  | 139°C | 130°C | 120°C | 118°C | 115°C |
| CH$_3$ | 3-NO$_2$ | — | 8.4 | — | — | — | 0.86 |
|  | 4-NO$_2$ | — | 9.2 | — | — | — | — |
|  | 2-NO$_2$ | — | 12.1 | — | — | — | — |
|  | 3,4-CH$_2$O$_2$ | — | 19.2 | — | — | — | 1.57 |
|  | 4-CH$_3$O | — | 26.5 | — | — | — | 1.95 |
|  | 4-(CH$_3$)$_2$N | — | 69.0 | 4.4 | — | — | 2.93 |
| C$_6$H$_5$ | 3-NO$_2$ | — | 31.0 | 8.8 | 3.4 | — | — |
|  | 4-NO$_2$ | — | — | 8.8 | — | — | — |
|  | 2-NO$_2$ | — | — | 13.0 | — | — | — |
|  | 3-CH$_3$O | — | — | — | 9.2 | — | 5.4 |
|  | H | — | 100 | — | 11.5 | — | 5.6 |
|  | 4-CH$_3$O | — | —[a] | — | 16.1 | — | 11.1 |
|  | 2-CH$_3$O | — | —[a] | — | 30.7 | — | 17.6 |
| CH$_2$Cl | 3-NO$_2$ | — | 27.0 | — | — | — | — |
|  | 4-NO$_2$ | — | 32.5 | — | — | — | — |
|  | 2-NO$_2$ | — | 32.5 | — | — | — | — |
|  | 3,4-CH$_2$O$_2$ | — | 92.0 | — | — | — | — |
| C$_2$H$_5$O | 3-NO$_2$ | — | 16.1 | — | — | — | — |
|  | 4-NO$_2$ | 17.7 | 17.0 | — | — | — | — |
|  | 2-NO$_2$ | 45.9 | 27.1 | — | — | — | — |
|  | 3-CH$_3$O | 50.9 | 46.2 | — | — | — | — |
|  | H | 51.0 | 50.6 | — | — | — | — |
|  | 4-CH$_3$O | 73.6 | 61.1 | — | — | — | — |
|  | 4-(CH$_3$)$_2$N | 156 | 112 | — | — | — | — |
|  | 2-CH$_3$O | 336 | 183 | — | — | — | — |
| (C$_6$H$_5$)$_2$N | 4-NO$_2$ | — | — | — | — | 36.7 | — |
|  | H | — | — | — | — | 100 | — |
| 4-NO$_2$C$_6$H$_4$ | 4-CH$_3$O | — | 22.4 | — | — | — | — |
| C$_6$H$_5$CH$_2$ | 4-CH$_3$O | — | —[a] | — | — | — | — |
|  | 3-NO$_2$ | — | 31.0 | 3.4 | — | — | — |
| 4-CH$_3$OC$_6$H$_4$ | 4-CH$_3$ | — | — | 12.4 | — | — | — |

[a] Too fast to measure.

Esters have also been found to be useful precursors for the thermal release of acids. Karabatsos and Krumel[17] examined the liquid-phase decomposition of oxalate esters. Dialkyl oxalates (**2-13**) of tertiary alcohols containing $\beta$-hydrogens thermally decomposed to release olefins and oxalic acid in 80–100% yields.

$$(R^1R^2CHCR^3R^4OCO)_2 \rightarrow HOOCCOOH + 2R^1R^2C{:}CR^3R^4$$

**2-13**

The decomposition occurred at relatively low temperatures—between 140 and 170°C—and the reaction, once started, was vigorously catalyzed by oxalic acid and proceeded to completion in about five minutes. These dialkyl oxalates were prepared as described by Karabatsos et al.[18] and were mostly oils or low-melting solids. The dialkyl oxalates of primary and secondary alcohols resisted decomposition even at temperatures of 325°C. Dicyclohexyl, dibornyl, and diisobornyl oxalates as well as oxalates of some tertiary alcohols decomposed at higher temperatures of 200–300°C, which resulted in lower yields of the oxalic acid. Table 2.2 contains examples reported in the literature by Karabatsos et al.[18]

A number of publications by Bailey et al.[19a,b] described the thermally triggered decomposition of ethyl esters (2-14) to release the parent acid and

*Table 2.2.* Thermal Release of Oxalic Acid from Oxalate Esters

| Oxalate ester | Release temperature (°C) | Percent yield (oxalic acid) |
|---|---|---|
| $[(CH_3)_3C\sim\!\!\langle\;\rangle\!\sim\!OCO]_2$ | | |
| trans | 140–145 | 80 |
| cis | 175–180 | 60 |
| $[C_6H_5CH(CH_3)OCO]_2$ | 150–170 | 67 |
| Di-t-butyl | 147–150 | 82 |
| Di-t-pentyl | 141–142 | 90 |
| Di-(1,1-dimethylbutyl) | 142–148 | 84 |
| Di-(1,1-diethylpropyl) | 142–148 | 90 |
| Di-(1-ethyl-1-methylpropyl) | 142 | 87 |
| Di-(1-ethyl-1-methylbutyl) | 147–150 | 100 |
| Di-(1-ethyl-1,2-dimethylpropyl) | 140–144 | 73 |
| Di-(1-cyclopentyl-1-methylethyl) | 140–145 | 89 |
| Di-(1-cyclohexyl-1-methylethyl) | 135–140 | 87 |
| Di-(trans-1,2-dimethylcyclohexyl) | 162–164 | 62 |
| Di-(1-methylcyclohexyl) | 159–162 | 86 |
| Di-(1-ethylcyclohexyl) | 156–165 | 73 |
| Di-(1-isopropylcyclohexyl) | 250–280 | 50 |
| Di-(1-methylcyclopentyl) | 134–138 | 91 |
| Di-(1-ethylcyclopentyl) | 173–190 | 91 |
| Di-(1-isopropylcyclopentyl) | 165–185 | 100 |
| Di-(1-methylcyclobutyl) | 220–250 | 39 |
| Di-(1-ethylcyclobutyl) | 197–205 | 57 |
| Dicyclohexyl | 245–250 | 81 |
| Di-1-bornyl | 300–301 | 88 |
| Di-1-isobornyl | 250–280 | 54 |

ethylene. The yields were acceptable for a pyrolysis reaction which proceeds at 470–550°C.

$$RCH_2CO_2CH_2CH_3 \xrightarrow{T°C} RCH_2COOH + CH_2=CH_2$$

**2-14**

Tertiary amyl acetates decompose at 150°C to release the parent acetic acid and the corresponding olefin.[19c]

Symmetrical diacetyl derivatives of alkylenediamines[20] (**2-15**) undergo thermally triggered ring closure to yield the corresponding imine, with the release of acetic acid.

$$(CH_2)_n \begin{matrix} NHCOCH_3 \\ \\ NHCOCH_3 \end{matrix} \xrightarrow[n=2,3,4]{T°C} (CH_2)_n \begin{matrix} N \\ \\ N \end{matrix} CCH_3 + CH_3COOH$$

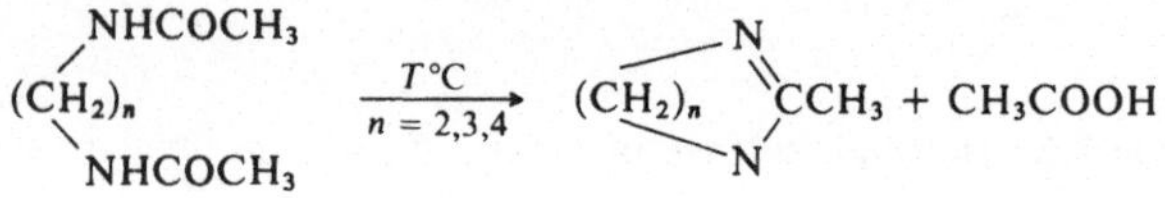

**2-15**

## 2.2.1.2. Nitric Acid Derivatives from Nitroalkyl Derivatives

Ethyl dinitroacetate (**2-16**) undergoes a cyclization reaction under heating and in certain cases at room temperature to yield ethyl furoxandicarboxylate (**2-17**) with the release of nitric acid.[21]

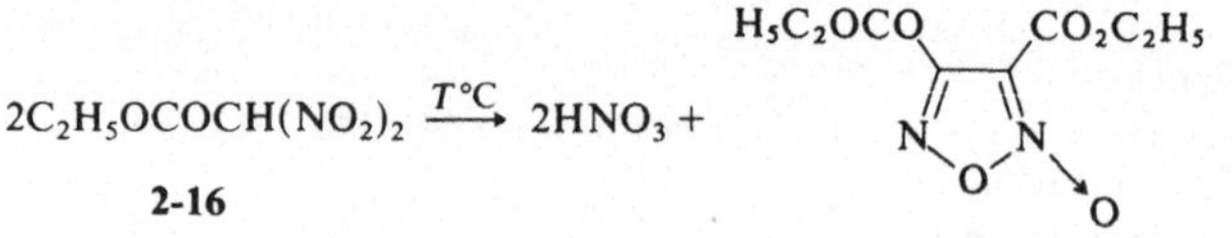

$$2C_2H_5OCOCH(NO_2)_2 \xrightarrow{T°C} 2HNO_3 +$$

**2-16**

H_5C_2OCO, CO_2C_2H_5

**2-17**

1-Cyano-1-(4′-bromophenyl)-3,4-dinitro-4-phenylbut-1-ene (**2-18**) melts at 128°C with the rapid release of nitrous acid[22] and the formation of 1-cyano-1-(4′-bromo)phenyl-3-nitro-4-phenylbuta-1,3-diene (**2-19**).

$$4\text{-}BrC_6H_4C(CN)=CHCH(NO_2)CH(NO_2)C_6H_5$$

**2-18**

$$\xrightarrow{T°C} HNO_2 + 4\text{-}BrC_6H_4C(CN)=CHC(NO_2)=CHC_6H_5$$

**2-19**

Ethyl nitrolic acid (**2-20**) on heating at 86°C undergoes a triggered decomposition with the release of acetic acid, nitrogen, and nitrogen dioxide[23] gas. This reaction may be potentially explosive and should be treated with care.

$$3CH_3C(NO_2)=NOH \xrightarrow{[O]} 3CH_3COOH + 2NO_2 + 2N_2$$

**2-20**

## 2.2.1.3. Ammonia and Amines from Amide Derivatives

Several decades ago Holley and Holley[24] introduced the o-amino-phenoxyacetyl moiety (**2-21**) as an amino protecting group for peptide synthesis. This type of protecting group can be removed thermally via an intramolecular cyclization reaction with the formation of a lactam ring (**2-22**) and the release of the amine at temperatures of about 100°C.

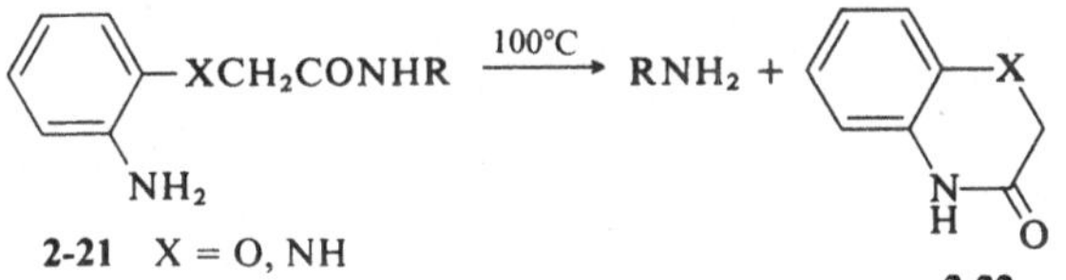

**2-21**  X = O, NH

**2-22**

| Released amine | X | Percent yield (2-22) |
|---|---|---|
| Glycylglycine | O (N) | 73 (76) |
| Glycylglycylglycine | O (N) | 76 (68) |
| glycyl-L-alanyl-L-leucine | O (N) | 65 (59) |
| L-Phenylalanyl-L-leucine | O (N) | 70 (31) |

Amides of diacids in most cases release ammonia under thermal triggering by an intramolecular cyclization reaction to form the corresponding imide. For example, methyl diglyconamide (**2-23**) cyclizes to the corresponding ether imide at about 150°C with the release of ammonia.[25] In the same way, glutarimide and succinimide are formed from the corresponding diamides by heating, with the release of ammonia.[26] Carbamide and diaminomalonamide (**2-24**) similarly release ammonia at 90–100°C.[27]

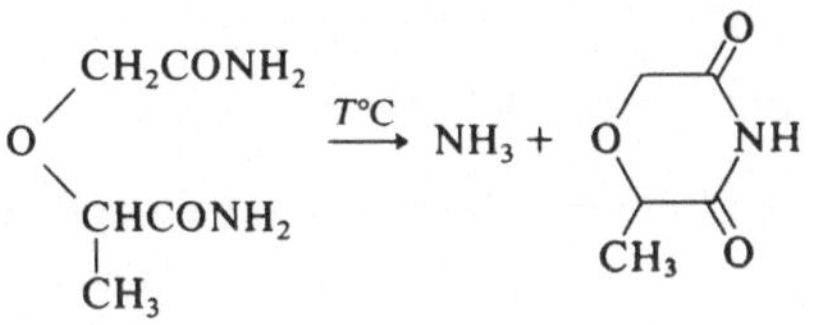

**2-23**

$$(NH_2)_2C(CONH_2)_2 \rightarrow NH_3 + NH=C(CONH_2)_2$$

**2-24**

Reilly and Brown[39] have described tris(perfluoroalkyl)-s-triazines (2-25b) as by-products accompanying the thermal release of ammonia from perfluoroalkylamidines (2-25a). Perfluoroadipodiamidine (2-25c) and perfluoroglutaramidine (2-25d) when heated above their melting points released ammonia rapidly and became viscous and eventually set hard into an infusible amber resin, as the result of a condensation cross-linking reaction. The perfluoroalkylamidines were prepared from the condensation reaction of perfluoroalkylnitrile or dinitrile and ammonia.[40a,40b]

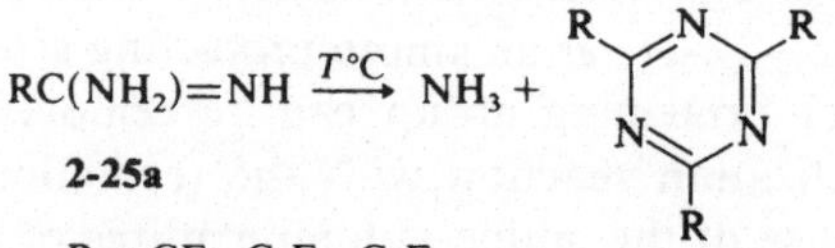

$$RC(NH_2)=NH \xrightarrow{T\,°C} NH_3 +$$

2-25a

$$R = CF_3, C_2F_5, C_3F_7$$

2-25b

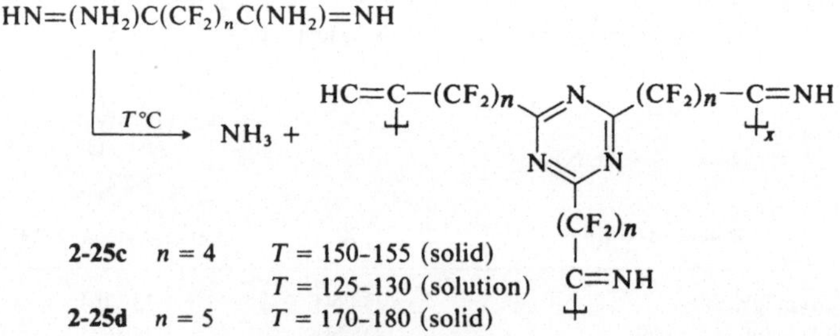

$$HN=(NH_2)C(CF_2)_nC(NH_2)=NH$$

$$\xrightarrow{T\,°C} NH_3 +$$

2-25c   $n = 4$   $T = 150-155$ (solid)
                  $T = 125-130$ (solution)
2-25d   $n = 5$   $T = 170-180$ (solid)
                  $T = 160-165$ (solution)

In a similar reaction, 1,3-dimethyl-5-hydroxyhydantoyl amide (2-26) thermally decomposes with formation of cholesterophane and the release of ammonia and carbon monoxide.[28]

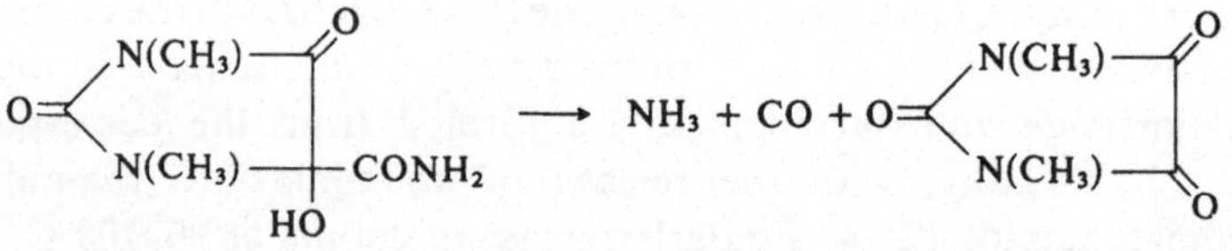

$$\longrightarrow NH_3 + CO +$$

2-26

## 2.2.1.4. Ammonia from Urea Derivatives

Ammonia has been known to be released via the thermal decomposition of urea and its derivatives. Unsubstituted ureas dissociate into cyanic acid and ammonia above 160°C, as reported by Davies and Underwood[29] and Werner.[30]

Alkyl-substituted ureas, such as ethylurea, yield ammonia and the corresponding alkyl cyanurate.[30] Aryl-substituted ureas, such as phenylurea, when heated at about 150°C yield ammonia and aryl isocyanates, as described by Hofmann.[31] Thiourea derivatives follow the same path[32] and yield phenyl isothiocyanates and ammonia at 160°C.

Arylureas with a nucleophilic group at the ortho position of the aryl ring, such as hydroxyl or an amino group, can undergo thermally triggered intramolecualr cyclization with the release of ammonia. For example, 2-aminophenylurea[33] (**2-27**) as well as 2-hydroxy-4-nitrophenylurea[34] (**2-28**) cyclize at 150°C with the release of ammonia.

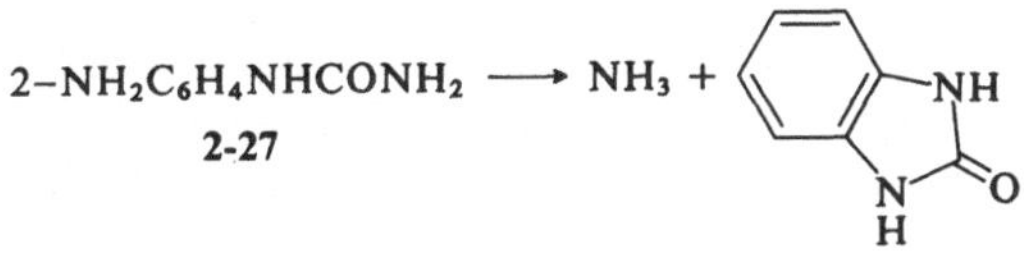

$$2-NH_2C_6H_4NHCONH_2 \longrightarrow NH_3 +$$

**2-27**

$$O_2N \!-\!\!\langle \;\; \rangle \!-\! NHCONH_2 \longrightarrow NH_3 + O_2N$$

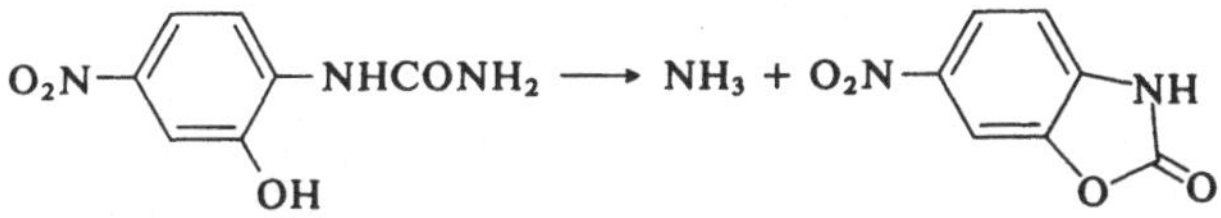

**2-28**

Alkyl-substituted ureas with a nucleophilic group on the alkyl carbon chain whose configuration allows cyclization into a stable ring structure, for example, β-hydroxyethylurea[35] (**2-29**), also exhibit thermally triggered expulsion of ammonia.

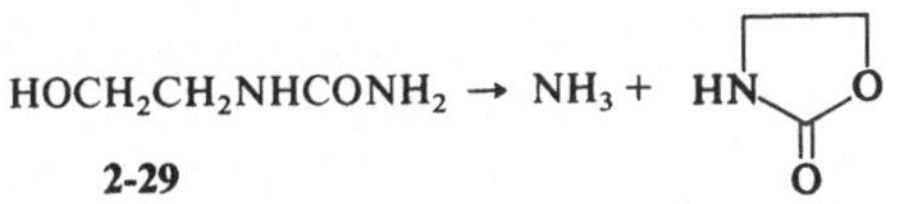

$$HOCH_2CH_2NHCONH_2 \rightarrow NH_3 + HN \qquad O$$

**2-29**

Intermolecular cyclization is also observed in some urea derivatives, accompanying the thermally triggered release of ammonia. For example, aminoguanidine bicarbonate (**2-30a**) upon heating decomposes spontaneously yielding the free base which in turn quickly releases ammonia with the formation of 3,6-diamino-1,2,4,5-tetrazine (**2-30b**) as described by Busch and Ulmer.[36]

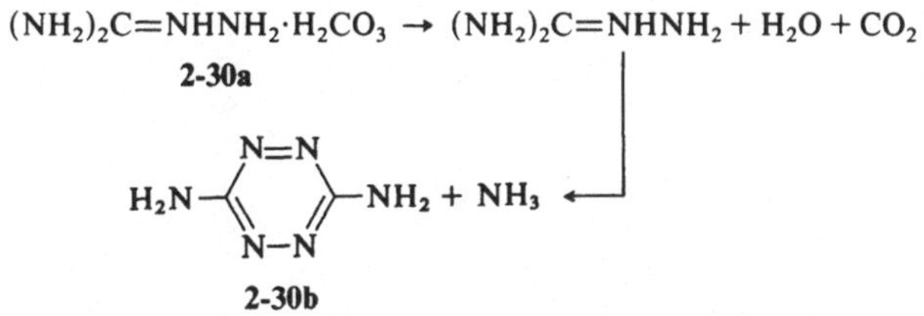

$$(NH_2)_2C\!=\!NHNH_2 \cdot H_2CO_3 \rightarrow (NH_2)_2C\!=\!NHNH_2 + H_2O + CO_2$$

**2-30a**

$$H_2N \!-\!\!\langle \overset{N=N}{\underset{N-N}{}} \rangle \!-\! NH_2 + NH_3 \longleftarrow$$

**2-30b**

Biuret is also unstable to heat and undergoes two types of reactions: one is a deammonation[30] and the other a dehydration reaction.[37] The former reaction involves the release of the corresponding cyanic acid and ammonia. The formation of biuret can be via the thermal decomposition of urea hydrochloride.[38]

## 2.2.1.5. Radicals from Peroxide and Azo Derivatives

A major number of compounds have been examined as sources for the thermal release of radicals. The most useful are peroxides, azo compounds, tri- and tetrahaloalkanes.

Peroxides are known to decompose by homolytic fission to release radicals. They are also known to detonate, thus making them hazardous to handle. The most commonly known peroxides are alkyl peroxides (2-31), acyl peroxides (2-32), hydroperoxides (2-33), peracids (2-34), and peresters (2-35) including percarbonates (2-36) and peroxalates (2-37).

| ROOR | RCOOOCOR | ROOH | RCOOOH | RCOOOR' | ROOCOOR' | ROOCOCOOOR' |
|------|----------|------|--------|---------|----------|-------------|
| 2-31 | 2-32 | 2-33 | 2-34 | 2-35 | 2-36 | 2-37 |

$$(CH_3)_3COOCC(CH_3)_3 \rightarrow 2(CH_3)_3CO^{\cdot}$$

**2-31a**

The most intensively studied alkyl peroxide is di-$t$-butyl peroxide (2-31a), which is commonly used as an initiator in radical reactions. This

*Table 2.3.* Radical Release by $t$-Butyl Peroxide
in Various Solvents[a,b]

| Solvent | Percent yield (radical release) |
|---------|----------------------------------|
| Decalin | 42 |
| Dimethylamine | 47 |
| Tri-$n$-butylamine | 48 |
| Anisole | 48 |
| $t$-Butylbenzene | 49 |
| Ethylbenzene | 54 |
| 4-Chlorotoluene | 59 |
| Benzyl chloride | 60 |
| Benzyl acetate | 60 |
| Benzylamine | 62 |
| Benzaldehyde | 68 |
| Chlorobenzene | 69 |
| Carbon tetrachloride | 71 |
| Tetrachloroethylene | 83 |
| Benzyl alcohol | 85 |
| $\alpha$-Methylbenzyl alcohol | 88 |

[a] At 110°C for 72 hours.
[b] Ref. 41.

follows from its relatively safe properties as compared to other more hazardous homologues. The rate of its thermally triggered decomposition and the release of radicals is a function of reaction medium, as shown in Table 2.3, and temperature.

Alkyl peroxides undergo decomposition with the release of radicals at similar rates, independent of the alkyl chain.[42] The activation energies required for a series of ethyl through $t$-butyl peroxides are between 34 and 37 kcal/mol.

Acyl peroxides are another class frequently studied. The acetyl (**2-32a**) and benzoyl (**2-32b**) are the most reported derivatives. The decomposition mechanism of benzoyl peroxide has been reported[43] to be very complex. Acetyl peroxides undergo homolytic cleavage to form acetoxy radicals, followed by the loss of carbon dioxide and the release of the corresponding alkyl radicals.[43] The rate of decomposition of acetyl peroxides shows a dependence on the nature of the reaction solvent: for example, the rates are higher in alcohols than in hydrocarbons.[44,45] The most commonly used is the benzoyl derivative because of its stability[46]—about 30 minutes at 100°C. Benzoyl peroxide decomposes in a similar manner to its acetyl analogue.

$$RCOOOCOR \rightarrow 2RCOO^. \rightarrow 2CO_2 + 2R^.$$

**2-32**

**2-32a**   $R = CH_3$

**2-32b**   $R = C_6H_5$

The hydrocarbon substituents of acyl peroxides (**2-32**) influence the rate of the release of the radicals as shown in Table 2.4; for example, electron-donating groups enhance the dissociation rates.[47]

Table 2.4. Relative Rate Constants for Radical
Release from Acyl Peroxides (**2-32**)

| R | Relative rate constants | | Reference |
|---|---|---|---|
| | 60°C | 70°C | |
| Methyl | 1.0 | 5.0 | 48, 51 |
| Ethyl | 1.0 | — | 48 |
| $n$-Propyl | 1.0 | — | 48 |
| $t$-Butyl | 100.0 | — | 48 |
| Phenyl | — | 1.0 | 49, 46 |
| Benzyl | — | very fast | 50 |
| Cyclopropyl | — | 0.4 | 49 |
| Cyclopentyl | — | 32 | 49 |
| Cyclohexyl | — | 60 | 49 |

Table 2.5. Half-Lives of t-Butyl
Peresters (RCOOOBu$^t$)

| R | Half-life at 60°C (min)$^a$ |
|---|---|
| CH$_3$ | $5 \times 10^5$ |
| C$_6$H$_5$ | $3 \times 10^5$ |
| C$_6$H$_5$CH$_2$ | $1.7 \times 10^3$ |
| (C$_6$H$_5$)$_2$CH | 26 |
| C$_6$H$_5$(CH$_3$)$_2$C | 12 |
| CH$_3$(C$_6$H$_5$)$_2$C | 6 |
| Cl$_3$C | 970 |
| 4-NO$_2$C$_6$H$_4$CH$_2$ | 4700 |
| 4-CH$_3$C$_6$H$_4$CH$_2$ | 880 |

$^a$ Ref. 52.

Peracids (**2-34**) and peresters (**2-35**) also release radicals. Peracids are formed from the acid chlorides or anhydrides by the action of hydroperoxides. The half-lives of peracids and peresters are influenced by the nature of the chain substituents, as shown in Table 2.5 for the case of t-butyl peresters.

The half-life kinetic studies revealed that thermal decomposition and the release of the radicals is faster with those groups which tend to stabilize the released radical; for example, the rate is higher for benzyl than for methyl.

t-Butyl[53] and cumyl[54] hydroperoxides are reported to release radicals either in the pure form or in inert solvents such as chlorobenzene. The latter hydroperoxide can be heated to 140°C to yield hydroxy radicals.

Azo compounds are also a source for radicals. They have been known since 1896 when Thiele and Heuser[55] reported azobisisobutyronitrile. It was not until 1949 that it was recognized that its thermal reaction can induce polymerization.

Table 2.6. Rates of Decomposition of Some Azo Compounds

| Azo compound | Solvent | $T$(°C) | $k \times 10^4$ (s$^{-1}$) |
|---|---|---|---|
| CH$_3$N=NCH$_3$ | Gas | 300 | 5.6 |
| (CH$_3$)$_3$CN=NC(CH$_3$)$_3$ | Gas | 250 | 4.8 |
| C$_6$H$_5$CH(CH$_3$)N=NCH(CH$_3$)C$_6$H$_5$ | Toluene | 110 | 1.69 |
| (CH$_3$)$_2$C(CN)N=NC(CN)(CH$_3$)$_2$ | Toluene | 80 | 1.55 |
| (C$_6$H$_5$)$_2$CHN=NCH(C$_6$H$_5$)$_2$ | Toluene | 64 | 3.4 |
| [(CH$_3$)$_2$C(COOC$_2$H$_5$)N=]$_2$ | Nitrobenzene | 80 | 1.57 |
| C$_6$H$_5$N=NC$_6$H$_5$ | Decahydronaphthalene | 144 | 2.69 |
| C$_6$H$_5$N=NC(C$_6$H$_5$)$_3$ | Toluene | 53 | 2.25 |

*Table 2.7.* Effect of Substituents
on Reactivity of Azo Compounds

| X | $(CH_3)_3CN=NC(X)(CH_3)_2$<br>$T(°C)$ for $t_{1/2} = 10$ h |
|---|---|
| $CH_3$ | 160 |
| RO | 143–154 |
| RS | 111–128 |
| $H_2NCO$ | 110 |
| $C_6H_5$ | 84 |
| CN | 79 |

Azo alkanes (**2-38**) constitute a useful group of radical initiators. These compounds are sensitive to the nature of the carbon chain substituents. For example, azomethane decomposes with an activation energy of about 51 kcal/mol and undergoes thermal homolysis at useful rates only at temperatures of about 400°C, while phenylazotriphenylmethane decomposes with an activation energy of about 27 kcal/mol and is useful at temperatures of 45–80°C. The decomposition rates of some azo compounds are presented in Table 2.6, while the effects of substituents on reactivity are shown in Tables 2.7–2.9.

$$R-N=N-R^1 \xrightarrow{\Delta} R^· + R^{1·} + N_2$$

**2-38**

| R | $R^1$ | $E_a$ (kcal/mol) | Reference |
|---|---|---|---|
| $CH_3$ | $CH_3$ | 51 | 56 |
| $(CH_3)_3C$ | $(CH_3)_3C$ | 43 | 57 |
| $(C_6H_5)_2CH$ | $(C_6H_5)_2CH$ | 27 | 58 |
| $C_6H_5$ | $(C_6H_5)_3C$ | 27 | 59 |
| $CH_3(CN)_2C$ | $CH_3(CN)_2C$ | 31 | 60 |

*Table 2.8.* Effect of Substituent Size on
Reactivity of Azo Compounds

| R | $(CH_3)_3CN=NCR(CN)(CH_3)$<br>$T(°C)$ for $t_{1/2} = 10$ h |
|---|---|
| $CH_3$ | 79 |
| $CH_3CH_2$ | 82 |
| $(CH_3)_2CH$ | 83 |
| $(CH_3)_3C$ | 80 |
| $(CH_3)_2CHCH_2$ | 70 |
| $(CH_3)_3CCH_2$ | 52 |

Table 2.9. Effect of Substituent
Ring Size on Reactivity of
Azo Compounds

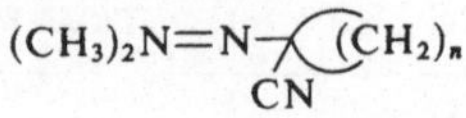

$$(CH_3)_2N=N-\underset{CN}{\overset{}{\diagup}}(CH_2)_n$$

| $n$ | $T(°C)$ for $t_{1/2} = 10$ h |
|-----|------------------------------|
| 5 | 96 |
| 4 | 78 |
| 6 | 68 |
| 7 | 55 |

Azoalkanes are surveyed in Chapters 1 and 4 as sources for the release of nitrogen gas and hydrocarbons, respectively, and will not be considered further in this section.

The thermally triggered decomposition of methylbenzenediazo sulfone (2-39) and the release of radicals has been examined by Kice and Gabrielsen.[63] The principal products are rationalized as the result of homolytic fission. The rate of decomposition increases with time due to an autocatalytic effect, and thus is significantly affected by the solvent, and is directly proportional to the triggering temperature. The release of radicals in solution is accompanied by a color change from yellow-orange to deep red.

$$C_6H_5N=NSO_2R \rightarrow C_6H_5^{·} + R^{·} + SO_2 + N_2$$

2-39

| R | Release temperature (°C) | $k_0 \times 10^9$ (Ms)$^{-1}$ at release temperature $^a$ |
|---|--------------------------|-----------------------------------------------------------|
| $CH_3$ | 80 | 19 |
| | 70 | 7.3 |
| | 59.9 | 1.9 |
| | 54.8 | 0.89 |
| $C_6H_5CH_2$ | 80 | 24 |
| | 70 | 12 |
| | 65 | 12 |
| | 60 | 6.2 |
| | 54.8 | 1.2 |

$^a$ $k_0$ = zero-order rate constant for the decomposition via the disappearance of $\alpha,\gamma$-bis(biphenylene)-$\beta$-phenylallyl used as a radical scavenger and monitor.

### 2.2.1.6. Nitrenes and Carbenes from Azide and Diazo Derivatives

Nitrenes were first observed by Tiemann[84] in 1891 as short-lived intermediates in the Lossen rearrangement. Later Stieglitz[85] proposed nitrenes as intermediates in the Curtius rearrangement. Their formation is most pronounced in the thermal decomposition of alkyl and aryl azides (**2-40**). These and related reactions, such as the rearrangement of tritylhydroxyl-amines[86] (**2-41**) and $N$-haloimines[86] (**2-42**), are sources for the thermal release of nitrenes and are usually termed Stieglitz rearrangements.[86b]

$$(C_6H_5)_3CN_3 \longrightarrow (C_6H_5)_3CN: + N_2$$

**2-40**

$$(C_6H_5)_3CNHOH \xrightarrow[H^+]{} (C_6H_5)_3CN: + H_2O$$

**2-41**

$$(C_6H_5)_3CNHX \longrightarrow (C_6H_5)_3CN: + HX$$

**2-42**

$$(C_6)H_5)_2C=NC_6H_5$$

Alkyl azides (**2-40**) provide a valuable source of nitrenes. These azides are generally stable at room temperature but above 100°C decompose with the release of nitrenes and nitrogen gas.[87,88]

Substituents have a pronounced effect on the direction of the reaction and the decomposition products. Saunders and Ware[89] examined the thermal decomposition of a range of triaryl ethyl azides (**2-43**) and correlated the effects of the substituents with the yields of the released nitrenes and their relative rates of formation.

$$4\text{-}RC_6H_4C(C_2H_5)_2N_3 \xrightarrow{\Delta} 4\text{-}RC_6H_4C(C_2H_5)_2N: + N_2$$

**2-43**

$$4\text{-}RC_6H_4N=C(C_2H_5)_2$$

**2-44**

| R | Percent yield (**2-44**) | $k$ (relative) |
|---|---|---|
| $(CH_3)_2N$ | 70 | 2.50 |
| $CH_3O$ | 70 | 1.53 |
| $CH_3$ | 66 | 1.08 |
| H | 75 | 1.00 |
| Cl | 43 | 1.17 |
| $NO_2$ | 75 | 1.07 |

Aryl nitrenes are generally formed by the loss of nitrogen from aryl azides (**2-40b**). These mostly decompose smoothly in solution at temperatures between 140 and 200°C. 2-Nitrophenyl azide tends to decompose at lower temperatures.[90] The thermal decomposition of aryl azides has been studied neat[91] but due to their explosive nature, their decomposition is not recommended as a facile route to the release of aryl nitrenes.

Carbonyl nitrenes (2-45) are the best-known class of nitrenes and exist in two forms, (2-45a) and (2-45b), of similar properties. Both can be generated from the corresponding acyl azide (2-46). Ethoxycarbonyl nitrene was first reported as an intermediate in the decomposion of ethylazido formate.[92] The nature of the carbon chain and its substituents influences the reaction rates, which are also a function of temperature (Table 2.10).

$$R_3CON: \qquad ROCON:$$
$$\text{2-45a} \qquad\quad \text{2-45b}$$

Sulfonyl nitrenes (2-47) are released by the thermal decomposition of sulfonyl azides (2-48). This was first reported by Curtius and Lorenzen[95] in 1898, and Curtius later[96] defined those sulfonyl azides capable of thermal release of the corresponding nitrenes without rearrangement.

$$ROCON_3$$
$$\text{2-46}$$

$$RSO_2N_3 \;\rightarrow\; RSO_2N: + N_2$$
$$\text{2-48} \qquad\quad \text{2-47}$$

A number of reviews[97] on the chemistry of sulfonyl azides appeared between the 1930s and the 1960s. A well-studied example is tosyl azide. Breslow *et al.*[98] examined a variety of aliphatic sulfonyl azides in diphenyl ether as a solvent and in each case observed a first-order rate, as calculated from the volume of the released nitrogen gas. The results of their kinetic studies are presented in Tables 2.11 and 2.12.

Cyano nitrenes are the youngest members of the nitrene family. The molecule was first reported in 1964 by Herzberg and Travis.[99] In the same

*Table 2.10.* Thermal Decomposition of Alkylazido Formates in Diphenyl Ether[a]

| Alkylazido formate | $T(°C)$ | $k \times 10^4 \; (s^{-1})$ |
|---|---|---|
| n-Octadecyl | 133.3 | 9.0 |
|  | 120.0 | 2.4 |
|  | 100.0 | 0.27 |
| Tetramethylenebis | 133.3 | 9.0 |
|  | 120.0 | 2.34 |
|  | 100.0 | 0.26 |
| n-Propyl | 130.0 | 6.0 |

[a] Refs. 80 and 81.

Table 2.11. First Order Rates for Release of Nitrenes
from Sulfonyl Azide Derivatives[a]

| Substituted sulfonyl azide | $T(°C)$ | $k \times 10^4 \ (s^{-1})$ |
|---|---|---|
| 1-Pentane | 166 | 4.46 |
| 1,4-Butanedi | 163 | 5.02 |
| 1,6-Hexanedi | 163 | 5.02 |
| 1,9-Nonanedi | 160 | 2.25 |
| 1,10-Decanedi | 163 | 4.45 |
| 1,4-Dimethylcyclohexane-$\alpha,\alpha'$-di | 163 | 4.82 |
| 3-Xylene-$\alpha,\alpha'$-di | 163 | 6.09 |
| 4-Xylene-$\alpha,\alpha'$-di | 163 | 5.78 |

[a] Ref. 98.

year cyanogen azide was reported[100] as its precursor in thermal[101] or photochemical[102] reactions. Unlike other organic azides, cyanogen azides required
a triggering temperature in excess of 100°C for the release of the corresponding nitrenes.

The thermal formation of carbenes has been examined extensively.
A carbene can be defined as a divalent carbon intermediate having two
covalent bonds and two nonbonding electrons. Carbenes can be formed by
chemical, photochemical, or thermal reactions. For example, they can be
released from diazo compounds by thermal or photochemical methods.[81,82]

Alkyl carbenes are generally formed from ketenes and undergo rapid
polymerization. Aryl carbenes are thermally released from diazo compounds.[83] Ethoxycarbonyl carbenes are relatively more stable. For example,
the thermal decomposition of ethyl diazoacetate proceeds at about 150°C;
the presence of a catalyst allows the decomposition to proceed at 90–100°C.

Table 2.12. Rates for Release of Sulfonyl Nitrene from
4-Toluenesulfonyl Azide in Various Solvents[a]

| Solvent | $T(°C)$ | $k \times 10^4 \ (s^{-1})$ |
|---|---|---|
| Diphenyl ether | 155 | 3.43 |
| Tetradecane | 155 | 3.80 |
| Nitrobenzene | 155 | 3.97 |
| 1-Octanol | 155 | 3.63 |
| Dimethyl terephthalate | 155 | 3.23 |
| n-Hexanoic acid | 155 | 2.97 |
| Diphenyl ether | 145 | 1.44 |
| 1,4-Dichlorobutane | 145 | 1.20 |

[a] Ref. 98.

Halocarbenes are known but are commonly generated by the action of reagents such as strong bases.

The reactivity of the released carbenes is dependent on the nature of the substituents and not on the mode of release, which requires cleavage of two bonds to a carbon atom in the precursor.

The thermal elimination of nitrogen gas from diazo compounds provides a major route to the release of carbenes.[64]

$$R^1R^2CN_2 \rightarrow R^1R^2C: + N_2$$

Diazo compounds can be prepared by a number of methods, most conveniently *in situ.* The most used method is the Bamford–Stevens reaction.[66,67]

Cycloeliminations[68-70] triggered thermally or photochemically also provide useful routes to carbenes. For example, the triggered fragmentation of three-membered rings (**2-49a**–**2-49c**) releases carbenes. Cyclopropanes generally do not give high yields of carbenes but yields are better with norcaradienes (**2-49d**) and (**2-49e**). Larger rings such as norbornadienes (**2-50a**) eliminate dialkoxy- and dichlorocarbenes under relatively mild conditions, as do 1,3-dioxalanes (**2-50b**) and cyclic carbonates (**2-50c**).

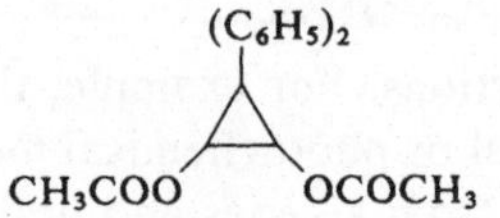

2-49a 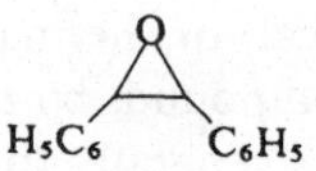          2-49b 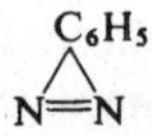          2-49c

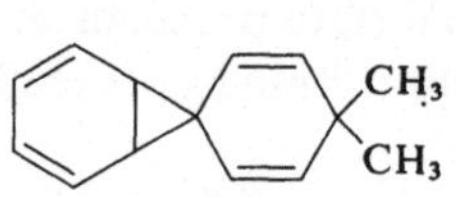

2-49d 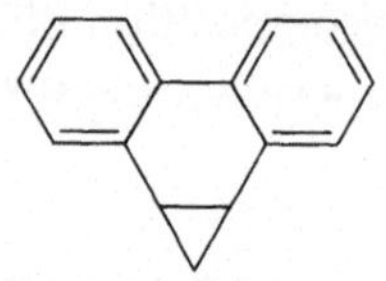                    2-49e

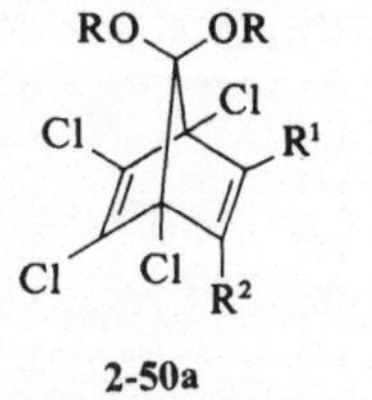

2-50a 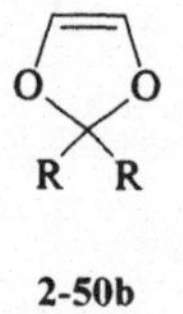          2-50b 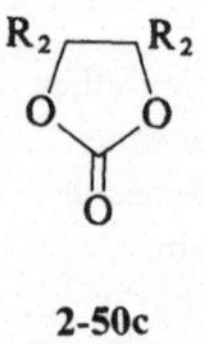          2-50c

Hoffmann and Hauser[72a] reported a number of norbornadiene derivatives (**2-50a**) which under mild conditions of about 100–150°C release the corresponding carbenes.

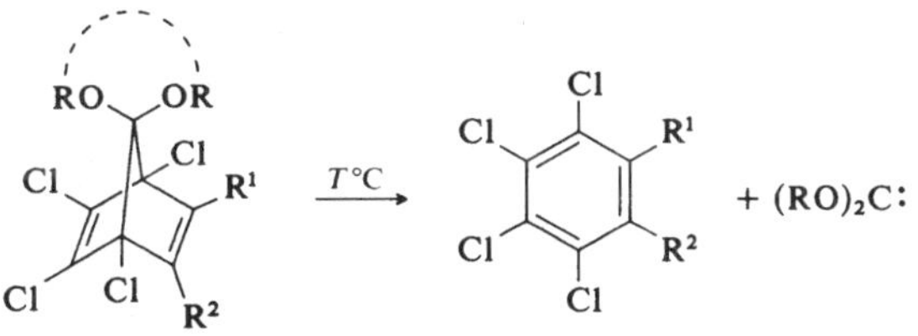

**2-50d**

| R | R$^1$ | R$^2$ | $T$(°C) | Percent yield | Reference |
|---|---|---|---|---|---|
| CH$_3$ | H | H | 85 | 38 | 72c |
| CH$_3$ | H | Br | 130 | 27 | 72a |
| CH$_3$ | H | CH$_3$ | 160 | 84 | 72b |
| CH$_3$ | CH$_2$OCOCH$_3$ | CH$_2$OCOCH$_3$ | 140 | 60 | 72d, 71 |
| CH$_3$ | CH$_2$OCOCH$_3$ | CH(CH$_3$)-OCOCH$_3$ | 140 | 70 | 72d |
| C$_6$H$_5$CH$_2$ | H | C$_6$H$_5$ | 160–195 | — | 72b |
| CH$_2$CH$_2$ | H | H | 120 | 94 | 72c |
| CH$_2$CH$_2$ | H | CH$_3$ | 160 | 100 | 72b |
| CH$_2$CH$_2$ | H | C$_6$H$_5$ | 130–170 | 95 | 72b |
| CH(CH$_3$)-CH(CH$_3$) | H | H | 105 | 72 | 72c |
| CH(CH$_3$)-CH(CH$_3$) | H | C$_6$H$_5$ | 140–150 | 80 | 72c |

Norcaradienes (**2-51b**), formed from 1,7-dimethylaminocyclohep-tatriene (**2-51a**), release[73] carbenes at 110°C. Similarly, heating 1,6-methano[10]annulene (**2-51c**) at temperatures between 60 and 250°C releases the corresponding carbene.[74]

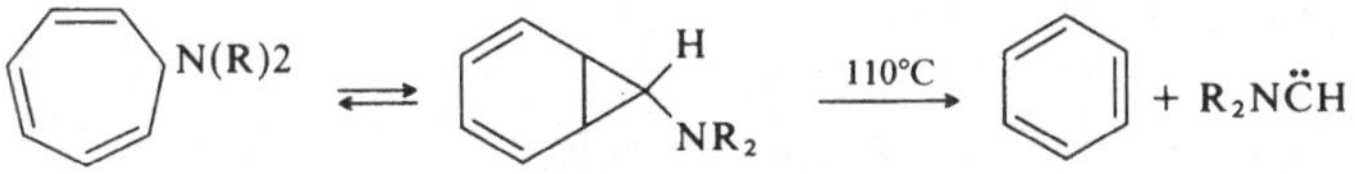

**2-51a**          **2-51b**

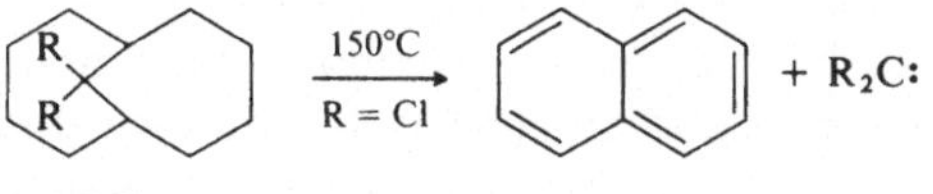

**2-51c**

The release of carbenes from cyclopropanes is usually observed with the fluorinated derivatives[75-80] such as **2-52a** at 160–270°C.

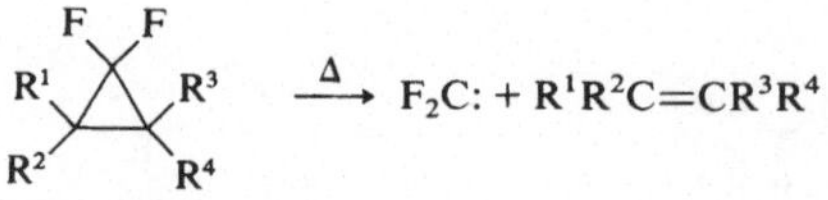

$$\underset{\substack{R^2 \quad\quad R^4 \\ \textbf{2-52a}}}{\overset{F \quad F}{R^1 \triangle R^3}} \xrightarrow{\Delta} F_2C: + R^1R^2C{=}CR^3R^4$$

| R¹ | R² | R³ | R⁴ |
|----|----|----|----|
| F | Cl | F | F |
| Cl | Cl | F | F |
| Cl | Cl | Cl | F |
| Cl | Cl | Cl | Cl |
| F | F | F | $CHF_2$ |
| F | F | F | $CH_2CF{=}CF_2$ |
| H | H | F | F |

Similarly, perhalogenated oxiranes and aziridines thermally release the corresponding carbenes.

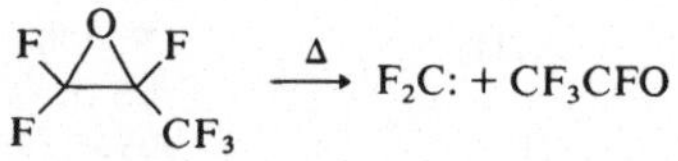

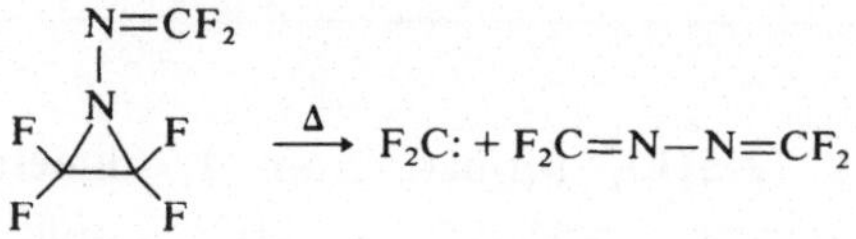

$$\underset{\substack{F \quad\quad CF_3 \\ \textbf{2-52b}}}{\overset{F \overset{O}{\triangle} F}{}} \xrightarrow{\Delta} F_2C: + CF_3CFO$$

$$\underset{\substack{F \quad\quad F \\ \textbf{2-52d}}}{\overset{\substack{N{=}CF_2 \\ | \\ F \overset{N}{\triangle} F}}{}} \xrightarrow{\Delta} F_2C: + F_2C{=}N{-}N{=}CF_2$$

## 2.2.2. Photochemically Triggered Release

### 2.2.2.1. Lewis Acids from Diazonium Salts

A common and versatile source of Lewis acids is diazonium salts. These have a variety of properties, some of which are touched upon in Chapters 1 and 4. Aryl diazonium salts are photosensitive and undergo a photolysis reaction known as the Schiemann reaction to form aryl halides.[103] When an aryl diazonium fluoroborate (2-53) is photolyzed, an aryl fluoride, nitrogen gas, and a Lewis acid are generated. Similarly, when other salts containing hexafluorophosphate, hexafluoroarsenate, hexafluoroantimonate, tetrachloroferrate, and tetrachloroantimonate anions are photolyzed, they release phosphorus pentafluoride, arsenic pentafluoride, antimony pentafluoride, iron trichloride, and antimony trichloride, respectively.

$$ArN_2^+BF_4^- \rightarrow ArF + N_2 + BF_3$$

**2-53**

Compounds (**2-54**) have been reported to have a broad variation in their spectral absorption characteristics throughout the ultraviolet and visible regions.[97]

**2-54**

| R | $X^-$ | $\lambda_{max}$ (nm)[a,b] |
|---|---|---|
| 2,6-Dimethoxy-4-( p-tolylmercapto) | BF$_4$ | 355, 391 |
| 4-NO$_2$ | BF$_4$ | 258, 311 |
| 2,4-Cl$_2$ | BF$_4$ | 285, 325 |
| 4-( N-morpholino) | BF$_4$ | 275–375 |
| 2,4-Cl$_2$ | Hexachlorostannate(IV) | 285 |
| 4-NO$_4$ | Hexachlorostannate(IV) | 258, 310 |
| 2,4-Cl$_2$ | Tetrachloroferrate | 259, 258, 360 |
| 4-NO$_2$ | Tetrachloroferrate | 243, 257, 310, 360 |
| 4-( N-morpholino) | Tetrachloroferrate | 240, 267, 313, 364 |
| 4-Cl | Hexafluorophosphate | 273 |
| 4-NO$_2$ | Hexafluorophosphate | 258, 310 |
| 4-( N-morpholino) | Hexafluorophosphate | 377 |
| 2-Cl-4-(CH$_3$)$_2$N-5-CH$_3$O | Hexafluorophosphate | 273 |
| 2.5-Diethoxy-4-( p-phenetole) | Hexafluorophosphate | 265, 415 |
| 2,5-Diethoxy-4-( p-tolylmercapto) | Hexafluorophosphate | 247, 359, 397 |
| 2,5-(CH$_3$O)$_2$-4-Cl | Hexafluorophosphate | 243, 287, 392 |
| 2,5-Dimethoxy-4-( N-morpholino) | Hexafluorophosphate | 266, 396 |
| 2,5-Dimethoxy-4-( p-tolyl) | Hexafluorophosphate | 405 |
| 2,4-(CH$_3$)$_2$-6-NO$_2$ | Hexafluorophosphate | 237, 290 |
| 2-NO$_2$-4-CH$_3$ | Hexafluorophosphate | 286 |
| 2-CH$_3$-4-NO$_2$ | Hexafluorophosphate | 262, 319 |
| 2,4,6-Br$_3$ | Hexafluorophosphate | 306 |

[a] In acetonitrile solution.
[b] Ref. 104a.

The photosensitivity of these salts depends both on the structure of the cationic and anionic portions of the molecule. deMoria and Murphy[105] reported salts having difluorophosphate, phosphotungstate, silicotungstate, and molybdosilicate anions. Other aryl diazonium salts have also been reported by Schlessinger[104b] and Tsunoda and Yamaoka.[106] The latter reported salts (**2-55**) with broad spectral characteristics and listed their quantum yields. Spectral sensitization of some diazonium salts, to extend their absorption into the visible region, has also been reported.[107]

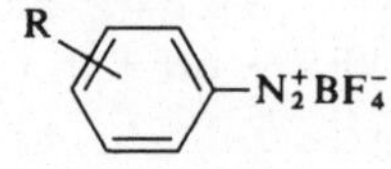

**2-55**

| R | $\lambda_{max}$ (nm) | Quantum yield[a] |
|---|---|---|
| H | 264 | 0.38 |
| 4-Cl | 283 | 0.29 |
| 3-Cl | 266 | 0.41 |
| 2-Cl | 269 | 0.18 |
| 4-Br | 293 | 0.35 |
| 4-F | 269 | 0.50 |
| 4-CH$_3$ | 279 | 0.23 |
| 3-CH$_3$ | 279 | 0.35 |
| 2-CH$_3$ | 269 | 0.30 |
| 4-CH$_3$O | 315 | 0.52 |
| 3-CH$_3$O | 275 | 0.24 |
| 2-CH$_3$O | 355 | 0.18 |
| 4-OH | 350 | 0.54 |
| 4-(CH$_3$)$_2$N | 385 | 0.58 |
| 3-NO$_2$ | 232 | 0.17 |
| 2-NO$_2$ | 280 | 0.10 |
| 4-C$_6$H$_5$ | 340 | 0.25 |
| 4-C$_6$H$_5$N=N | 352 | 0.18 |
| 2,3- 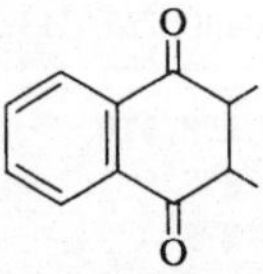 | 249 | 0.18 |
| 2,3-(C$_2$H$_5$O)$_2$-4-morpholino | 405 | 0.43 |
| 2,3- 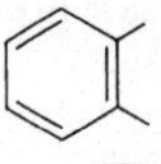 -4-(CH$_3$)$_2$N | 420 | 0.42 |
| 4-$\overset{+}{N}_2$ 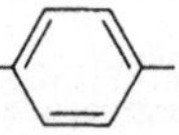 NH | 420, 520 | 0.27, 0.13 |
| 3-CH$_3$O-4-(CH$_3$)$_2$N | 380 | 0.74 |
| 3,4- 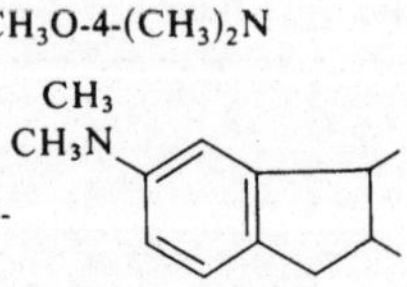 | 470 | 0.34 |
| 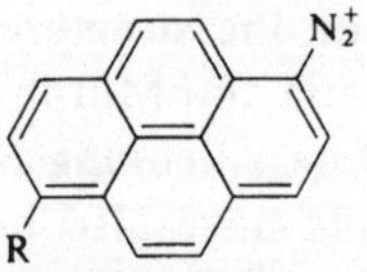 | | |
| R=H | 460 | 0.32 |
| R=CH$_3$ 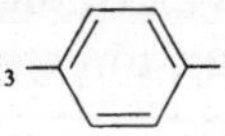 SO$_2$NH | 630 | 0.18 |

[a] Ref. 106.

Heterocyclic diazonium salts are known to a lesser extent than their carbocyclic analogues. For example, diazopyrazoles were reported in 1896 by Knorr and Stolz[108] and later examined by Morgan and Reilly.[109] 3(5)-Diazopyrazoles are noticeably less stable than their 4-diazo analogues, but both are sensitive to light and both couple with phenols to form dyes. Other stable five-membered heterocyclic rings have been easily converted into the corresponding diazonium salts, for example, pyrroles (**2-56**).[110] A review on the chemistry of heterocyclic diazonium salts was published by Tedder[111] in 1967.

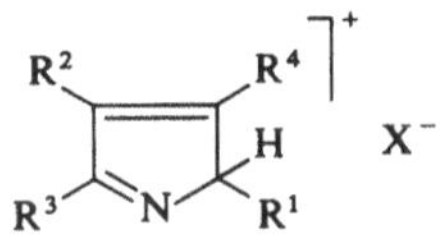

**2-56**

| $R^1$ | $R^2$ | $R^3$ | $R^4$ | $\lambda_{max}$ (nm)$^a$ |
|---|---|---|---|---|
| $C_6H_5$ | H | $C_6H_5$ | $N_2$ | 345 |
| $N_2$ | $NO_2$ | $C_6H_5$ | $C_6H_5$ | 339 |
| $CH_3CO$ | $C_6H_5$ | $CH_3$ | $N_2$ | 350 |
| $N_2$ | H | $C_6H_5$ | $C_6H_5$ | 387 |
| $N_2$ | H | $CH_3$ | $C_6H_5$ | 333 |
| $N_2$ | $-C_4H_4-$ | | $C_6H_5$ | 350 |
| $N_2$ | $-C_4H_4-$ | $-C_4H_4-$ | $4-CH_3OC_6H_4$ | 347 |
| 2-Diazo-5-methoxy-2-(4'-methoxyphenyl) | | | | 346 |
| 3-Diazo-2-phenyl-4,5-benzindole | | | | 359 |

$^a$ Refs. 105 and 106.

### 2.2.2.2. Brønsted Acids from Iodonium, Sulfonium, and Selenium Salts

The release of strong Brønsted acids from iodonium, sulfonium, and selenium salts is reported to be feasible.

Diaryliodonium salts (**2-57**) were first prepared by Hartmann and Meyer[112] in 1894 by the condensation of two molecules of iodosobenzene (**2-58**). Since then a few general synthetic procedures have been developed, making the choice of the preparative method dependent on the desired iodonium salt and its substituents.[113]

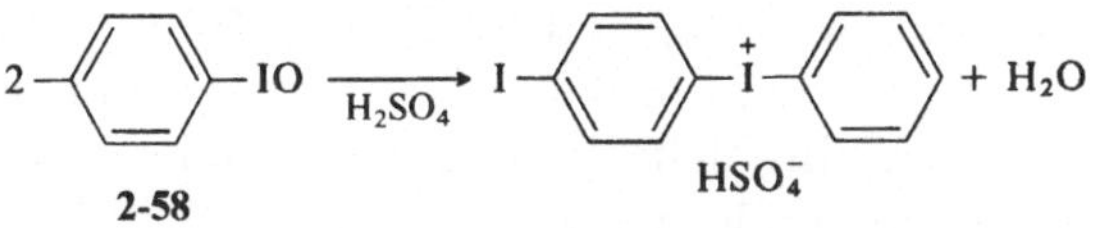

**2-58**

Crivello and Lam[114] have reported iodonium salts with metal complex anions having high photosensitivity and capable of photochemically releasing strong acids used in cationic polymerization. Parallel work in this area has also been reported by Smith[115] of the 3M Company and investigators at Imperial Chemical Industries.[116] Iodonium salts are mostly crystalline, either white or pale yellow in color. They can be prepared with a variety of simple and complex counterions such as tetrafluoroborate, pentafluorophosphate, perchlorate, and trifluoroacetate.

2-57

| $R^1$ | $R^2$ | $X^-$ | $\lambda_{max}$ (nm) ($\varepsilon \times 10^4$),[a,b] |
|---|---|---|---|
| H | H | $BF_4$ | 227 (17.8) |
| 4-$CH_3O$ | H | $BF_4$ | 246 (15.4) |
| 3-$NO_2$ | 3-$NO_2$ | $AsF_6$ | 215 (35.0), 245 (17.0) |
| 4-$CH_3$ | 4-$CH_3$ | $BF_4$ | 236 (18.0) |
| 4-$CH_3$ | 4-$CH_3$ | $PF_6$ | 237 (18.2) |
| 4-$(CH_3)_3C$ | 4-$(CH_3)_3C$ | $BF_4$ | 238 (20.0) |
| 4-$(CH_3)_3C$ | 4-$(CH_3)_3C$ | $PF_6$ | 238 (20.0) |
| 4-$(CH_3)_3C$ | 4-$(CH_3)_3C$ | $AsF_6$ | 238 (20.7) |
| 4-$(CH_3)_3C$ | 4-$(CH_3)_3C$ | $SbF_6$ | 238 (21.2) |
| 4-Cl | 4-Cl | $AsF_6$ | 240 (23.0) |
| 4-$CH_3CONH$ | 4-$CH_3CONH$ | $AsF_6$ | 275 (30.0) |
| | | $AsF_6$ | 264 (17.3) |

[a] In methanol solution.
[b] Ref. 116.

The mechanism of photolysis of the diaryliodonium salts has been examined in some detail and has been shown to proceed via the following overall reaction.[117]

$$Ar_2I^+X^- \xrightarrow[SH]{h\nu} ArI + ArS + HX$$

2-57

This explains the use of iodonium salts as good sources for the release of strong Brønsted acids of the type $HBF_4$, $HAsF_6$, $HSbF_6$, etc.

There is a relation between the rate of photolysis of the salt and its sensitivity to the triggering light.[114b] Dye sensitization of diaryliodonium salts to energies in the visible spectrum is possible.[115,116]

A variety of dialkyliodonium salts have been reported.[116] These are characterized by spectral properties in the ultraviolet region only and are active alkylating agents for nucleophiles.

Diarylchloronium and diarylbromonium salts have also been reported, although their synthetic yields are generally low.[117b,118] These compounds are stable under ambient conditions, and only those with complex metal anions are photosensitive.

Dialkylhalonium fluoroantimonate salts, reported[119,120] to be prepared in quantitative yields from secondary alkyl halides and $SbF_5$–$SO_2$ solution, have found most application as alkylating agents for heterocyclic derivatives.

Analogously, triarylsulfonium salts (**2-59**) bearing non-nucleophilic complex metal halide anions are highly photosensitive and undergo photochemically triggered release of strong acids. Triarylsulfonium salts were

$$R^1R^2R^3S^+X^-$$

**2-59**

| $R^1$ | $R^2$ | $R^3$ | $X^-$ | $\lambda_{max}$ (nm)($\varepsilon \times 10^2$) |
|---|---|---|---|---|
| $C_6H_5$ | $C_6H_5$ | $C_6H_5$ | $BF_4$ | 230 (170) |
| $C_6H_5$ | $C_6H_5$ | $C_6H_5$ | $AsF_6$ | 230 (170) |
| $C_6H_5$ | $C_6H_5$ | 4-$(CH_3)_3CC_6H_4$ | $PF_6$ | 237 (200) |
| 4-$CH_3OC_6H_4$ | 4-$CH_3OC_6H_4$ | 4-$CH_3OC_6H_4$ | $AsF_6$ | 225 (210), 280 (100) |
| 4-$CH_3C_6H_4$ | 4-$CH_3C_6H_4$ | 4-$CH_3C_6H_4$ | $BF_4$ | 243 (240), 278 (49) |
| 2,5-$(CH_3)_2C_6H_3$ | $C_6H_5$ | $C_6H_5$ | $AsF_6$ | 275 (420), 287 (360), 307 (240) |
| 4-Hydroxy-3,5-dimethylphenyl ($R^2 = R^3 = C_6H_5$) | | | $AsF_6$ | 263 (250), 280 (220), 316 (77) |

*Fused ring compounds*

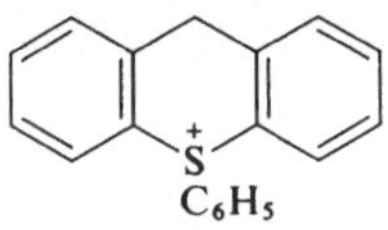

BF₄    232 (310), 277 (250)

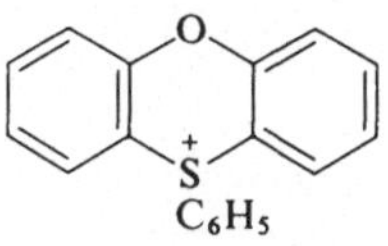

AsF₆    238 (200), 292 (52)

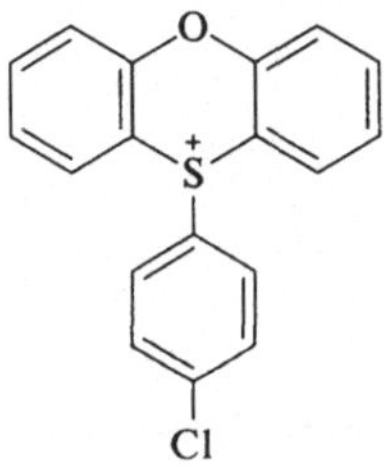

AsF₆    238 (220)

first reported[121] as early as 1891, and recently workers at GE,[122,123] 3M,[124] and ICI[125] reported a variety of sulfonium salts bearing complex anions. The photolysis of triarylsulfonium salts appears to be dependent on both the cationic and anionic portions of the molecule and is analogous to that of iodonium salts in terms of the acid release mechanism and quantum yields. Salts with the same cationic portion have their reactivity determined by the nature of the anion.

These salts, like their iodonium counterparts, absorb in the ultraviolet region and can be spectrally sensitized[126-128,124] to different regions of the visible spectrum using anthracene, benzophenones, thioxanthone, and phenothiazine dyes.[127]

Triarylsulfonium salts are readily prepared in moderate yields by a variety of methods,[129-133] such as the reaction of diaryl thioethers and diaryliodonium salts reported by Crivello and Lam.[129]

$$Ar_2I^+X^- + (C_6H_5)_2S \rightarrow Ar(C_6H_5)_2S^+X^- + ArI$$

| 2-57 | | 2-59 | |
|------|------|------|------|

| Ar | X⁻ | Percent yield of sulfonium salt[a] |
|----|----|-----|
| $C_6H_5$ | $BF_4$ | 94 |
| $C_6H_5$ | $AsF_6$ | 97 |
| $4\text{-}(CH_3)_3CC_6H_4$ | $PF_6$ | 92 |
| $4\text{-}(CH_3)_3CC_6H_4$ | $SbF_6$ | 65 |
| $4\text{-}(CH_3)_2CHC_6H_4$ | $AsF_6$ | 88 |
| $4\text{-}CH_3C_6H_4$ | $PF_6$ | 87 |
| $4\text{-}CH_3CH_2C_6H_4$ | $AsF_6$ | 100 |
| $3,4\text{-}(CH_3)_2C_6H_3$ | $AsF_6$ | 59 |
| $4\text{-}ClC_6H_4$ | $AsF_6$ | 94 |

[a] Ref. 130.

| Ar | Y | Percent yield of sulfonium salt[a] |
|----|----|-----|
| $C_6H_5$ | O | 79 |
| $4\text{-}(CH_3)_2CHC_6H_4$ | O | 41 |
| $4\text{-}CH_3C_6H_4$ | $CH_2$ | 95 |
| $4\text{-}ClC_6H_4$ | O | 99 |
| $4\text{-}(CH_3)_3CC_6H_4$ | $CH_2$ | 67 |
| $C_6H_5$ | $CH_2$ | 87 |

Sulfonium salts containing a thiophenoxy substituent (2-60) on the cationic portion of the molecule have been reported[134-136] to be more reactive than the unsubstituted parent compound. For example, diphenyl-4-thiophenoxyphenylsulfonium salts were described by Crivello and Lam[136] as new efficient acid precursors. This was attributed to the extended conjugation resulting from the 4-thiophenoxy group. The photolysis reactions were studied using a Hanovia 450-W medium-pressure mercury lamp with an output between 200 and 300 nm and a quartz cooling jacket.

**2-60**

| $R^1$ | $R^2$ | $X^-$ | $\lambda_{max}$ (nm) $(\varepsilon \times 10^3)^{a,b}$ |
|---|---|---|---|
| H | H | $AsF_6$ | 227 (21.0) |
| $4\text{-}C_6H_5S$ | H | $AsF_6$ | 225 (23.4) |
| $4\text{-}C_6H_5S$ | H | $PF_6$ | 227 (22.24), 300 (18.0) |
| $3\text{-}C_6H_5S$ | H | $AsF_6$ | 230 (24.33) |
| $2\text{-}C_6H_5S$ | H | $AsF_6$ | 230 (25.25) |
| $4\text{-}C_6H_5S$ | $3,4\text{-}(CH_3)_2$ | $AsF_6$ | 235 (21.6), 300 (19.6) |
| $4\text{-}C_6H_5S$ | $4\text{-}C(CH_3)_3$ | $AsF_6$ | 230 (23.3), 300 (19.6) |
| $4\text{-}C_6H_5SO$ | H | $AsF_6$ | 230 (22.3), 300 (19.3) |
| $4\text{-}C_6H_5SO_2$ | H | $AsF_6$ | 243 (25.2) |

[a] In methanol solution.
[b] Refs. 134–136.

Other substituted sulfonium salts were also examined and found to be useful acid precursors, for example, dialkylphenacylsulfonium salts[137] (**2-62**) and dialkyl-4-hydroxyphenylsulfonium salts[138] (**2-61**). These were

**2-61**

| $R^1$ | $R^2$ | $R^3$ | $X^-$ | $\lambda_{max}$ (nm) $(\varepsilon \times 10^2)^{a,b}$ |
|---|---|---|---|---|
| $CH_3$ | $CH_3$ | H | $BF_4$ | 252 (93), 279 (40), 284 (41), 300 (30) |
| $CH_3$ | $CH_3$ | H | $AsF_6$ | 252 (93), 279 (40), 284 (41), 300 (30) |
| H | $CH_3$ | H | $AsF_6$ | 250 (106), 275 (49), 283 (45), 295 (25) |
| H | H | $CH_3$ | $AsF_6$ | 252 (99), 275 (64), 284 (64) |
| $CH_3O$ | $CH_3O$ | H | $AsF_6$ | 220 (247), 275 (62), 315 (42) |
| H | $-\alpha$-Naphthyl$-$ | | $BF_4$ | 248 (164), 317 (77), 326 (81), 353 (65), 366 (36) |
| $C_6H_5$ | $C_6H_5$ | H | $AsF_6$ | 317 (81), 327 (82), 366 (50) |

[a] In methanol solution.
[b] Ref. 138

prepared in good yields and proceed to release the corresponding strong acids by mechanisms analogous to those for the parent sulfonium salts.

**2-62**

| R—R¹ | R² | R³ | X⁻ | $\lambda_{max}$ (nm) $(\varepsilon \times 10^2)^{a,b}$ |
|------|-----|-----|------|------|
| $CH_3$ | H | H | $BF_4$ | 290 (41) |
| $(CH_2)_4$ | H | H | $BF_4$ | 248 (103), 300 (47) |
| $(CH_2)_4$ | H | H | $PF_6$ | 248 (102), 300 (47) |
| $(CH_2)_4$ | H | H | $AsF_6$ | 247 (102), 300 (47) |
| $(CH_2)_4$ | H | H | $SbF_6$ | 248 (102), 300 (47) |
| $(CH_2)_4$ | $NO_2$ | H | $BF_4$ | 263 (159), 355 (38), 360 (38) |
| $(CH_2)_4$ | Br | H | $BF_4$ | 270 (152), 310 (45), 360 (350) |
| $(CH_2)_5$ | H | H | $AsF_6$ | 250 (111), 293 (43) |
| $(CH_2)_4$ | $-\beta$-Naphthyl$-$ | | $BF_4$ | 300 (219), 360 (10) |
| $(CH_2)_4$ | $C_6H_5$ | H | $BF_4$ | 259 (115), 288 (115), 296 (43), 360 (19) |

$^a$ In methanol solution.
$^b$ Ref. 137.

Trialkylsulfonium salts are strong alkylating agents,[139,140] transparent in the spectral region 250–700 nm and capable of being sensitized to other regions of the spectrum.

Triarylselenium salts (**2-63**) prepared by Leicester and Bergstrom[141] via a synthetic route involving an oxidation followed by a Friedel-Crafts reaction show photosensitivities comparable to the sulfonium salts. The anions can be varied and may include metal complex anions.

**2-63**

| R | X⁻ | $\lambda_{max}$ (nm) $(\varepsilon \times 10^2)^{a,b}$ |
|------|------|------|
| H | $BF_4$ | 258 (108), 266 (28), 275 (21) |
| H | $AsF_6$ | 258 (109), 266 (28), 275 (21) |
| H | $SbF_6$ | 258 (109), 266 (28), 275 (21) |
| 4-$(CH_3)_3C$ | $AsF_6$ | 266 (210), 264 (37), 372 (23) |
| 2-$CH_3$ | $AsF_6$ | 265 (30), 272 (32) |
| 4-$CH_3$ | $AsF_6$ | 225 (220), 265 (37), 272 (25) |

$^a$ In methanol solution.
$^b$ Ref. 141.

2.2.2.3. Brønsted Acids from Nitrophenyl Phosphate and Sulfate Esters

The release of Brønsted acids from nitrophenyl phosphate (**2-64**) and sulfate (**2-65**) esters has been reported by Havinga et al.[142] and Kirby and Varvogolis.[143] These esters are stable in aqueous solution over a wide pH range and undergo photochemically triggered hydrolysis with the release of phosphoric or sulfuric acids, together with the corresponding nitrophenol.

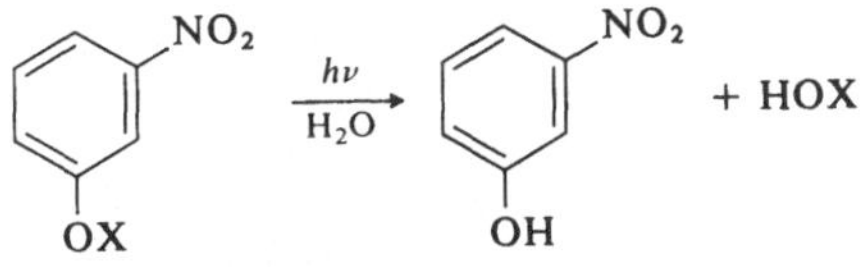

2-64   X = $H_2PO_3$, $HPO_2$
2-65   X = $HSO_3$

Data for the release of nitrophenol from (**2-64**) as a function of irradiation time are shown in Table 2.13 for (**2-64**). The reaction is unaffected by temperature variations but substituents at the meta position of the phenyl ring caused an increase in the release rate. Such an acceleration is more pronounced for meta than for ortho and para substituents. The half-life for an aqueous solution of 3,5-dinitrophenyl phosphate under irradiation using a 125-W medium-pressure mercury lamp fell to five minutes from a half-life of several months in the dark. The sulfate esters (**2-65**) are less affected by substituents on the phenyl ring than their phosphate counterparts.[144]

| R | Concentration of the released phenolate ion ($\times 10^4$ $M$) | |
|---|---|---|
|  | X = $HSO_3$ | X = $H_2PO_3$ |
| 4-$NO_2$ | 0.024 | 0.21 |
| 3-$NO_2$ | 7.08 | 6.14 |
| 2-$NO_2$ | 0.021 | 0.18 |

The kinetics of the photochemically triggered release of strong acids from both the phosphate and sulfate esters was examined by Kirby and Jenks[145] and found to be dependent on the $pK_a$ of the leaving aryl group. The preparation of such esters is easily accomplished through a room-temperature hydrolysis of phenolphosphorooxydichloroiodate.[146]

Table 2.13. Photochemically Triggered Release of the
Phenolate Ion from the Corresponding Phosphate Ester

| Irradiation time of (2-64)<br>(minutes at 25°C) | Concentration of released<br>phenolate ion ($\times 10^3$ $M$) |
|---|---|
| 0 | 0.00 |
| 90 | 0.27 |
| 180 | 0.49 |
| 240 | 0.59 |
| 330 | 0.75 |

### 2.2.2.4. Brønsted Acids from Intramolecular Cyclization

The photochemically triggered cyclization of certain heterocyclic compounds via the displacement of a hydrogen halide molecule is a potential route for the release of acids. For example, hydrogen chloride is released in the cyclization of chlorobenzo[b]thiophene (2-66) to form the corresponding pyrimidine derivative (2-67).[147] Similarly, the formation of benzazepine derivatives (2-69) from iodobenzene derivatives (2-68) allows the release of hydroiodic acid.[148] Furthermore, N-(chloroacetyl)benzylamine (2-70) upon photochemical irradiation forms two isomers of isoquinoline derivatives (2-71) with the release of hydrochloric acid.[149] 1-[3-(Chloroacetylamino)propyl]-3-methylindole (2-72) is converted into three heterocyclic isomers with the release of hydrochloric acid.[150]

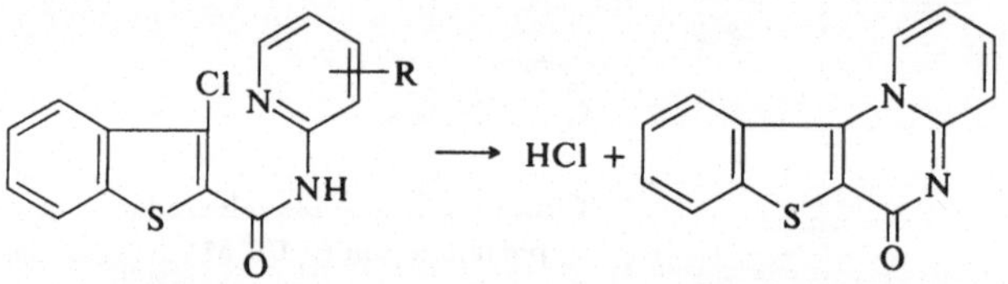

2-66                                2-67

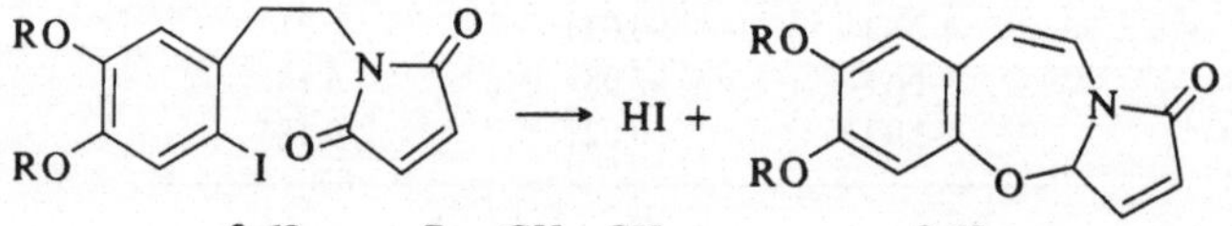

2-68          R = CH$_2$, CH$_3$          2-69

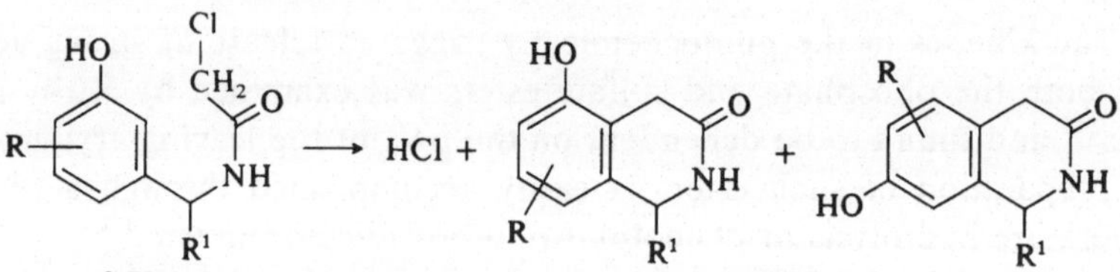

2-70                                2-71

| R | R$^1$ | Percent yield |
|---|---|---|
| 3-OH | H | 49 |
| 3-OH | CH$_3$ | — |
| 3,5-(OH)$_2$ | H | — |
| 3-OH | C$_6$H$_5$CH$_2$ | — |
| 3-OH | 4-CH$_3$OC$_6$H$_4$CH$_2$ | 41 |
| 3-OH | 3,4,5-(CH$_3$O)$_3$C$_6$H$_2$CH$_2$ | — |
| 3,5-(OH)$_2$ | H | — |
| 3,5-(OH)$_2$ | 3,4,5-(CH$_3$O)$_3$C$_6$H$_2$CH$_2$ | 60 |

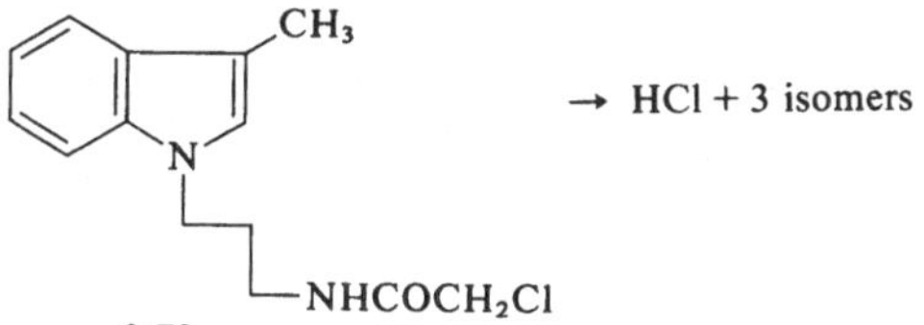

→ HCl + 3 isomers

**2-72**

The reaction of N-chloroacetyl-3-hydroxylbenzylamine derivatives (**2-70**) was facilitated by the presence of electron-donating groups on the aryl ring.[151]

Similarly, the irradiation of 4-phenoxy-3-iodocoumarin (**2-73**) gave the corresponding tetracyclic compound (**2-74**) with the release of hydroiodic acid.[152] The irradiation of 2-chlorophenylnicotinate ester (**2-75**) gave 4-azanthone (**2-76**) with the release of hydrochloric acid.[153]

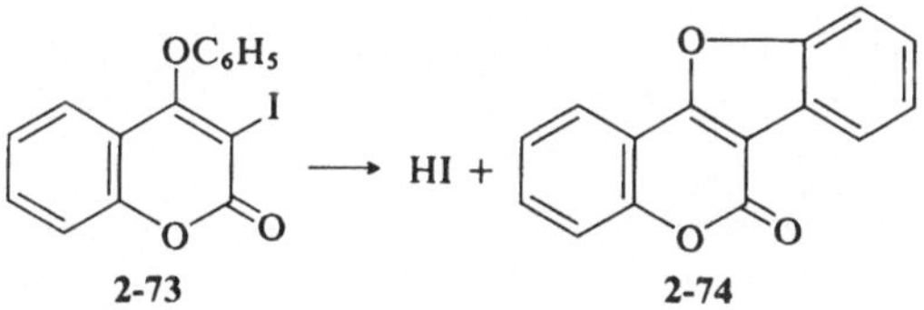

**2-73**      → HI +      **2-74**

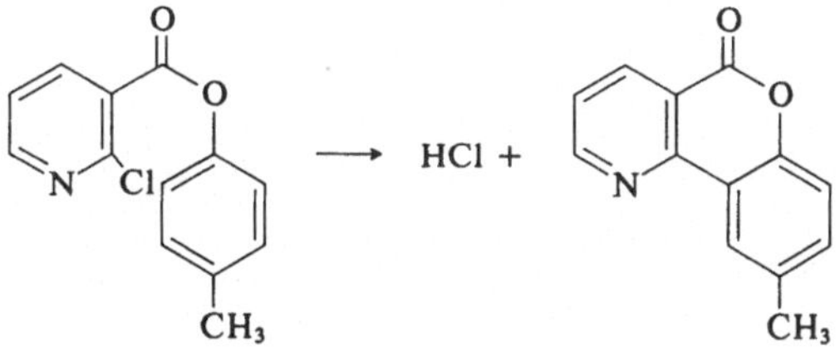

**2-75**      → HCl +      **2-76**

### 2.2.2.5. Carboxylic Acids from Benzyl Esters

Ciamician and Silber[154] observed that photochemical irradiation of 2-nitrobenzaldehyde (**2-77**) caused the release of nitrosobenzoic acid (**2-78**). The reaction, however, was limited to the ortho nitro derivatives only. Some related reactions include the triggered release of nitrocinnamic acid derivatives and other acids.[155-167] This reaction was later used by Patchornik et al.,[168,169] Barltrop et al.,[170] and Pillai[171] as a route for the photochemically triggered release of carboxylic acids. The reaction is usually initiated by irradiation with light of wavelength 300 nm or longer using, for example, a RPR-100 Rayonet apparatus. The yields are generally high to quantitative, making the reaction a valuable synthetic one.

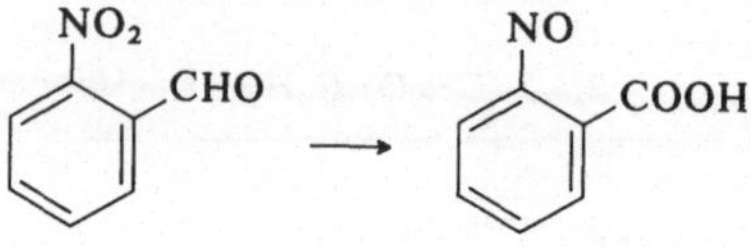

2-77                    2-78

Sachs and Hilpert[172] examined 2-nitrobenzyl ester derivatives (2-79) and came to the conclusion that such structures are also photosensitive.

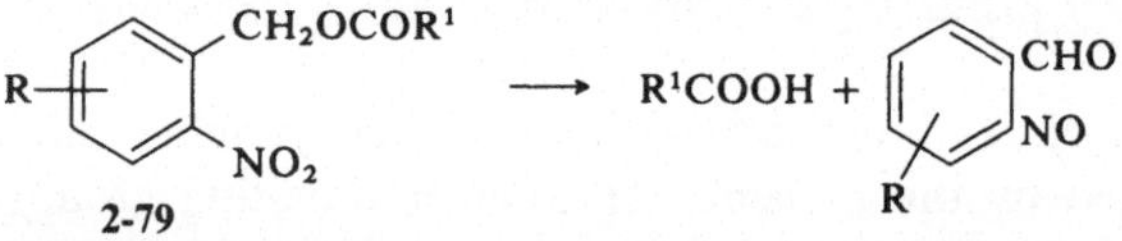

2-79

Electron withdrawing groups on the benzylic carbon atom (2-79) tend to facilitate the release of the corresponding acid as shown from the reaction yields. Electron-rich groups on the benzene ring facilitate such a release reaction. The reaction was also extended to release amines from the corre-

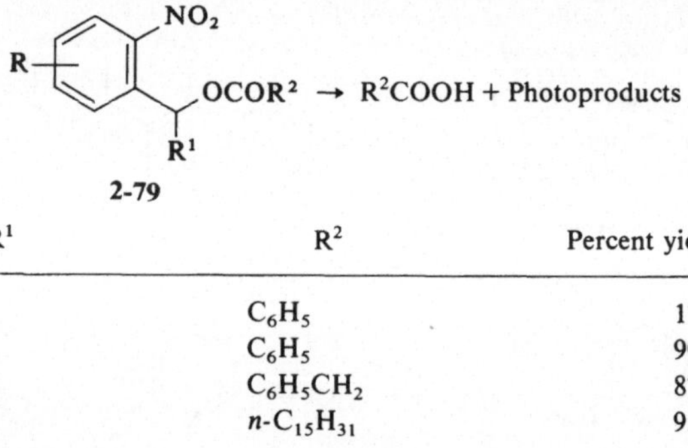

2-79

| $R^1$ | $R^2$ | Percent yield (acid)[a] |
|---|---|---|
| H | $C_6H_5$ | 17 |
| $C_6H_5$ | $C_6H_5$ | 90 |
| $C_6H_5$ | $C_6H_5CH_2$ | 87 |
| $C_6H_5$ | $n\text{-}C_{15}H_{31}$ | 95 |
| $C_6H_5$ | 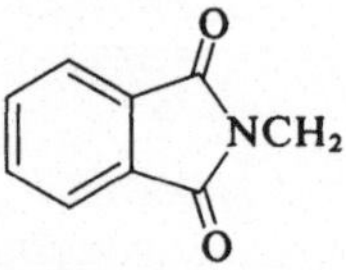 | 75 |
| $o\text{-}NO_2C_6H_4$ | $C_6H_5$ | 100 |
| $o\text{-}NO_2C_6H_4$ | $C_6H_5CH_2$ | 100 |
| $o\text{-}NO_2C_6H_4$ | $1\text{-}CH_2C_{10}H_6$ | 100 |
| $o\text{-}NO_2C_6H_4$ | L-BocNHCHCH$_3$[b] | 100 |
| $o\text{-}NO_2C_6H_4$ | L-BocNHCHCH$_2$C$_6$H$_5$[b] | 100 |

[a] Refs. 168 and 170.
[b] Boc = benzyloxycarbonyl.

sponding amides. This latter reaction will be included under the amine release section of this chapter.

Benzyloxycarbonyl esters (2-80) have also been shown[173] to release the corresponding carboxylic acid under photochemical triggering. Electron-rich benzylic derivatives more readily release the acid than their electron-poor analogues. The mechanism of such a reaction is believed to proceed by a heterolytic fission of the benzyl carbon–oxygen bond to give the carboxylic acid and a benzylic cation which adds water or the reaction solvent. The quantum yields, measured in ethanol solution under irradiation from a medium-pressure mercury lamp, reinforce the proposed mechanism. This reaction has been found useful in the triggered release of amines as well.

$$R^2CH_2OCOR \xrightarrow[ROH]{h\nu} R^1COOH + R^2CH_2OR$$

2-80

### 2.2.2.6. Carboxylic Acids from Phenacyl, Sulfenyl, and Benzoin Esters

Phenacyl esters of carboxylic acids (2-81) have been reported by Sheehan and Umezawa[174] to release carboxylic acid under photochemical irradiation in the 300-nm region.

$$\xrightarrow[R^1OH]{h\nu} RCOOH +$$

2-81

| $R^a$ | $R^1$ | $R^2$ | Percent yield (acid)$^b$ |
|---|---|---|---|
| $C_6H_5$ | $CH_3O$ | H | 60 |
| Boc—L-Ala$^c$ | $CH_3O$ | H | 71 |
| Z-D,L-Ala | $CH_3O$ | H | 85 |
| Boc—L-Phe$^c$ | $CH_3O$ | H | 80 |
| Trityl-Gly | $CH_3O$ | H | 73 |
| Z-L-Trp | $CH_3O$ | H | 82 |
| Phthaloyl-Gly | $CH_3O$ | H | 30 |
| Z-Gly—Gly | $CH_3O$ | H | 67 |
| Z-L-Asp(OBzl)—L-Ser | $CH_3O$ | H | 75 |
| $C_6H_5$ | H | $CH_3$ | 76 |
| Phthaloyl-Gly | H | $CH_3$ | 76 |
| Boc—Gly$^c$ | H | $CH_3$ | 91 |
| Boc—L-Ala$^c$ | H | $CH_3$ | 61 |
| Boc—L-Phe$^c$ | H | $CH_3$ | 80 |

$^a$ Amino acid abbreviations follow IUPAC-IUB rules; see *J. Biol. Chem.* **247**, 997 (1972).
$^b$ Ref. 174.
$^c$ Boc = benzyloxycarbonyl.

Similarly, 2,4-dinitrobenzenesulfenyl esters of carboxylic acids (2-82) have been shown by Barton *et al.*[175-177] to produce carboxylic acids in high yields under photochemical triggering. The reaction rates were enhanced with electron-rich groups on the aryl ring as well as by nucleophilic solvents which tend to stabilize the 2,4-dinitrophenylsulfenium cation and thus drive the yields higher. Irradiation is generally with a 125-W medium- to high-pressure mercury lamp.

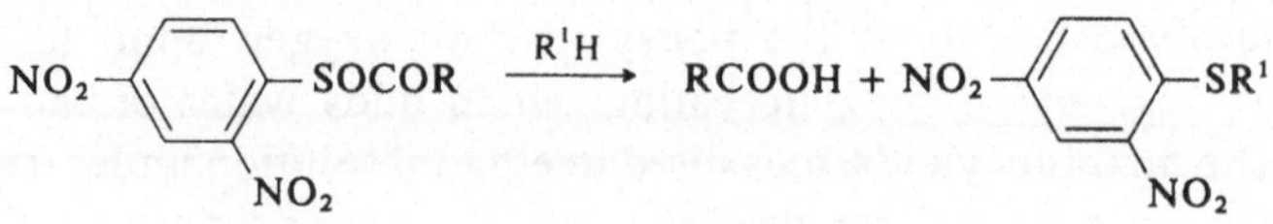

$$NO_2-\bigcirc\!\!\!\!\!\bigcirc-SOCOR \xrightarrow{R^1H} RCOOH + NO_2-\bigcirc\!\!\!\!\!\bigcirc-SR^1$$

2-82

| R | Percent yield (RCOOH)[a] |
|---|---|
| $CH_3$ | 73 |
| $C_5H_{11}$ | 98 |
| $C_6H_5CH_2$ | 88 |

[a] Ref. 211.

Sulfenyl carboxylates are prepared by the reaction of a sulfenyl chloride and the sodium salt of the carboxylic acid.[178] The thermal stability of these esters has been studied by Putman and Sharkey.[179]

Analogous 2,5-dinitrophenyl phosphate esters of carboxylic acids are known to be extremely photosensitive and have a half-life of five minutes under irradiation and of many months in the dark.

Unsymmetrical benzoin acetates (**2-83**) or their carboxylic analogues were shown by Sheehan et al.[180] to be photosensitive precursors for the triggered release of carboxylic acids.

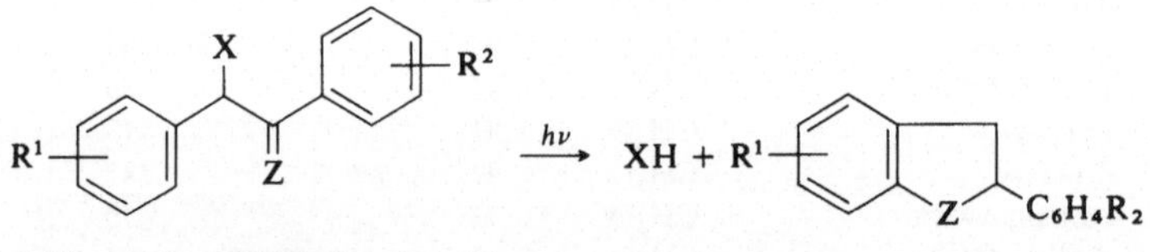

2-83    X = RCOO, Z = O or S

| R¹ | R² | R | Z |
|---|---|---|---|
| H | H | $CH_3$ | O |
| 4-$CH_3$O | 4-$CH_3$O | $CH_3$ | O |
| 3-$CH_3$O | 3-$CH_3$O | $CH_3$ | O |

The yields of the reactions were found to depend on the nature of the acid molecule released and its properties as a leaving group: the better the leaving group, the higher the reaction yield. Irradiation is generally conducted using a 125-W high-pressure mercury lamp with a Pyrex filter. The reaction was also shown to be sensitive to monochromatic light sources such as the 337.1-nm line of a nitrogen laser and the 351.1- and 365.8-nm lines of an argon-ion laser.

### 2.2.2.7. Carboxylic Acids from Amide Derivatives

In 1937 Reidal and Mitchell[181] showed that stearic anilide (**2-84**) irradiated as a monolayer with light of wavelength 235–240 nm decomposed to release stearic acid and aniline.

Later, in 1940, Carpenter[182] extended the reaction to include $\beta$-phenylethyl stearylamide (**2-86**) and benzyl stearylamide (**2-85**).

$$C_{17}H_{35}CONHC_6H_5 \qquad\qquad C_{17}H_{35}CONHOH_2C_6H_5 \qquad\qquad C_{17}H_{35}CONH(CH_2)_2C_6H_5$$

$$\textbf{2-84} \qquad\qquad\qquad\qquad \textbf{2-85} \qquad\qquad\qquad\qquad \textbf{2-86}$$

In later years Amit et al.[183] extended previous observations and examined the photochemical properties of N-acyl-1,2,3,4-tetrahydro-8-nitroquinoline (**2-87**) and its derivatives. Irradiation studies showed that the position of the nitro group had a great influence on the photosensitivity of the amide. When the nitro group was in the 6- or 7-position, the corresponding amides were very stable to light. In contrast, 8-nitroanilides were extremely photosensitive and released the corresponding acids in high yields.[183a] These nitro derivatives of tetrahydroquinolines were prepared by nitration followed by acylation as described by Amit et al.[183b] Irradiation reactions were carried out in solution using a Rayonet apparatus with light output above 305 nm.

$$R = 5\text{-}NO_2,\ 6\text{-}NO_2,\ 7\text{-}NO_2; \qquad R^1 = C_6H_5,\ 3,4\text{-}Cl_2C_6H_3$$

| R = 8-NO$_2$ | R$^1$ | Solvent | Percent yield (R$^1$COOH)$^a$ |
|---|---|---|---|
| | $C_6H_5$ | $CH_3OH$ | 91 |
| | $C_6H_5$ | $(CH_3)_2CHOH$ | 85 |
| | $C_6H_5$ | $C_6H_6$ | 80 |
| | $3,4\text{-}Cl_2C_6H_3$ | $CH_3OH$ | 95 |
| | $3,4\text{-}Cl_2C_6H_3$ | $C_2H_5OH$ | 95 |
| | $3,4\text{-}Cl_2C_6H_3$ | $C_6H_6$ | 80 |

$^a$ Ref. 183b.

Indoline amide derivatives (**2-88**) also showed photosensitivity toward triggered release of the corresponding acid.[183] Irradiation conditions for the indoline derivatives were similar to those for the quinoline derivatives.

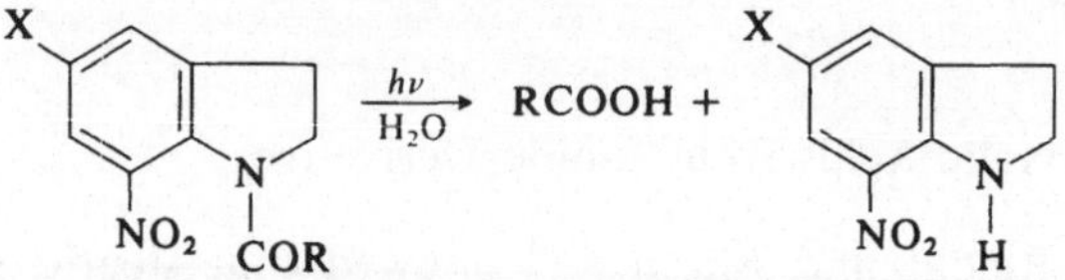

| X = Br | R | Percent yield (RCOOH)[a] |
|---|---|---|
| | $C_6H_5$ | 100 |
| | $4\text{-}CH_3OC_6H_4$ | 100 |
| | $3\text{-}NO_2C_6H_4$ | 87 |
| | $3,4\text{-}Cl_2C_6H_3$ | 87 |
| | 2-Naphthyl | 100 |
| | 2-Furyl | 90 |
| | $n\text{-}C_7H_{15}$ | 100 |
| | $C_6H_5$ | 100 |

[a] Ref. 183b.

Amit and Patchornik[184] have extended the acid release reactions of photosensitive anilides to include acyclic anilides (**2-89**). For example,

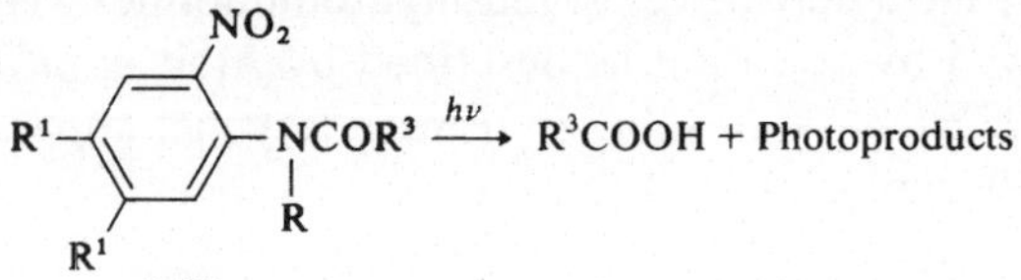

| R | $R^1$ | $R^3$ | Percent yield ($R^3$COOH) |
|---|---|---|---|
| H | H | $C_6H_5$ | 0 |
| H | H | $CH_3$ | 90 |
| $C_6H_5$ | H | $CH_3$ | 95 |
| $CH_3$ | H | $CH_3$ | 95 |
| $CH_3$ | H | $C_6H_5$ | 90 |
| $CH_3$ | H | 2-Naphthyl | 90 |
| Cyclohexyl | H | 2-Naphthyl | 80 |
| Cyclohexyl | $CH_3O$ | $CH_3$ | 90 |
| $CH_3(CH_2)_3$ | $CH_3O$ | $CH_3$ | 70 |
| $C_6H_5CH_2$ | $CH_3O$ | $CH_3$ | 90 |
| $C_6H_5CH_2$ | $CH_3O$ | $3,4\text{-}Cl_2C_6H_3$ | 85 |
| Cyclohexyl | $CH_3O$ | $3,4\text{-}Cl_2C_6H_3$ | 93 |
| $CH_3$ | $CH_3O$ | $2\text{-}ClC_6H_4$ | 93 |
| $CH_3(CH_2)_3$ | $CH_3O$ | 2-Naphthyl | 80 |

N-n-butyl-4',5'-dimethoxy-2'-nitro-2-naphthanilide exposed to sunlight through a GWV filter transmitting light above 370 nm gave naphthoic acid in 80% yield. The yields are highest in polar solvents such as methanol, ethanol, and water. The mechanism of such a cleavage involves the ortho nitro group, and the amide nitrogen substituent has a key influence on the sensitivity of the anilides.

Organic compounds (**2-90**) where three electron-withdrawing groups are attached to the same carbon atom are sensitive toward photochemically triggered decomposition with the release of the corresponding acid.[185] Examples are $\alpha$-phenylsulfonylketoanilides, which are prepared from $\alpha$-chlorophenylketoanilides with the sodium salt of an aryl sulfinic acid.

$$\begin{array}{c} X \\ | \\ RA\overset{}{C}BR^1 \\ | \\ DR^2 \end{array}$$

**2-90**

A, B, D = CO, CONH, SO$_2$, CN, NO$_2$

X = H or alkyl

R, R$^1$, R$^2$ = aryl, alkyl, heterocyclic

$$RCOCH(SO_2R^2)CONHR^1 \xrightarrow[H_2O]{h\nu} RCOOH + R^2SO_2CH_2CONHR^1$$

| R | R$^1$ | R$^2$ | $\lambda_{max}$ (nm)$^a$ |
|---|---|---|---|
| C$_6$H$_5$ | C$_6$H$_5$ | C$_6$H$_5$ | 250 |
| C$_6$H$_5$ | C$_6$H$_5$ | 2-C$_{16}$H$_{33}$SO$_2$C$_6$H$_4$ | 290 |
| C$_6$H$_5$ | C$_6$H$_5$ | 2-C$_{16}$H$_{33}$OC$_6$H$_5$ | 228, 259 |
| 2-Furyl | C$_6$H$_5$ | C$_6$H$_5$ | 278 |
| C$_6$H$_5$ | C$_6$H$_5$ | 2-CH$_3$O-5-C$_{16}$H$_{33}$SC$_6$H$_{13}$ | 258 |
| C$_6$H$_5$ | C$_6$H$_5$ | 4-NO$_2$C$_6$H$_4$ | 305 |
| 4-CH$_3$C$_6$H$_4$ | C$_6$H$_5$ | C$_6$H$_5$ | 297 |
| C$_6$H$_5$ | C$_6$H$_5$ | 1-Naphthyl | 293 |
| 4-NO$_2$C$_6$H$_4$ | C$_6$H$_5$ | C$_6$H$_5$ | 227, 261 |
| C$_6$H$_5$ | C$_6$H$_5$ | C$_6$H$_5$ | 243 |

$^a$ Ref. 185.

## 2.2.2.8. Carboxylic Acids from Quinone Diazides, Aryl Azides, and Unsaturated Carbonyl Compounds

The photochemically triggered release of carboxylic acids by the ring contraction of o-quinone diazides and related derivatives was observed as early as 1944 by Sus[186] during the study of *o*-diazophenols (**2-91**). The reaction corresponds to the Arndt-Eistert rearrangement of aliphatic diazo ketones.

Sus *et al.*,[187-190] Horner *et al.*,[191] and Cava *et al.*[192] later described a number of related compounds which display the same photochemical sensitivity and labeled the mechanism as a photochemical Wolff rearrangement.

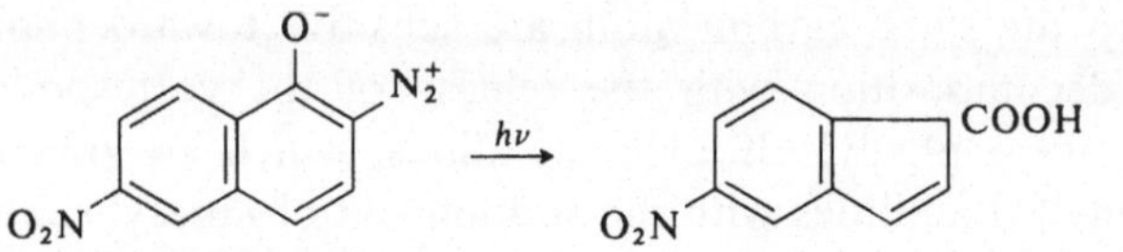

(Ref. 186)

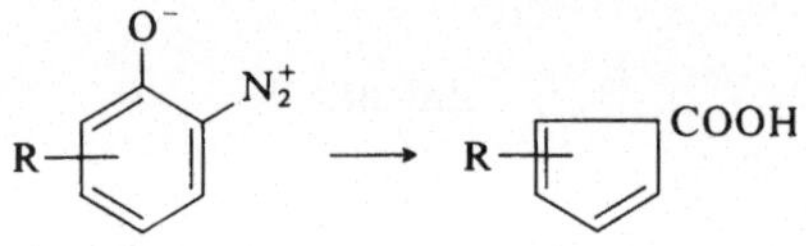

**2-91**

(Ref. 186)

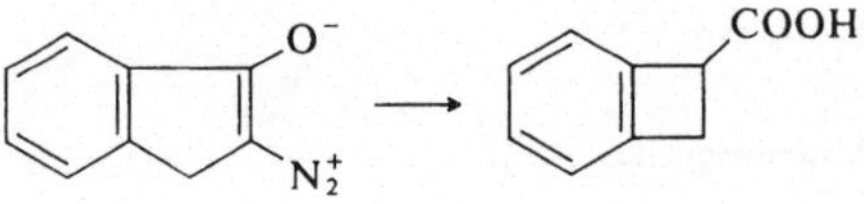

(Ref. 187)

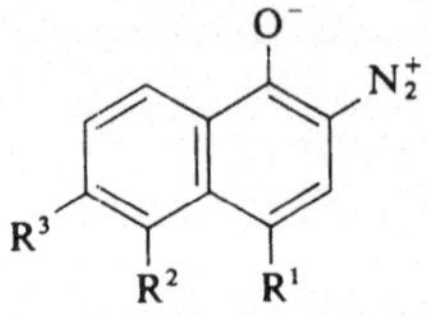

| $R^1$ | $R^2$ | $R^3$ |
|-------|-------|-------|
| H     | H     | H     |
| CN    | H     | H     |
| H     | CN    | CN    |
| H     | H     | CN    |
| OH    | H     | H     |

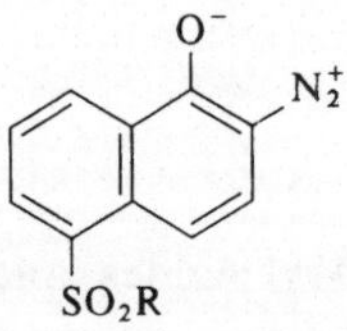

| R | $\lambda_{max}$ (nm)($\varepsilon \times 10^2$)[a] |
|---|---|
| 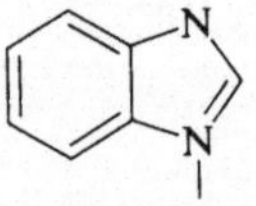 (benzimidazolyl) | 352 (107) |
| $N_3$ | 349 (90) |
| $C_6H_5(CH_3)N$ | 341 (75) |
| $C_6H_5O$ | 343 (77) |
| $CH_3(CH_3O)N$ | 337 (67) |
| $C_8H_{17}O$ | 334 (69) |
| $(CH_3)_2N$ | 332 (63) |

[a] In methanol solution.

This reaction was later extended to a number of other molecules which are capable of releasing carboxylic acids via the same route, including 16-diazo-3$\beta$-hydroxyandrostan-17-one, reported by Mateos and Chao,[193] diazocamphor (**2-92**), reported by Wiberg et al.,[194] and 10-diazo-9(10$H$)-phenanthrone (**2-93**) and 6-diazo-5(6$H$)-chrysenone (**2-94**), both reported by Sus.[195] Quinone diazides of heterocyclic derivatives are also known, but to a lesser extent. Mustafa[196] reported 3-diazo-2(3$H$)-pyridone (**2-95**), and Sus and Moller[188,189] described 3-diazo-4(3$H$)-quinolone (**2-96**) and 3-diazo-1,7-naphthyridin-4(3$H$)-one (**2-97**).

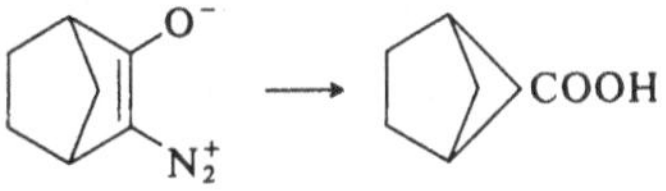

2-92

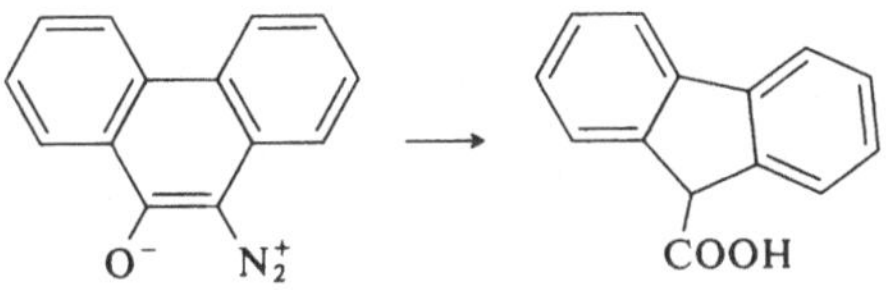

2-93

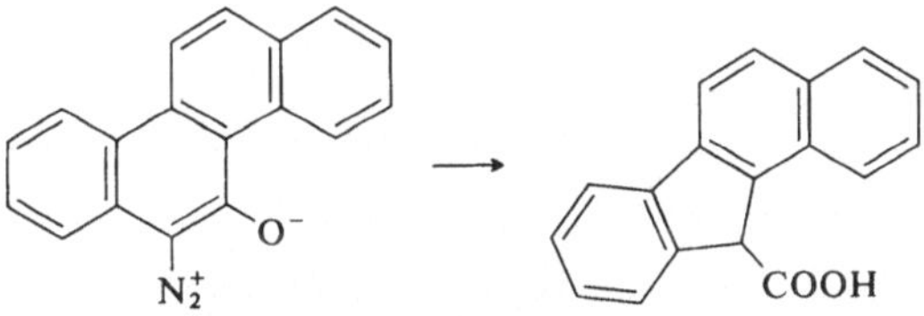

2-94

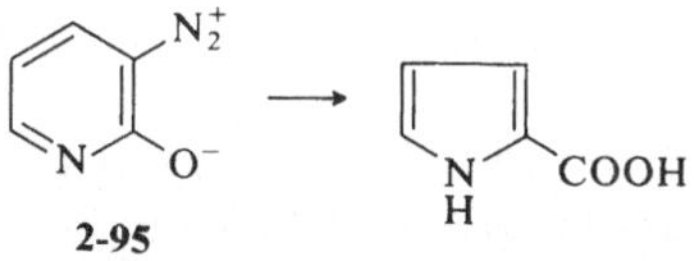

2-95

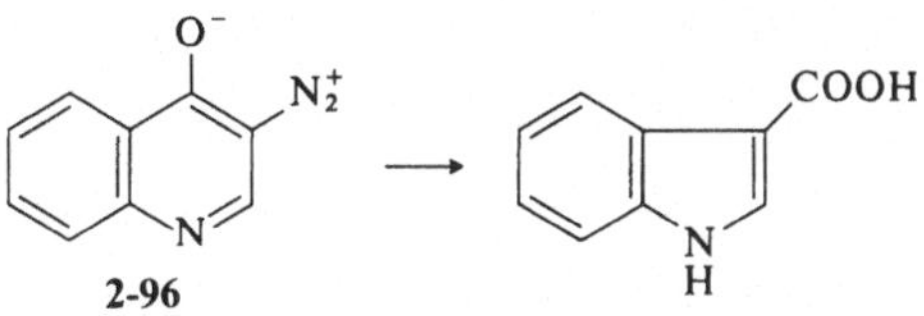

2-96

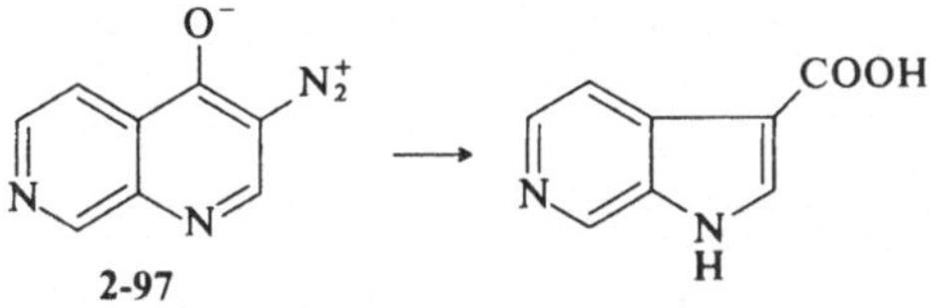

2-97

Poot *et al.*[197] reported that dibenzenesulfonyldiazomethane (**2-98**) or benzenesulfonylbenzoyl diazomethane (**2-99**) derivatives are photochemically sensitive and release sulfinic acids together with sulfonic and carboxylic acid derivatives, respectively. These compounds are prepared[198] from the reaction of *p*-toluenesulfonazide and the corresponding disulfone or benzoyl sulfone derivatives. Irradiation was accomplished using a HP 125-W lamp for 10 minutes at a distance of 15 centimeters.[197]

$$\underset{\substack{\| \\ N_2}}{RSO_2CSO_2R} \xrightarrow[-CO]{h\nu/2H_2O} RSO_2H + RSO_3H + N_2 + 2H^+$$

**2-98**

$$\underset{\substack{\| \\ N_2}}{R^1SO_2CCOR^1} \xrightarrow[-CO]{h\nu/5H_2O} R^1SO_2H + R^1SO_3H + RCOOH + N_2 + 7H^+$$

**2-99**

| R = $C_6H_5$ | $R^1$ | $\lambda_{max}$ (nm)$^a$ |
|---|---|---|
| | 4-CH$_3$C$_6$H$_4$ | 379 |
| | 4-NO$_2$C$_6$H$_4$ | — |
| | 4-BrC$_6$H$_4$ | 370 |
| | 4-IC$_6$H$_4$ | 370 |
| | 4-FC$_6$H$_4$ | 370 |
| | 4-CH$_3$OC$_6$H$_4$ | 385 |
| | $\alpha$-C$_{10}$H$_7$ | 300 |

$^a$ In methanol solution.

Aryl azides have been used as precursors of carboxylic acids. The azide yields a nitrene and nitrogen upon irradiation. The former can be used in a subsequent cyclization reaction to release a carboxylic acid moiety. Derivatives (**2-100**) and (**2-101**) have been prepared and used by Barton *et al.*[199] as synthetic precursors for carboxylic acids. The reaction is commonly conducted in solution using a variety of solvents. The light source used is a medium-pressure mercury lamp with a Pyrex filter.

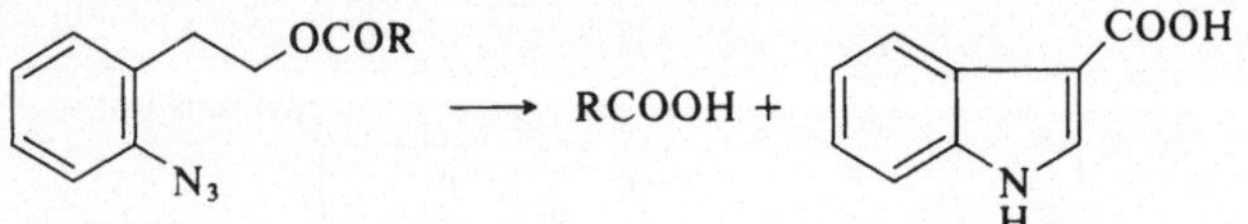

**2-100**

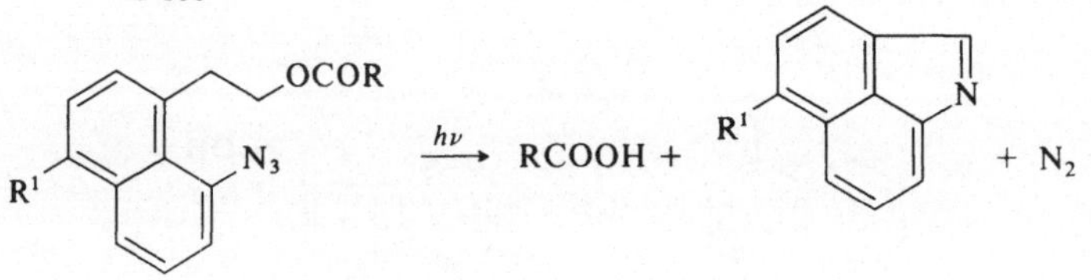

**2-101**

R = C$_6$H$_5$'
R = CH$_3$O, yield = 60–70%

Irradiation of unsaturated cyclic ketones in aqueous ether yields the release of a carboxylic acid, the result of a ring opening reaction. Barton and Quinkert[200,201] examined 6,6-disubstituted cyclohexadienones (**2-102**) which gave the corresponding diene-acids (**2-103**). Similarly, the irradiation of α-pyrone (**2-104**) released the corresponding acid[202] (**2-105**). The cyclohexadienones are prepared from the corresponding phenol and lead tetraacetate.[203]

$$\rightarrow R^1R^2C{:}CHCR^2{:}CHCHR^4COOH$$

**2-103**

**2-102**

| $R^1$ | $R^2$ | $R^3$ | $R^4$ | Percent yield (**2-103**) |
|-------|-------|-------|-------|---------------------------|
| $CH_3$ | $CH_3COO$ | H | H | 79 |
| $CH_3$ | $CH_3$ | H | H | 64 |
| $CH_3COO$ | $CH_3COO$ | $CH_3$ | H | 54 |
| $CH_3COO$ | $CH_3$ | H | $CH_3$ | 50 |

$$\xrightarrow[CH_3OH]{h\nu/H_2O} CH_3COCH{=}C(CH_3)CH_2COOH$$

**2-105**

**2-104**

The photochemically triggered oxidation of perfluoroalkanes which contain di- or trihalo-substituted carbons provides a route to perfluorocarboxylic acids. For example, Hazeldine[204] reported the release of 3,4-dichloropentafluorobutyric acid (**2-107**) from 1,3,4-trichlorohexafluoro-1-iodobutane (**2-106**) in 63% yield upon irradiation in the presence of sodium hydroxide using an ultraviolet light source.

$$ClF_2CCFClCF_2CFClI \xrightarrow[H_2O]{h\nu} ClF_2CCFClCF_2COOH$$

**2-106**                    **2-107**

## 2.2.2.9. Amines from Nitrobenzyl, Benzyloxycarbonyl, Phenyloxycarbonyl, Phenacyl, and Sulfonamide Derivatives

2-Nitrobenzyl derivatives of amines (**2-108**) have been found to undergo photochemically triggered release of amines[205] analogous to the production of acids discussed in Section 2.2.5. The main difference between these reactions is that amines are incorporated into the nitrobenzyl structure as the carbamates, which decarboxylate to the free amine.

A general preparative route to the 2-nitrobenzyloxycarbonyl derivatives (2-108) is from 2-nitrobenzyl alcohols (2-109) and phosgene followed by reaction with the required amine.[205b,205c] The corresponding dimethoxy derivatives, studied by Amit et al.,[206] have photosensitive properties similar to the unsubstituted parent compounds. These were prepared from the chloroformates and amines and were cleaved by 320-nm irradiation at 25°C to give the free amine in 70–90% yield.

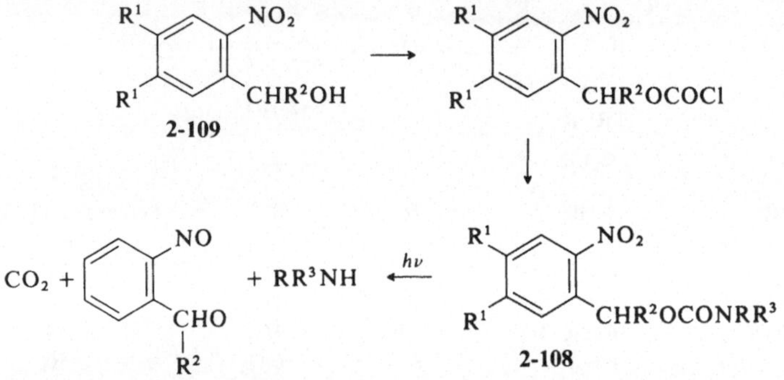

| Released amine[a] | Percent yield ($RR^3NH$) |
|---|---|
| N-Boc—Gly | 17 |
| N-Boc—L-Ala | 35 |
| N-Boc—L-Phe | 17 |
| N-Voc—L-Val | 42 |
| N-Voc—L-Pro | 42 |
| N-Voc—L-Met | 41 |
| N-Voc—L-Ala | 80 |
| N-Voc—L-Phe | 35 |
| N-Voc—L-Trp | 40 |
| N-Voc—L-Phe—Gly | 40 |
| ND Moc—Gly | 94 |
| ND Moc—L-Leu | 70 |
| ND Moc—L-Leu—Gly | 80 |
| ND Boc—L-Phe | 70 |
| ND Boc—L-Ala | 95 |

[a] Amino acid abbreviations follow IUPAC-IUB rules; see J. Biol. Chem. 247, 997 (1972).

The 2-nitrophenyloxycarbonyl derivatives (2-110) and benzyloxycarbonyl derivatives (2-111) have been also reported.[207,208] The methoxy substitution at the aromatic ring was found to enhance the release reaction.[209] The benzyloxy derivatives were examined more closely than their phenyloxy counterparts. Chamberlain[210] reported that this reaction liberates amines in 40–85% yields. The amines are bonded in the molecular structure as a carbamate which can be prepared from the corresponding chloroformates and the appropriate amine.[210,211] The reactions of other benzyloxycarbonyl derivatives, such as the $\alpha$-dimethyl-3,5-dimethoxybenzoyloxycarbonyl derivative, are known to proceed in quantitative yield.[212,213]

$$\text{ROCONR}^1\text{R}^2 \xrightarrow{h\nu} \text{R}^1\text{R}^2\text{NH} + \text{CO}_2 + \text{ROH}$$

**2-110**  R = 2-NO$_2$C$_6$H$_4$
**2-111**  R = 3,5-(CH$_3$O)$_2$C$_6$H$_3$CH$_2$

| R$^1$R$^2$NH | Percent yield (R$^1$R$^2$NH) |
|---|---|
| Glycine | 85 |
| D,L-Methionine | 60 |
| D-Phenylglycine | 66 |
| L-Serine | 72 |
| t-Butyloxycarbonyl-L-Lysine | 42 |
| D-Phenylglycylglycine | 65 |

Phenacyl derivatives (**2-112**) have been shown by Sheehan and Umezawa[214] to release amines as well as carboxylic acids under photochemical triggering conditions. These derivatives are prepared from the phenacyl halides (**2-113**) and the corresponding amine.

| R$^3$ | R$^4$ | R$^1$R$^2$NH | Percent yield (R$^1$R$^2$NH)$^a$ |
|---|---|---|---|
| H | 4-CH$_3$O | Boc—L-Ala | 82 |
| H | 4-CH$_3$O | Z-D,L-Ala | 84 |
| CH$_3$ | H | Z-Gly—Gly | 77 |
| CH$_3$ | H | Boc—Gly | 87 |

Benzylsulfonamide derivatives (**2-114**) have been reported by Pincock et al.[215] to be photosensitive and capable of releasing amines in yields of 10-90%. The reaction proceeds by the cleavage of the C—S bond to release the amines with the loss of sulfur dioxide.

Aromatic amines were found to be the most suitable for such a process, under irradiation using a medium-pressure mercury lamp with a Vycor filter. These sulfonamide derivatives can be prepared as described by Hendrickson and Berberon[216] and include p-toluene[217] and unsubstituted benzene[218] derivatives. The following are a few examples.

$$RC_6H_5CH_2SO_2NR^1R^2 \xrightarrow[R^3OH]{h\nu} R^1R^2NH + SO_2 + RC_6H_4CH_2OR^3$$

**2-114**

| R | $R^1$ | $R^2$ | $R^3$ | Percent yield ($R^1R^2NH$) |
|---|---|---|---|---|
| H | $C_4H_9$ | $C_4H_9$ | $CH_3$ | 61 |
| H | $C_4H_9$ | $C_4H_9$ | $(CH_3)_2CH$ | 78 |
| H | $CH_3(CH_2)_8$ | H | $(CH_3)_2CH$ | 81 |
| H | $C_6H_{11}$ | H | $CH_3$ | 98 |
| H | $C_6H_5$ | H | $(CH_3)_2CH$ | 10 |
| 4-NO$_2$ | $C_6H_{11}$ | H | $(CH_3)_2CH$ | 0 |
| H | $C_4H_9$ | $C_4H_9$ | $CH_3$ | 63 |
| H | $C_4H_9$ | $C_4H_9$ | $(CH_3)_2CH$ | 80 |

### 2.2.2.10. Nitrenes from Azide Derivatives

Nitrenes are reactive nitrogen intermediates with four electrons. They can undergo a number of reactions including insertion into a C, H-single bond and addition to a C, C-double bond.

In general, the most convenient way to release nitrenes is by the decomposition of azides with the loss of nitrogen.

Alkyl azides (**2-115**) can be made to release nitrenes photochemically. The photodecomposition is a first-order reaction independent of temperature under most conditions. The reaction can be triggered by either 216- or 287-nm light. The quantum yields[219] are known to be high and independent of the solvent or azide concentration at the initial stages of the reaction.

$$RN_3 \xrightarrow{h\nu} [R\ddot{N}:] + N_2$$

**2-115**

| R | $\lambda_{max}$ (nm) | Solvent | $\phi_{219\,nm}$ |
|---|---|---|---|
| $C_6H_5$ | 253 | Methanol | 0.88 |
| | 313 | Methanol | 0.88 |
| | 313 | Diphenyl ether | 0.88 |
| $CH_3(CH_2)_2$ | 313 | Methanol | 0.83 |
| | 313 | Heptane | 0.79 |
| $CH_3(CH_2)_3$ | 313 | Hexane | 0.78 |
| $(CH_3)_2(CH_2)$ | 313 | Heptane | 0.79 |
| $CH_3(CH_2)_5$ | 253 | Methanol | 0.86 |
| | 313 | Methanol | 0.71 |
| | 313 | Diethyl ether | 0.71 |
| | 313 | Heptane | 0.69 |
| ⬡— | 313 | Heptane | 0.68 |

Alternatively, aryl nitrenes can be released by photochemical triggering from aryl azides (**2-116**) or arylisocyanates (**2-117**) with the loss of nitrogen or carbon monoxide, respectively. Smith and Brown[220] reported aryl nitrenes as intermediates in the formation of carbazole derivatives (**2-118**) from the corresponding 2-azidobiphenyls (**2-119**).

$$ArN_3 \xrightarrow{h\nu} [Ar\ddot{N}:] + N_2$$

**2-116**

$$ArNCO \xrightarrow{h\nu} [Ar\ddot{N}:] + CO$$

**2-117**

| Ar in **2-116** | Percent yield (nitrene)[a] |
|---|---|
| $C_6H_5$ | 5 |
| $4\text{-}CH_3OC_6H_4$ | 18 |
| $4\text{-}C_6H_4C_6H_5$ | 81 |
| $4\text{-}ClC_6H_5$ | 0 |

[a] Ref. 219; yield of the released nitrene is measured as the yield of ArN:NAr formed.

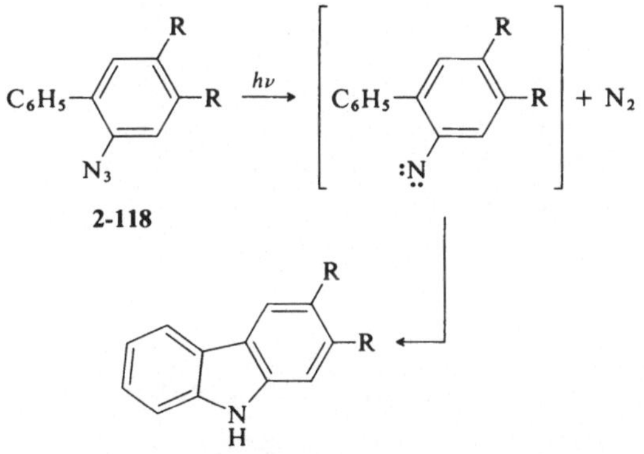

**2-118**

**2-119**

Photolyses of aryl azides have received less attention than their thermolytic reactions, but enough has been reported to demonstrate that it is a general reaction. Most aryl azides decompose under sunlight,[220,221] with the most effective radiation wavelength at about 250–310 nm.

Photolysis of aryl isocyanates has also received little attention and only few examples of the release of aryl nitrenes have been documented. For example, 2-isocyanatobiphenyl (**2-120**) was converted into carbazole in 15% yield by photolysis.[222]

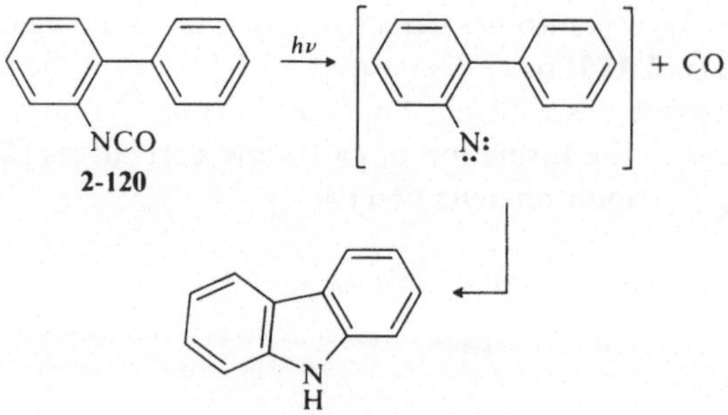

**2-120**

**2-121**

Photolyses of a group of nitrones (**2-122**) derived from *p*-quinone have led to release of azobenzenes in substantial amounts.[223] The two nitrone functions, under reaction conditions, react in sequence with the release of nitrenes as intermediates.[224]

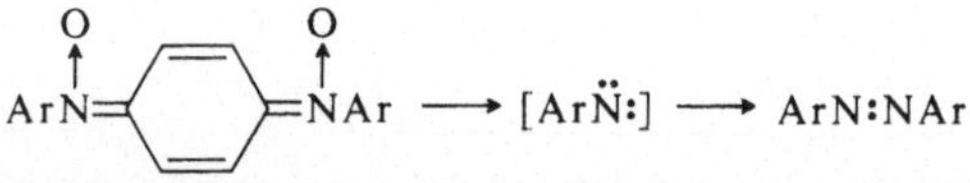

$$ArN \xrightarrow{\quad\quad} [Ar\ddot{N}:] \longrightarrow ArN:NAr$$

**2-122**

Carbonyl nitrenes are the best-known class of nitrenes. Two isomeric forms, (**2-123**) and (**2-124**), are known, both of similar properties; both can be released photochemically from the corresponding carbonyl azide.

Carbonyl nitrenes (**2-123**) were first postulated by Tiemann[225] in 1891 as intermediates in the Lossen rearrangement. Later, heterocyclic rings, such as compounds (**2-125**) and (**2-126**), were also found[226] to be useful in the release of carbonyl nitrenes.

$$RCON_3 \xrightarrow{h\nu} [RCO\ddot{N}:]$$

**2-123**

$$ROCON_3 \xrightarrow{h\nu} [ROCO\ddot{N}:]$$

**2-124**

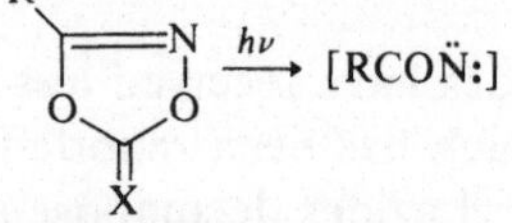

$$\xrightarrow{h\nu} [RCO\ddot{N}:]$$

**2-125**  X = O
**2-126**  X = S

R = $C_6H_5$, 4-$CH_3OC_6H_4$, 4-$NO_2C_6H_4$, 2-$CH_3OC_6H_4$,

2,4,6-$(CH_3)_3C_6H_2$, $C_6H_4CH:CH$, 3-$CH_3OC_6H_4$

The photolysis of ethyl azidoformate, for example, is carried out using a low-pressure mercury lamp with a Vycor 7212 filter.[226] The quantum yields of the photolysis reaction of ethyl azidoformate are reported to be about 0.2 in cyclohexane and about 1.0 in methanol,[227] with possible photosensitization of the reaction with acetophenone.

The photochemically induced Curtius rearrangement[228] of benzoyl azide (**2-127**) to yield phenyl isocyanate is postulated to proceed with the release of carbonyl nitrenes. Phenylacetyl azide (**2-128**) and adipic acid diazide[280] (**2-129**) follow the same reaction path.

$$RCON_3 \xrightarrow[-N_2]{h\nu} [RCO\ddot{N}:] \longrightarrow RNCO$$

**2-127**  $R = C_6H_5$
**2-128**  $R = C_6H_5CH_2$
**2-129**  $R = N_3CO(CH_2)_4$

Sulfonyl azides (**2-130**) are another source of nitrenes. Smolinsky *et al.*[229] and Horner and Christmann[230] reported the successful photolysis of sulfonyl azides by irradiation from a low-pressure mercury lamp at 253 nm with the release of the corresponding nitrenes.

$$RSO_2N_3 \xrightarrow{h\nu} [RSO_2\ddot{N}:] + N_2$$

**2-130**  $R = C_6H_5, 4\text{-}CH_3C_6H_4$

$$NCN_3 \xrightarrow{h\nu} [NC\ddot{N}:] + N_2$$

**2-131**

| | | Percent yield[a] | |
|---|---|---|---|
| Triggering radiation (nm) | Relative rate | *cis* | *trans* |
| 212 | — | 98 | 2 |
| 258 | — | 98 | 2 |
| 280 | — | 98 | 2 |
| 310–425 | 1 | 98 | 2 |
| 310–425 | 1.5[b] | 87 | 13 |
| 310–425 | 2.9[c] | 56 | 44 |

[a] Formation of the addition product with *cis*-1,2-dimethylcyclohexene.
[b] Fluorenone as a sensitizer.
[c] Benzophenone as a sensitizer.

## 2.2.2.11. Carbenes from Diazo Derivatives

Carbenes are the isoelectronic carbon analogues of nitrenes. As with the nitrenes, these are reactive intermediates which find applications involving insertion into a C, H-single bond, addition to a C, C-double bond, etc.

Methylene carbene (**2-132**), the parent compound of the series, has been studied extensively[236] by the photolysis of the corresponding ketene at wavelengths shorter than 280 nm.

$$CH_2:C:O \xrightarrow{h\nu} :CH_2 + CO$$

**2-132**

Aryl and diaryl carbenes can be released photochemically[239] from the corresponding diazo compounds, with quantum yields ranging between 0.65 and 0.58 under ultraviolet irradiation.[240]

Carboalkoxy carbenes are commonly released from diazo esters.[241] Wolf[241] studied the quantum yields in various solvents, under irradiation at different wavelengths. The quantum yields strongly decreased with increase in the irradiation wavelength. Whereas light of 260 nm appears more efficient in protic than hydrocarbon solvents, the reverse is true for radiation of greater than 300 nm wavelength.

Ketocarbenes (**2-133a**) are released via the decomposition of diazoketones (**2-133b**) by light, heat, and various catalysts. The relation between the structure and the quantum yields has been studied with a variety of diazoketones,[240] showing a decrease in quantum yields with increase in the polarization of the diazo group. The reaction is analogous to the Arndt-Eistert rearrangement.[243]

**2-133a**                              **2-133b**

| R | R$^1$ | $\phi^{a,b}$ |
|---|---|---|
| H | H | 0.46 |
| 4-CH$_3$ | H | 0.42 |
| 4-Cl | H | 0.41 |
| 4-C$_6$H$_5$ | H | 0.27 |
| 4-C$_6$H$_4$CO | H | 0.28 |
| 4-CH$_3$SO$_2$ | H | 0.92 |
| 4-NO$_2$ | H | 0.18 |
| H | CH$_3$CO$_2$ | 0.35 |
| H | CH$_3$CH$_2$CO | 0.31 |
| N$_2$CHCO | H | 0.15 |

[a] Irradiation in methanol at the UV $\lambda_{max}$.
[b] Ref. 240.

Halocarbenes, dihalocarbenes, and carbenes linked to a hetero atom have been reported in the literature but are generally released by alkaline chemical decomposition, which is beyond the scope of this survey.

Cyclopropanes, oxiranes, aziridines, diaziridines, $3H$-pyrazoles, and 1,3-dioxalanes have been reported to be stable precursors for the photochemical release of carbenes via a cycloelimination reaction. A review of this subject by Griffin[237] appeared in 1971.

Cyclopropane (**2-134a**) and its derivatives are the most studied precursors and yield a wide variety of substituted carbenes and olefins under mild reaction conditions.[238]

$$R^1 \text{---} \underset{R^2}{\overset{R^5 \quad R^6}{\triangle}} \text{---} R^3 \quad \overset{h\nu}{\longrightarrow} \quad R^6R^5C: + R^1R^2C:CR^3R^4$$

**2-134a**

| $R^1$ | $R^2$ | $R^3$ | $R^4$ | $R^5$ | $R^6$ | Percent yield (carbene) | Reference |
|---|---|---|---|---|---|---|---|
| H | H | H | H | H | H | — | 238a |
| H | H | H | H | H | $C_6H_5CH_2$ | — | 238b |
| $C_6H_5$ | H | H | H | H | H | — | 238c, d |
| $C_6H_5$ | H | H | H | Cl | Cl | 9–15 | 238e |
| $C_6H_5$ | H | H | H | H | $C_6H_5$ | 7–8 | 238f |
| $C_6H_5$ | H | $C_6H_5$ | H | $C_6H_5$ | H | 5 | 238f |
| H | H | $C_6H_5$ | $C_6H_5$ | $C_6H_5$ | $C_6H_5$ | 50 | 238g |

Similarly, oxiranes (**2-134b**) and diaziridines (**2-134c**) have been shown to release alkyl- and aryl-substituted carbenes in high yields.[242]

$$R^1 \text{---} \underset{R^2}{\overset{O}{\triangle}} \text{---} R^3 \quad \overset{h\nu}{\longrightarrow} \quad R^1R^2C: + R^3R^4CO$$

**2-134b**

| $R^1$ | $R^2$ | $R^3$ | $R^4$ | Percent yield (carbene) | Reference |
|---|---|---|---|---|---|
| H | $C_6H_5$ | $C_6H_5$ | H | 65–70 | 238f |
| H | $C_6H_5$ | H | $C_6H_5$ | — | 242 |
| $C_6H_5$ | $C_6H_5$ | H | $C_6H_5$ | 85 | 242 |
| $C_6H_5$ | $C_6H_5$ | $C_6H_5$ | $C_6H_5$ | 88–100 | 242 |
| $C_6H_5$ | H | $C_6H_5$ | CN | 62 | 242d |
| $C_6H_5$ | $C_6H_5$ | CN | $C_6H_5$ | 71 | 242d |

$$
\begin{array}{c}
\underset{\underset{2\text{-}134c}{N=N}}{\overset{R^1\quad R^2}{\diagdown\!\!\diagup}} \quad \longrightarrow \quad R^1R^2C\!: + N_2
\end{array}
$$

| $R^1$ | $R^{2a}$ |
| --- | --- |
| $C_2H_5$ | $C_2H_5$ |
| $CH_3$ | H |
| $CH_3$ | $C_2H_5$ |
| $3\text{-}C_3H_{17}$ | H |
| $3\text{-}C_3H_{17}$ | $CH_3$ |
| $(CH_3)_3C$ | H |
| $(CH_3)_3C$ | $CH_3$ |
| $n\text{-}C_5H_{11}$ | H |
| $C_6H_5$ | Br |

[a] Ref. 242d.

## 2.2.2.12. Radicals from Onium Salts and Other Compounds

The most useful and convenient route toward the triggered release of radicals is via the photodecomposition of onium salts. In general, the decomposition of diazonium, iodonium, sulfonium, and selenium salts produces acids, aryl halides, and gases by the generation of radical intermediates. These radicals have been trapped by a variety of reaction mechanisms. Since most of the reactions are common for the release of acids, much already has been covered in the section dealing with the release of acids.

Simple alkyl radicals are very reactive intermediates with short lifetimes. The stability of free radicals is in the order of tertiary greater than secondary and secondary greater than primary radicals. The stability is increased with the extension of conjugation, as, for example, in (2-135) and (2-136).

$$
\begin{array}{cc}
\text{CH}_3\text{CO} & \\
\text{(2-135)} & \text{(2-136)}
\end{array}
$$

CH$_3$CO

·C(C$_6$H$_5$)$_2$

2-135    2-136

The release of free radicals via homolytic fission has been reviewed by Walling.[244] Radicals can undergo four types of reactions: abstraction of another atom, addition to a double bond, decomposition, or rearrangement.

Some common free radicals are halogen atoms, alkyls, or aryls. Davidson[245] reported a correlation between the radical reactivity and activation energies for the abstraction of a hydrogen in an exchange reaction.

$$X^{\cdot} + C_2H_5 \rightarrow XH + [C_2H_4^{\cdot}]$$

| X | $E_a$ (kcal/mol) |
|---|---|
| F | 0.2 |
| Cl | 1.0 |
| CF$_3$ | 7.5 |
| H | 8.7 |
| CH$_3$ | 11.2 |
| Br | 13.3 |

Several types of compounds have been reported as sources for the photochemical release of radicals. The most common are benzoin ethers, $\alpha$-acyloxime esters, acetophenone derivatives, benzyl ketals, ketones, and aldehydes.

Benzoin ethers (**2-137**) dissociate from the excited triplet state to form free radicals by a Norrish type reaction.[246,247]

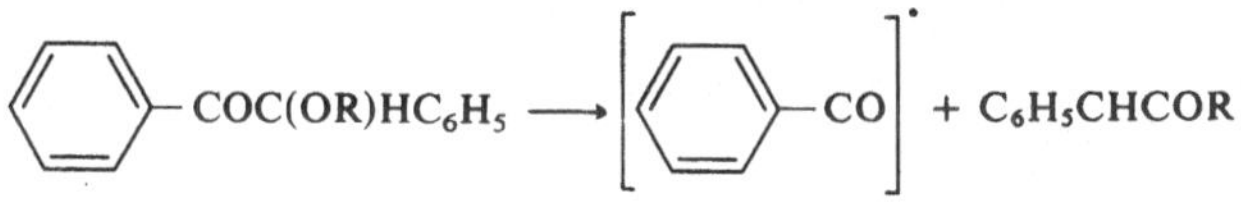

**2-137**

$\alpha$-Acyloxime esters (**2-138**), studied by Delzenne,[248a] also proceed via a Norrish type reaction with the release of two different radicals.

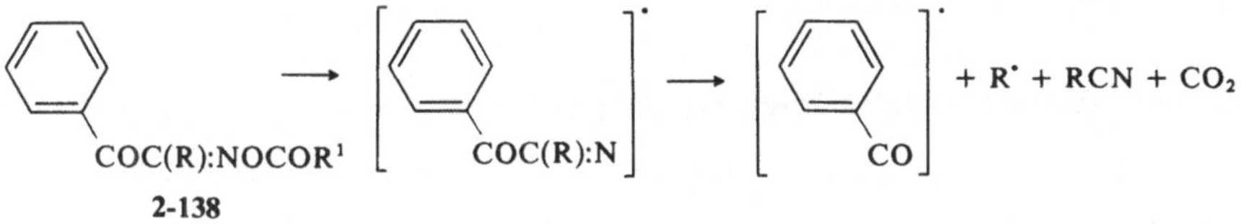

**2-138**

Di- and trihalo-substituted methanes such as di- and trichloroaceto-phenones (**2-139**) fragment photochemically with the release of acyl and trichloromethyl radicals.

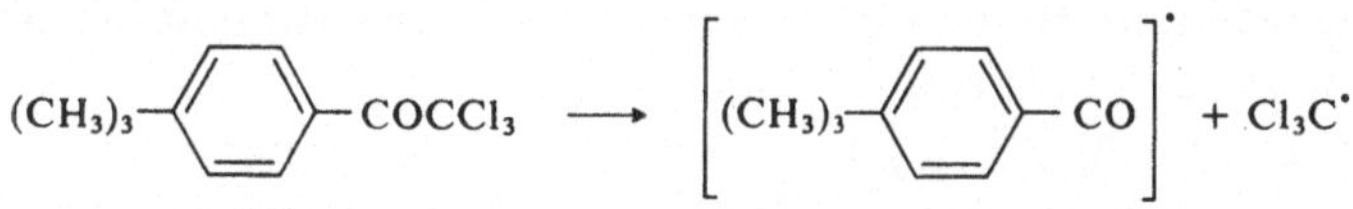

**2-139**

The sensitivity of haloalkanes such as carbon tetrabromide or iodoform toward photochemically triggered release of radicals has been known since the 1920s but was not fully exploited[248] until about the 1960s. Some other interesting compounds used as radical precursors are halosulfones,[248] such

Table 2.14. Photochemically Triggered Fragmentation of
Halogen-Containing Compounds

| Radical precursor | Dissociation energy (kJ/mol) |
|---|---|
| 4-Bromoacetanilide | 297 |
| 4-Bromodiphenyl | 297 |
| N-Bromosuccinimide | 192 |
| 2-Chloroanthraquinone | 218 |
| 1-Chloro-4-nitrobenzene | 293 |
| N-Chlorosuccinimide | 234 |
| 1,4-Dichlorobenzene | 310 |
| 2,4-Dichlorophenol | 289 |
| Hexachlorobenzene | 293 |
| Hexachloroethane | 297 |
| 1,2,3,4-Tetrabromobutane | 247 |
| Tetrabromo-o-cresol | 226 |
| Tetrabromomethane | 205 |
| Tetrabromophenolphthalein | 234 |
| 1,2,3,4-Tetrachlorobenzene | 284 |
| Tetrachloromethane | 284 |
| Tetrachlorotetrahydronaphthalene | 289 |
| Triiodomethane | 185 |

as tribromomethylphenyl, tribromomethyl-4-methylphenyl, tribromo-methyl-4-nitrophenyl, tribromomethylpyridyl, tribromomethyl-3-nitro-pyridyl, and trichloromethylphenyl sulfones. Many other examples have been reviewed by Lawton;[249] some of these are given in Table 2.14.

## 2.2.3. Electrochemically Triggered Release

The use of an electric current as a triggering agent for a chemical reaction has been known since the Kolbe reaction but has been exploited to a lesser extent than other forms of triggering. Generally, high potentials are required. The use of substituents or catalysts which can facilitate the transfer of the electrons is important. The mild reaction conditions and the ability to vary the electric current make electrolysis a viable method for triggering reactions. In principle, all reactions for the cathodic electroreduction of chemical bonds could be used.

Triphenylmethyl,[250] diphenylmethyl, benzyl, cinnamyl, and phenyl derivatives[251] are used as precursors for the release of amino and carboxylic compounds in 70–90% yields. Substituents on these derivatives which can increase the electron affinity of the compound enhance the reaction.

The choice of solvents can determine the site of the electrochemical reaction if two or more sites are available. A good review has been published by Maironovsky.[252]

The experimental setup, in general, consists of a vessel with two side and one center compartments. A mercury pool electrode in the center compartment acts as the cathode with a saturated calomel reference in one side compartment as the anode and a platinum sheet in the other compartment as a second anode. The reaction solution is added to the center compartment with electrolyte solutions in both side compartments. An external potentiostat provides the potential difference.

## 2.2.3.1. Acids and Bases from Ester and Amine Derivatives

The release of acids and amines by electrochemically triggered reduction of esters with 2-haloethyl groups (**2-140**) is dependent on the nature of the halogen and the substituents (Table 2.15).[253]

$$RXCH_2CR^1R^2R^3 \rightarrow RXH + CH_2{:}CR^1R^2 + R^3H$$

**2-140**   X = COO, NH

The release of amines from the corresponding *p*-toluenesulfonates (**2-141**) was found to be feasible. This reaction was first observed by Horner and Neumann,[254] with yields usually of about 70–90% in aprotic solvents. Reactivity is dependent on the nature of the amine substituents. The reaction has been used on a preparative scale of up to 10 grams.[255]

$$4\text{-}CH_3C_6H_4SO_2NR^1R^2 \rightarrow R^1R^2NH + 4\text{-}CH_3C_6H_4SO_2H$$

**2-141**

*Table 2.15.* Yields of Electrochemically Triggered Release of Acids and Amines

| Compound | Product | $E_{1/2}$ (volts) | Percent yield[a] |
|---|---|---|---|
| $C_6H_5CO_2CH_2CCl_3$ | $C_6H_5CO_2H$ | −1.65 | 87 |
| $C_6H_5CO_2CH_2CHCl_2$ | $C_6H_5CO_2H$ | −1.85 | 78 |
| $C_6H_5CO_2CH_2CBr_3$ | $C_6H_5CO_2H$ | −0.70 | 85 |
| $4\text{-}CH_3C_6H_4NHCOCH_2CCl_3$ | $4\text{-}CH_3C_6H_4NH_2$ | −1.70 | 88 |
| $4\text{-}CH_3C_6H_4NHCOCH_2CHCl_2$ | $4\text{-}CH_3C_6H_4NH_2$ | −2.15 | 47 |
| $C_6H_5CH_2OCO_2CH_2CCl_3$ | $C_6H_5CH_2OH$ | −1.50 | 70 |
| $C_6H_5CH_2SCO_2CH_2CCl_3$ | $C_6H_5CH_2SH$ | −1.50 | 90 |
| *N*-Acetyl-*S*-(2,2,2-trichloroethoxycarbonyl)cysteine | *N*-Acetylcysteine | −1.60 | 100 |
| *N*-Acetyl-*S*-(2,2,2-trichloroethoxycarbonyl)cysteine methyl ether | *N*-Acetylcysteine methyl ether | −1.50 | 88 |

[a] Ref. 253.

*Table 2.16.* Half-Wave Potentials of Compounds Useful in the Release of Acids and Amines

| | $E_{12}$ (volts)$^{a,b}$ | |
|---|---|---|
| Y | RNHY | RCOOY |
| Tosyl | 2.6 | — |
| Benzoyl | 2.5 | — |
| Trityl | NR | 2.6 |
| Benzyhydryl | NR | 2.6 |
| Phenyl | NR | 2.7 |
| Cinnamyl | NR | 2.2 |
| Benzyloxycarbonyl | 2.9 | — |
| Benzylidene | — | — |
| 2-Nitrophenylsulfenyl | 1.1 | — |
| 4-Nitrophenyl | — | 1.2 |
| 4-Nitrobenzyloxycarbonyl | 1.2 | — |
| 2,2,2-Trichloroethoxycarbonyl | 1.7 (methanol) | 1.5 (methanol) |
| 2,2-Dichloroethoxycarbonyl | 2.2 | — |

$^a$ $E_{1/2}$ versus SCE in dimethylformamide, with $(alkyl)_4N^+X^-$ as electrolyte.
$^b$ NR = no reaction.

The benzoyl derivatives of amines (**2-142**) were also found to be useful precursors of amines. The reaction was first reported in the 1940s with yields of 60–90% in methanol.[256,257]

$$C_6H_5CONR^1R^2 \rightarrow R^1R^2NH + C_6H_5CHO$$

**2-142**

Table 2.16 lists a variety of precursors useful in the release of acids and amines,[252] together with their reported polarographic half-wave potentials versus the standard calomel electrode.

The electrochemical release of radicals will not be discussed due to the size of the topic since a majority of electrochemical reactions proceed via radical mechanisms and their survey is beyond the scope of this book.

# REFERENCES

1. S. I. Schlessinger, *Polym. Sci. 14*, 513 (1974).
2. H. Ito and C. G. Wilson, *Tech. Papers, Soc. Plastics Eng. Meeting*, 331 (1982).
3. J. Winslow and K. G. Gatzke, U.S. Pat. 4,370,401 (1982).
4. J. Kosar, *Light Sensitive Systems: Chemistry and Applications of Nonsilver Halide Photographic Processes*, p. 218, J. Wiley and Sons, New York (1965).

5. (a) M. Morrison, U.S. Pat. 2,732,299 (1956); (b) R. J. Klimkowski, G. E. Beauchamp, and W. D. Bauer, Belgian Pat. 609,912 (1961); (c) R. J. Klimkowski, L. Amariti, and A. Janda, Belgian Pat. 625,554 (1962); (d) H. Goto, Japanese Pat. 37-17,734 (1962); (e) H. D. Murray, A. Tanenbaum, and R. P. Royer, British Pat. 818,912 (1954); (f) Bauchet & Cie, British Pat. 909,491 (1962).

6. E. Rouse, G. W. Sharp, and C. C. Hunt, British Pat. 859,781 (1961).

7. (a) W. Neugebauer, O. Sus, and H. R. Stumpf, U.S. Pat. 3,050,387 (1962); (b) W. Neugebauer, British Pat. 844,039 (1960).

8. (a) M. P. Schmidt and R. Zahn, U.S. Pat. 1,845,989 (1932); (b) M. K. Reichel and W. Neugebauer, U.S. Pat. 2,663,640 (1953).

9. (a) J. J. Sagura and J. A. van Allan, U.S. Pat. 3,062,650 (1965); (b) A. Schoen, U.S. Pat. 2,416,021 (1947).

10. (a) W. Neugebauer, U.S. Pat. 2,687,958 (1954); (b) Kodak Ltd., British Pat. 886,716 (1942); (c) A. G. Kalle, French Pat. 886,716 (1942); (d) M. K. Reichel and W. Neugebauer, German Pat. 1,123,204 (1959).

11. (a) S. Ruhemann and S. I. Levy, *J. Chem. Soc. 101*, 2542 (1912); (b) P. W. Neber, K. Hartung, and W. Ruopp, *Ber. 58b*, 1234 (1925).

12. (a) H. Lindemann & H. Thiele, *Ann. 449*, 63 (1926); (b) H. Lindemann, H. Konitzer, S. Romanoff, *Ann. 456*, 275 (1927).

13. K. von Auwers and E. Krese, *Ann. 450*, 273 (1926).

14. D. Ambrose and O. L. Brady, *J. Chem. Soc.*, 1243 (1950).

15. O. L. Brady and G. P. McHugh, *J. Chem. Soc., 127*, 2414 (1927).

16. O. L. Brady and F. P. Dunn, *J. Chem. Soc. 109*, 650 (1916).

17. G. J. Karabatsos and K. L. Krumel, *J. Am. Chem. Soc. 91*, 3324 (1969).

18. G. J. Karabatsos, J. M. Corbett, and K. L. Krumel, *J. Org. Chem. 30*, 689 (1965).

19. (a) W. J. Bailey and J. J. Daly, Jr., *J. Org. Chem.* 1249 (1964); (b) W. J. Bailey and W. G. Carpenter, *J. Org. Chem. 29*, 1252 (1964); (c) N. Menschutkin, *Ber. 15*, 2512 (1982).

20. (a) A. W. Hofmann, *Ber. 21*, 2332 (1888); (b) T. Haga and R. Majima, *Ber. 36*, 333 (1903).

21. A. Wahl, *Ann. Chim. 25*, 429 (1912).

22. P. W. Neber and S. Paeschke, *Ber. 59F*, 2140 (1926).

23. (a) V. Meyer, *Ann. 175*, 104 (1875); (b) R. Behrend and H. Tryller, *Ann. 283*, 209 (1895).

24. R. W. Holley and A. D. Holley, *J. Am. Chem. Soc. 74*, 3069 (1952).

25. E. Jungfleisch and M. Godchot, *Compt. Rend. 145*, 70 (1907).

26. A. Pinner, *Ber. 23*, 2943 (1890).

27. M. Conrad and C. Bruckner, *Ber. 24*, 2995 (1891).

28. H. Blitz, *Ber. 43*, 1589 (1910).

29. T. L. Davies and H. W. Underwood, Jr., *J. Am. Chem. Soc. 44*, 2597 (1922).

30. E. A. Werner, *J. Chem. Soc. 103*, 1010 (1913).

31. A. W. Hofmann, *Ber. 18*, 3228 (1885).

32. T. L. Davies and K. C. Blanchard, *J. Am. Chem. Soc. 45*, 1816 (1923).

33. C. Pellizari, *Gazz. Chim. Ital. 49*, 16 (1919).

34. L. Semper and L. Lichtenstadt, *Ann. 400*, 302 (1915).

35. L. Knorr and P. Rossler, *Ber. 36*, 1283 (1903).

36. *M. Busch and Th. Ulmer, Ber. 35*, 1716 (1902).

37. A. Hantzsch and H. Bauer, *Ber. 38*, 1005 (1905).

38. E. A. Werner, *J. Chem. Soc. 105*, 923-933 (1914).

39. W. L. Reilly and H. C. Brown, *J. Org. Chem. 22*, 698 (1957).

40. (a) W. L. Reilly and H. C. Brown, *J. Am. Chem. Soc. 78*, 6032 (1956); (b) H. C. Brown, *J. Polym. Sci. 44*, 9 (1960).

41. S. H. Goh and S. H. Ong, *J. Chem. Soc., B*, 870 (1970).

42. W. A. Pryor, D. M. Huston, T. R. Fiske, T. L. Pickering, and E. Ciuffarin, *J. Am. Chem. Soc.* **86**, 4237 (1964).
43. (a) R. M. Noyes, *J. Chem. Phys.* **18**, 999 (1950); (b) R. M. Noyes, *J. Chem. Phys.* **22**, 1349 (1954); (c) R. M. Noyes, *J. Phys. Chem.* **65**, 763 (1961).
44. M. S. Karasch, J. L. Rowe, and W. Murray, *J. Org. Chem.* **16**, 905 (1951).
45. E. S. Huyser and C. J. Bredeweg, *J. Am. Chem. Soc.* **86**, 2401 (1964).
46. S. Nozaki and P. D. Bartlett, *J. Am. Chem. Soc.* **68**, 1686 (1946).
47. W. Cooper, *J. Chem. Soc.*, 2408 (1952).
48. J. Smid and M. Szware, *J. Chem. Phys.* **29**, 432 (1958).
49. H. Hart and D. P. Wyman, *J. Am. Chem. Soc.* **81**, 4891 (1959).
50. P. D. Bartlett and J. E. Leffler, *J. Am. Chem. Soc.* **72**, 3030 (1950).
51. M. Levy, M. Steinberg, and M. Szwarc, *J. Am. Chem. Soc.* **76**, 5978 (1954).
52. P. D. Bartlett and R. R. Hiatt, *J. Am. Chem. Soc.* **80**, 1398 (1958).
53. E. R. Bell, J. H. Raley, F. F. Rust, F. H. Seubold, and W. E. Vaughan, *Disc. Faraday Soc.* **10**, 242 (1952).
54. M. S. Kharasch, A. Fono, and W. Nudenberg, *J. Org. Chem.* **16**, 113 (1951).
55. J. Thiele and K. Heuser, *Ann.* **290**, 1 (1896).
56. C. Stell and A. F. Trotman-Dickson, *J. Chem. Soc.* 975 (1959).
57. A. U. Blackham and N. L. Eatough, *J. Am. Chem. Soc.* **84**, 2922 (1962).
58. S. G. Cohen and C. H. Wang, *J. Am. Chem. Soc.* **77**, 2457 (1955).
59. M. G. Alder and J. E. Leffler, *J. Am. Chem. Soc.* **76**, 1425 (1954).
60. (a) C. E. H. Bawn and S. F. Mellish, *Trans. Faraday Soc.* **47**, 1216 (1951); (b) J. P. van Hook and A. V. Tobolsky, *J. Am. Chem. Soc.* **80**, 779 (1958).
61. H. Zollinger, *Azo and Diazo Compounds*, p. 266, Interscience Pub., New York (1961).
62. C. S. Sheppard, in: *Encyclopedia of Polymer Science and Engineering* (H. F. Mark, N. M. Bikales, C. G. Overberger, and G. Menges, eds.), Vol. 2, p. 143, Wiley-Interscience, New York (1985).
63. J. L. Kice and R. S. Gabrielsen, *J. Org. Chem.* **35**, 1004 (1970).
64. W. J. Baron, M. R. DeCamp, M. E. Hendrick, M. Jones, R. H. Levin, and M. B. Sohn, in: *Carbenes* (R. A. Moss and M. Jones, eds.), Vol. 1, p. 1, J. Wiley and Sons, New York (1973).
65. W. Kirmse, ed., *Carbene Chemistry*, 2nd ed., Academic Press, New York (1971).
66. J. Casanova and B. Waegell, *Bull. Soc. Chim. Fr.*, 22 (1975).
67. R. H. Shapiro, *Organic Reactions 23*, 405 (1976).
68. R. W. Hoffmann, *Angew. Chem. Intl. Ed. Engl.* **10**, 529 (1971).
69. G. W. Griffin and N. R. Bertoniere, in: *Carbenes* (R. A. Moss and M. Jones, eds.), Vol. 1, p. 305, J. Wiley and Sons, New York (1973).
70. S. S. Hixson, *J. Am. Chem. Soc.* **95**, 6144 (1973).
71. D. Seyferth and A. B. Evnin, *J. Am. Chem. Soc.* **89**, 1468 (1967).
72. (a) R. W. Hofmann and H. Hauser, *Tetrahedron 21*, 891 (1965); (b) D. M. Lemal, E. P. Gosselink, and S. D. McGregor, *J. Am. Chem. Soc.* **88**, 582 (1966); (c) K. Mackenzie, *J. Chem. Soc., Suppl. 1*, 5710 (1964); (d) H. Feichtinger and H. Linden, German Pat. 1,105,862 (1961).
73. A. P. terBorg, E. Razenberg, and H. Kloosterziel, *Recl. Trav. Chim. Pays-Bas 84*, 1230 (1965).
74. (a) E. Vogel, *Pure Appl. Chem. 20*, 237 (1969); (b) V. Rautenstrauch, H. J. Scholl, and E. Vogel, *Angew. Chem. Intl. Ed. Engl. 7*, 288 (1968).
75. (a) P. B. Sargeant, *J. Org. Chem. 35*, 678 (1970); (b) P. G. Gassman, T. J. Atkins, and F. J. Williams, *J. Org. Chem. 93*, 1812 (1971); (c) P. G. Gassman and F. J. Williams, *Tetrahedron Lett.*, 1409 (1971); (d) L. A. Paquette, R. P. Henzel, and S. E. Wilson, *J. Am. Chem. Soc. 93*, 2335 (1971).

76. R. A. Mitch, E. W. Nevar, and P. H. Ogden, *J. Heterocycl. Chem. 4*, 389 (1967).
77. W. R. Brasen, H. N. Cripps, C. G. Bottom Ley, M. W. Farlow, and C. G. Krepspan, *J. Chem. Soc. 30*, 4188 (1965).
78. J. M. Birchall, R. N. Haszeldine, and D. W. Roberts, *J. Chem. Soc., Chem. Commun.*, 287 (1967).
79. R. N. Haszeldine and J. G. Speight, *J. Chem. Soc., Chem. Commun.*, 995 (1967).
80. R. A. Mitch and E. W. Nevar, *J. Phys. Chem. 70*, 546 (1966).
81. W. E. V. Doering and L. H. Knox, *J. Am. Chem. Soc. 78*, 4947 (1956).
82. H. Meerwein, H. Rathjen, and H. Werner, *Ber. 75*, 1610 (1942).
83. H. Staudinger and O. Kupfer, *Ber. Dtsch. Chem. Ges. 44*, 2197 (1911).
84. F. Tiemann, *Ber. 24*, 4162 (1891).
85. J. Stieglitz, *Am. Chem. J. 18*, 751 (1896).
86. (a) J. Stieglitz and P. N. Leech, *Chem. Ber. 46*, 2147 (1913); (b) J. K. Senior, *J. Am. Chem. Soc. 38*, 2718 (1916).
87. J. A. Leermakers, *J. Am. Chem. Soc. 55*, 2719, 3098 (1933).
88. G. Geiseler and W. Konig, *Z. Phys. Chem. 227*, 81 (1964).
89. W. H. Saunders, Jr., and J. C. Ware, *J. Am. Chem. Soc. 80*, 3328 (1958).
90. T. F. Fagley, J. R. Sutter, and R. L. Oglukian, *J. Am. Chem. Soc. 78*, 5567 (1956).
91. R. Kwok and P. Pranc, *J. Org. Chem. 33*, 2880 (1968).
92. W. Lwowski and T. W. Mattingly, Jr., *Tetrahedron Lett.*, 277 (1962).
93. D. S. Breslow and E. I. Edwards, *Tetrahedron Lett.*, 2123 (1967).
94. R. Huisgen and H. Blaschke, *Chem. Ber. 98*, 2985 (1965).
95. T. Curtius and F. Lorenzen, *J. Prakt. Chem. 58*, 160 (1898).
96. T. Curtius, *Z. Angew. Chem. 27*, 213 (1914).
97. (a) O. C. Dermer and M. T. Edmison, *Chem. Rev. 57*, 99 (1957); (b) L. Horner and A. Chirtmann, *Angew. Chem. Intl. Ed. Engl. 2*, 599 (1963); (c) R. A. Abramovitch and B. A. Davis, *Chem. Rev. 64*, 149 (1964).
98. D. S. Breslow, M. F. Sloan, N. R. Newburg, and W. B. Renfrow, *J. Am. Chem. Soc. 91*, 2293 (1969).
99. G. Herzberg and D. N. Travis, *Can. J. Phys. 42*, 1658 (1964).
100. F. D. March and M. E. Hermes, *J. Am. Chem. Soc. 86*, 4506 (1964).
101. A. G. Anastassiou, H. E. Simmons, and F. D. March, *J. Am. Chem. Soc. 87*, 2296 (1965).
102. (a) G. J. Pontrelli and A. G. Anastassiou, *J. Chem. Phys. 42*, 3735 (1965); (b) E. Wasserman, L. Barash, and W. A. Yager, *J. Am. Chem. Soc. 87*, 2075 (1965).
103. (a) A. Roe, *Org. Reactions 5*, 193 (1949); (b) K. H. Saunders, *The Aromatic Diazo Compounds*, 2nd ed., E. Arnold and Co., London (1949); (c) J. Calvert and J. N. Pitts, *Photchemistry*, p. 473, J. Wiley and Sons, New York (1966).
104. (a) S. I. Schlessinger, *Photogr. Sci. Eng. 18*, 387 (1974); (b) S. I. Schlessinger, *Polym. Eng. Sci.*, *14*, 513 (1974).
105. P. P. deMoria and J. P. Murphy, U.S. Pat. 3,930,856 (1976).
106. T. Tsunoda and T. Yamaoka, *Nippon Shasin Gakkaishi, 92*, 27 (1966).
107. M. P. Schmidt and G. von Poser, German Pat. 763,388 (1952).
108. L. Knorr and F. Stolz, *Ann. Chem. 293*, 68 (1896).
109. (a) G. T. Morgan and J. Reilly, *J. Chem. Soc. 103*, 808 (1913); (b) G. T. Morgan and J. Reilly, *J. Chem. Soc.*, 439 (1914).
110. (a) J. M. Tedder and B. Webster, *J. Chem. Soc.*, 3270 (1960); (b) J. M. Tedder and B. Webster, *J. Chem. Soc.*, 1638 (1962); (c) H. P. Patel and J. M. Tedder, *J. Chem. Soc.*, 4593 (1963).
111. J. M. Tedder, in: *Advances in Heterocyclic Chemistry* (A. R. Katritzky and A. J. Boulton, eds.), Vol. 8, p. 1, Academic Press, London (1967).
112. C. Hartmann and V. Meyer, *Ber. 27*, 426 (1894).

113. (a) I. Masson, *Nature 139*, 150 (1937); (b) I. Masson and E. Race, *J. Chem. Soc.*, 1718 (1937); (c) I. Masson and W. E. Hanby, *J. Chem. Soc.*, 1699 (1938); (d) I. Masson and C. Arugurment, *J. Chem. Soc.*, 1703 (1938); (e) D. A. Berry, R. W. Greene, W. C. Ellis, and M. M. Baldwin, 72nd Intl. Congr. Pure and Appl. Chem., New York, Sept. 1950, abstract p. 465; (f) R. B. Sandin, *Chem. Rev. 32*, 249 (1943); (g) M. C. Casserio, D. L. Glusker, and J. D. Roberts, *J. Am. Chem. Soc. 81*, 336 (1959).

114. (a) J. V. Crivello and J. H. W. Lam, 4th Intl. Symp. on Cationic Polym., Akron, Ohio, June 1976; *J. Polym. Sci., Symp. Issue No. 56*, 1 (1976); (b) J. V. Crivello and L. H. W. Lam, *Macromolecules 10*, 1307 (1977).

115. G. H. Smith, Belgian Pat. 828,841 (1975).

116. ICI, Ltd., Belgian Pat. 837,782 (1972).

117. (a) G. A. Olah and J. R. DeMember, *J. Am. Chem. Soc. 92*, 2562 (1970); (b) R. B. Sandin and A. S. Hay, *J. Am. Chem. Soc. 92*, 2562 (1970).

118. J. V. Crivello, Belgian Pat. 828,670 (1974).

119. G. A. Olah and J. R. DeMember, *J. Am. Chem. Soc. 92*, 718 (1970).

120. G. A. Olah and J. R. DeMember, *J. Am. Chem. Soc. 91*, 2113 (1969).

121. A. Michaelis and E. Godcheux, *Ber. 24*, 757 (1891).

122. J. V. Crivello and J. H. W. Lam, *J. Polym. Sci., Chem. Ed. 17*, 977 (1979).

123. J. V. Crivello and J. H. W. Lam, *J. Rad. Curing 5*, 2 (1978).

124. G. H. Smith, U.S. Pat. 4,069,054 (1978).

125. ICI, Ltd., Belgian Pat. 833,372 (1976).

126. J. V. Crivello, in: *UV Curing: Science & Technology* (S. P. Pappas, ed.), Technology Marketing Corp., Stamford, Connecticut (1978).

127. S. P. Pappas and J. H. Jilek, *Photogr. Sci. Eng. 23*, 141 (1979).

128. J. V. Crivello and J. H. W. Lam, *J. Polym. Sci., Chem. Ed. 17*, 1059 (1979).

129. J. V. Crivello and J. H. W. Lam, *Synth. Commun. 9*, 151 (1979).

130. J. V. Crivello and J. H. W. Lam, *J. Org. Chem. 43*, 3055 (1978).

131. H. M. Pitt, U.S. Pat. 2,807,648 (1958).

132. W. Hahn and R. Stroh, U.S. Pat. 2,833,827 (1958).

133. P. Manya, A. Sekera, and P. Rumpf, *Bull. Soc. Chim. Fr. 1*, 286 (1971).

134. J. V. Crivello and J. H. W. Lam, *J. Polym. Sci., Chem. Ed. 18*, 2677 (1980).

135. G. H. Smith, U.S. Pat. 4,173,476 (1979).

136. J. V. Crivello and J. H. W. Lam, *J. Polym. Sci., Chem. Ed. 18*, 2679 (1980).

137. J. V. Crivello and J. H. W. Lam, *J. Polym. Sci., Chem. Ed. 17*, 2877 (1979).

138. J. V. Crivello and J. H. W. Lam, *J. Polym. Sci., Chem. Ed. 18*, 1021 (1980).

139. Toyo Rayon, K. K., French Pat. 1,377,261 (1964).

140. W. J. Hogsed, U.S. Pat. 3,418,289 (1968).

141. H. M. Leicester and F. W. Bergstrom, *J. Am. Chem. Soc. 51*, 3587 (1929).

142. E. Havinga, R. O. Dejonghe, and W. Dorst, *Recl. Trav. Chim. Pays-Bas 75*, 278 (1956).

143. A. J. Kirby and A. G. Varvogolis, *J. Am. Chem. Soc. 89*, 415 (1967).

144. E. Havinga, R. O. DeJonghe, and W. Dorst, *Recl. Trav. Chim. Pays-Bas 75*, 378 (1956).

145. A. J. Kirby and W. O. Jenks, *J. Am. Chem. Soc. 87*, 3209 (1965).

146. M. Y. Kraft and V. V. Katshkina, *Dokl. Akad. Nauk SSSR 86*, 725 (1952); *Chem. Abstr. 47*, 8032 (1953).

147. M. Terashima, K. Seti, K. Itoh, and Y. Kanaoka, *Heterocycles 8*, 421 (1977).

148. I. Tse and V. Snieckus, *J. Chem. Soc., Chem. Commun.*, 505 (1976).

149. M. Ikeda, K. Hirao, Y. Okuno, N. Numao, and O. Yonemitsu, *Tetrahedron 33*, 489 (1971).

150. O. Schindler, P. Nilaus, U. Strauss, and H. P. Harter, *Helv. Chim. Acta 59*, 2704 (1976).

151. Y. Okuno, M. Kawamori, K. Hirao, and O. Yonemitsu, *Chem. Pharm. Bull. (Tokyo) 23*, 2584 (1975).

152. T. Kappe, G. Korbuly, and W. Stadlbauer, *Chem. Ber. 111*, 3857 (1978).

153. K. Beelitz and K. Praefcke, *Justus Liebigs Ann. Chem.*, 1081 (1979).
154. G. Ciamician and P. Silber, *Chem. Ber. 34*, 2040 (1901).
155. J. A. Berson and E. Brown, *J. Am. Chem. Soc. 77*, 447 (1955).
156. F. Sachs and R. Kempf, *Ber. 35*, 2704 (1902).
157. S. Secareanu and I. Lupas, *Bull. Soc. Chim. Fr.*, 1161 (1936).
158. H. Suida, *J. Prakt. Chem. 48*, 829 (1911).
159. I. Tanasecu, *Bull. Soc. Chim. Fr.*, 1443 (1926).
160. K. Dimroth, M. Bohlmann, and F. Bohlmann, *Angew. Chem. 59*, 176 (1947).
161. F. Krohnke, G. Krohnke, and I. Vogt, *Chem. Ber. 86*, 1500 (1953).
162. P. Pfeiffer, *Ann. 411*, 72 (1916).
163. P. Pfeiffer, *Ber. 45*, 1819 (1912).
164. W. Reid and M. Wilk, *Ann. 590*, 91 (1954).
165. J. S. Splitter and M. Calvin, *J. Org. Chem. 20*, 1086 (1955).
166. R. Stoermer and H. Oehelert, *Ber. 55*, 1232 (1922).
167. I. Tanasecu, *Bull. Soc. Chim. Fr.*, 1074 (1927).
168. A. Patchornik, B. Amit, and R. B. Woodward, *J. Am. Chem. Soc. 92*, 6333 (1970).
169. A. Patchornik, *Pharmacology of Hormonal Polypeptides and Proteins*, p. 11, Plenum Pub. Corp., New York (1968).
170. J. A. Barltrop, P. J. Plant, and P. Schofield, *J. Chem. Soc., Chem. Commun.*, 822 (1966).
171. N. N. R. Pillai, *Synthesis* (1980).
172. F. Sachs and S. Hilpert, *Chem. Ber. 37*, 3425 (1904).
173. J. A. Barltrop and P. Schofield, *Tetrahedron Lett.*, 697 (1962).
174. J. C. Sheehan and K. Umezawa, *J. Org. Chem. 34*, 3771 (1973).
175. D. H. R. Barton, Y. L. Chow, A. Cox, and G. S. Kirby, *J. Chem. Soc.*, 3571 (1965).
176. D. H. R. Barton and W. K. Harasch, *J. Am. Chem. Soc. 78*, 1207 (1956).
177. D. H. R. Barton, T. Nakano, and P. G. Sammes, *J. Chem. Soc., Sect. C*, 322 (1968).
178. A. Poot, G. Delzenne, R. Pollet, and U. Laridon, *J. Photogr. Sci. 19*, 88 (1971).
179. R. E. Putman and N. H. Sharkey, *J. Am. Chem. Soc. 79*, 6526 (1957).
180. (a) J. C. Sheehan, R. M. Wilson, and A. W. Oxford, *J. Am. Chem. Soc. 93*, 7222 (1971); (b) J. C. Sheehan and R. M. Wilson, *J. Am. Chem. Soc. 86*, 5277 (1964).
181. E. K. Rideal and Mitchell, *Proc. Roy. Soc. (London) 159A*, 206 (1937).
182. D. C. Carpenter, *J. Am. Chem. Soc. 76*, 289 (1940).
183. (a) B. Amit, D. A. Ben-Efraim, and A. A. Patchornik, *J. Chem. Soc., Perkin Trans. 1*, 57 (1976); (b) B. Amit, D. A. Ben-Efraim, and A. Patchornik, *J. Am. Chem. Soc. 98*, 843 (1976).
184. B. Amit and A. Patchornik, *Tetrahedron Lett.*, 2205 (1973).
185. A. Poot, G. Delzenne, R. Pollet, and U. Laridon, *J. Photogr. Sci. 19*, 88 (1971).
186. O. Sus, *Ann. 556*, 65 (1944).
187. O. Sus, M. Glos, K. Moller, and H. D. Eberhardt, *Ann. 583*, 150 (1953).
188. O. Sus and K. Moller, *Ann. 599*, 233 (1956).
189. O. Sus, *Ann. 579*, 133 (1953).
190. O. Sus and K. Moller, *Ann. 593*, 91 (1955).
191. L. Horner, E. Spietschika, and E. Becker, *Chem. Ber. 88*, 934 (1955).
192. M. P. Cava, R. L. Litlie, and D. R. Napier, *J. Am. Chem. Soc. 80*, 2257 (1958).
193. J. L. Mateos and O. Chao, *Biol. Inst. Quim. Univ. Nat. Auton. Mex. 13*, 3 (1961).
194. K. B. Wiberg, R. L. Lowry, and T. H. Colby, *J. Am. Chem. Soc. 83*, 3998 (1961).
195. (a) O. Sus, *Liebigs Ann. Chem. 556*, 65 (1944); (b) O. Sus, *Liebigs Ann. Chem. 556*, 85 (1944).
196. A. Mustafa, *Advan. Photochem. 2*, 63 (1964).
197. A. Poot, G. Delzenne, R. Pollet, and V. Laridon, *J. Photogr. Sci. 19*, 88 (1971).
198. F. Klages and K. Bott, *Chem. Ber. 97*, 735 (1964).

199. D. H. R. Barton, P. G. Sammes, and G. G. Weingarten, *J. Chem. Soc., Sect. C,* 721 (1971).
200. D. H. R. Barton and G. Quinkert, *Proc. Chem. Soc., London,* 197 (1958).
201. D. H. R. Barton, *Helv. Chim. Acta 42,* 2604 (1959).
202. P. DeMayo, in: *Advances in Organic Chemistry* (R. A. Raphael, E. C. Taylor, and H. Wynberg, eds.), p. 394, Interscience Pub. Inc., New York (1960).
203. F. Wessely and F. Sinwel, *Monatsh Chem. 81,* 1055 (1950).
204. R. N. Hazeldine, *J. Chem. Soc.,* 4291 (1951).
205. (a) A. Patchornik, B. Amit, and R. B. Woodward, *J. Am. Chem. Soc. 2,* 6333 (1970); (b) J. A. Baltrop and P. Schofield, *J. Chem. Soc.,* 4758 (1963); (c) J. A. Baltrop, P. J. Plant and P. Schofield, *J. Chem. Soc., Chem. Commun.,* 822 (1966).
206. B. Amit, U. Zahavi, and A. Patchornik, *J. Org. Chem. 39,* 192 (1974).
207. T. Wieland, C. Lamperstoreer, and C. Birr, *Makromol. Chem. 92,* 279 (1966).
208. J. A. Barltrop and P. Schofield, *Tetrahedron Lett.,* 697 (1962).
209. H. E. Zimmermann and V. R. Sandel, *J. Am. Chem. Soc. 85,* 915 (1963).
210. J. W. Chamberlain, *J. Org. Chem. 31,* 1658 (1966).
211. G. W. Anderson and and A. C. McGregor, *J. Am. Chem. Soc. Soc. 79,* 6180 (1957).
212. C. Birr, W. Lochinger, G. Stahnkf, and P. Lang, *Liebigs Ann. Chem. 763,* 162 (1972).
213. C. Birr, F. Flor, P. Fleckenstein, and T. Wieland, *Peptides, Proceedings of the European Peptide Symposium,* p. 175, North Holland Pub. Co., Amsterdam (1973).
214. J. C. Sheehan and K. Umezawa, *J. Org. Chem.,* 38, 3771 (1973).
215. (a) J. A. Pincock and A. Jurgens, *Tetrahedron Lett.,* 1029 (1979); (b) R. F. Langler, Z. Marini, and J. A. Pincock, *Can. J. Chem. 56,* 903 (1978).
216. J. B. Hendrickson and R. Berberon, *Tetrahedron Lett.,* 345 (1970).
217. A. Abad, D. Mellier, J. P. Pete, and C. Portella, *Tetrahedron Lett.,* 4555 (1971).
218. H. B. Milne and C. H. Peng, *J. Am. Chem. Soc. 79,* 639 (1957).
219. (a) E. Kock, *Tetrahedron 23,* 1747 (1967); (b) F. D. Lewis and W. H. Saunders, *J. Am. Chem. Soc. 90,* 3828 (1969); (c) W. E. Doering and R. A. Odum, *Tetrahedron 22,* 81 (1966); (d) L. Horner, A. Christmann, and A. Gross, *Chem. Ber. 96,* 399 (1963).
220. P. A. S. Smith and B. B. Brown, *J. Am. Chem. Soc. 73,* 2435 (1951).
221. J. H. Boyer and F. C. Canter, *Chem. Ber. 54,* 1 (1954).
222. J. S. Swenton, *Tetrahedron Lett.,* 2855 (1967).
223. C. J. Pedersen, *J. Am. Chem. Soc. 79,* 5014 (1957).
224. W. Kirmse, *Angew. Chem. 71,* 537 (1959).
225. F. Tiemann, *Ber. 24,* 4162 (1891).
226. (a) W. Lwowski and T. W. Mattingly, Jr., *Tetrahedron Lett.,* 277 (1962); (b) W. Lwowski and T. W. Mattingly, Jr., *J. Am. Chem. Soc. 87,* 1947 (1965).
227. W. Lwowski, in: *Nitrenes* (W. Lwowski, ed.), p. 188, Interscience Pub., New York (1970).
228. L. Horner, E. Spietschka, and A. Gross, *Liebigs Ann. Chem. 573,* 17 (1951).
229. G. Smolinsky, E. Wasserman, and W. A. Yager, *J. Am. Chem. Soc. 84,* 3220 (1962).
230. (a) L. Horner and A. Christmann, *Chem. Ber. 96,* 388 (1963); (b) L. Horner and A. Christmann, *Angew. Chem. Intl. Ed. Engl. 2,* 599 (1963).
231. G. Herzberg and D. N. Travis, *Can. J. Phys. 42,* 1658 (1964).
232. G. J. Pontrelli and A. G. Anatassiou, *J. Chem. Phys. 42,* 3735 (1965).
233. E. Wasserman, L. Barash, and W. A. Yager, *J. Am. Chem. Soc. 87,* 2075 (1965).
234. A. G. Anatassiou and J. N. Shepelavy, *J. Am. Chem. Soc. 90,* 492 (1963).
235. A. G. Anatassiou, H. E. Simmons, and F. A. March, in: *Nitrenes* (W. Lwowski, ed.), p. 334, Interscience Pub., New York (1970).
236. (a) A. N. Strachan and W. A. Noyes, *J. Am. Chem. Soc. 76,* 3258 (1954); (b) W. A. Noyes, G. B. Porter, and I. E. Jolley, *Chem. Rev. 56,* 49 (1956); (c) G. B. Porter, *J. Am. Chem. Soc. 79,* 827 (1957).
237. G. W. Griffin, *Angew. Chem. Intl. Ed. Engl. 10,* 537 (1971).

238. (a) C. L. Currie, H. Okabe, and J. R. McNesby, *J. Phys. Chem.* 67, 1494 (1963); (b) P. A. Leermakers and G. F. Vesely, *J. Org. Chem.* 30, 539 (1965); (c) P. A. Leermakers and M. E. Moss, *J. Org. Chem.* 31, 301 (1966); (d) D. B. Richardson, L. R. Durett, J. M. Martin, Jr., W. E. Putmann, S. C. Slaymaker, and I. Dvoretsky, *J. Am. Chem. Soc.* 87, 2763 (1965); (e) M. Jones, Jr., W. H. Sachs, A. Kulczycki, Jr., and F. T. Waller, *J. Am. Chem. Soc.* 88, 3167 (1966); (f) H. Dietrich, G. W. Griffin, and R. C. Petterson, *Tetrahedron Lett.*, 153 (1968); (g) H. Kristinsson, K. N. Mehrotra, G. W. Griffin, R. C. Petterson, and C. S. Irving, *Chem. Ind. (London),* 1562 (1966).

239. H. Staudinger and O. Kupfer, *Ber. Dtsch. Chem. Ges.* 44, 2197 (1911).

240. W. Kirmse and L. Horner, *Ann.* 625, 34 (1959).

241. E. Wolf, *Z. Physik. Chem.* B17, 46 (1932).

242. (a) R. S. Becker, J. Kolc, R. O. Bost, H. Dietrich, P. Petrellis, and G. W. Griffin, *J. Am. Chem. Soc.* 90, 3292 (1968); (b) R. S. Becker, R. O. Bost, J. Kolc, N. R. Bertoniere, R. L. Smith, and G. W. Griffin. *J. Am. Chem. Soc.* 92, 1302 (1970); (c) P. Petrellis, H. Dietrich, E. Meyer, and G. W. Griffin, *J. Am. Chem. Soc.* 89, 1967 (1967); (d) A. M. Mansoor and I. D. R. Stevens, *Tetrahedron Lett.*, 1733 (1966).

243. (a) K. B. Wiberg and T. W. Hutton, *J. Am. Chem. Soc.* 74, 5367 (1954); (b) W. Kirmse, *Angew. Chem.* 71, 537 (1959).

244. C. Walling, *Free Radicals in Solution*, pp. 467–563, J. Wiley and Sons, New York (1957).

245. R. S. Davidson, *Quart. Rev. (London)* 21, 249 (1967).

246. H. G. Hew, *Tetrahedron Lett.*, 4755 (1972).

247. S. P. Pappas and A. Chattopadhyay, *J. Am. Chem. Soc.* 95, 6484 (1973).

248. (a) Delzenne, U.S. Pat. 3,042,515 (1962); (b) E. Wainer, U.S. Pat. 3,056,673 (1962); (c) K. Itano, M. Nakano, S. Hoshino, and A. Kato, U.S. Pat. 3,502,476 (1970); (d) H. Hori, M. Tagushi, A. Kato, and K. Yumiki, U.S. Pat. 3,560,216 (1971).

249. W. R. Lawton, SPSE Seminar, Novel Imaging Systems, 1969, p. 63.

250. V. G. Mairanovsky, A. Ja. Veinberg, and G. I. Samokhvalov, *Avt. Svid. N222*, 353 (1966); *Zh. Obsch. Khim.* 38, 666 (1968).

251. V. G. Mairanovsky and I. G. Samokhvalov, *Electrokhimiya 2*, 717 (1966).

252. V. G. Mairanovsky, *Angew. Chem. Intl. Ed. Engl.* 15, 281 (1976).

253. M. F. Semmelhack and G. E. Heinsohn, *J. Am. Chem. Soc.* 94, 5139 (1972).

254. L. Horner and H. Neuman, *Chem. Ber.* 98, 1715, 3462 (1965).

255. T. Iwasaki, K. Matsumoto, M. Matsuoka, T. Takahashi, and K. Okumura, *Bull. Chem. Soc. Japan* 46, 852 (1973).

256. S. Wawzonek, H. Laitinen, and S. Kwiatkowski, *J. Am. Chem. Soc.* 66, 830 (1944).

257. M. von Stackelberg and W. Stracke, *Z. Electrochem. Angew. Phys. Chem.* 53, 118 (1949).

# 3

# Triggered Release of Other Monomeric Species

## 3.1. PROCESSES UTILIZING THE TRIGGERED RELEASE OF MONOMERIC SPECIES

A number of patented applications utilize the release of monomeric species under specific conditions. An example is found in the Kodak patent[1] describing the use of the photochemically triggered release of phenanthrene (3-2) from stilbene (3-1).

$$C_6H_5CH:CHC_6H_5 \xrightarrow{h\nu}$$

**3-1**

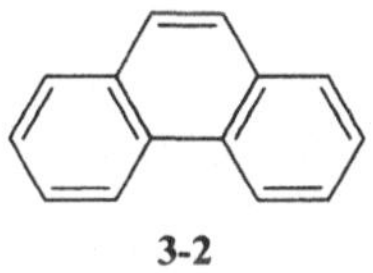

**3-2**

The phenanthrene product serves as a sensitizer for a photopolymerization process used in presensitized printing plates.

Another example is the dry-developed, photothermographic diazo-imaging process which was made possible by the thermal release of a base from urea or its derivatives, such as ethylurea,[2] guanidine,[3] or biuret.[3]

Ultraviolet-sensitive nitrones (3-3), known[4] to undergo triggered reversible cyclization to the isomeric oxaziranes (3-4), were also utilized as a basis for a thermally developed imaging system.[5] The thermal development step, following the light exposure step, causes the decomposition of the released oxazirane (3-4) into an azo dye (3-5) and an aldehyde. This results in the formation of a yellow-colored image.

The photochemical generation of the kinetically active *cis* isomer from the more thermodynamically stable *trans* isomer is a well-known feature of olefins: an example is the maleic to fumaric acid isomerization.[6] Each

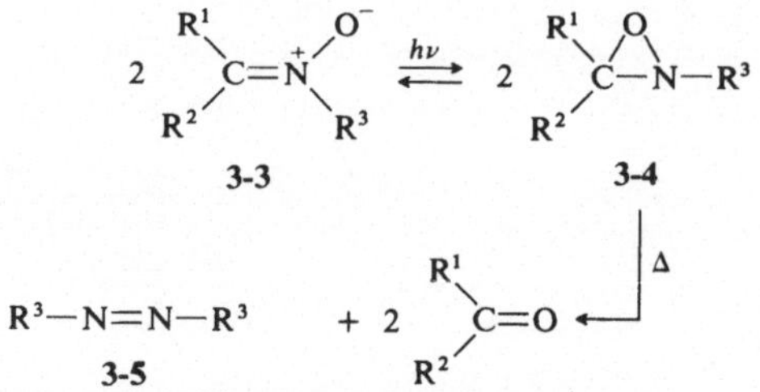

$$2 \underset{R^2}{\overset{R^1}{\diagup}} C = \overset{+}{N} \underset{R^3}{\overset{O^-}{\diagdown}} \underset{hv}{\overset{hv}{\rightleftharpoons}} 2 \underset{R^2}{\overset{R^1}{\diagup}} \overset{O}{\overset{\diagdown}{C}} - N - R^3$$

3-3                                3-4

$$R^3 - N = N - R^3 + 2 \underset{R^2}{\overset{R^1}{\diagup}} C = O \longleftarrow \overset{\Delta}{\rfloor}$$

3-5

isomer has its own characteristic chemical and physical properties which are frequently used in applied chemistry. Kodak utilized such an approach in their early photoresist chemistry.[7] They employed polymeric cinnamates (**3-6**) which can only cross-link via 1,2-cycloaddition reactions between the C,C-double bonds in the *cis* form (**3-7**), which is predominantly generated during the light imaging exposure.

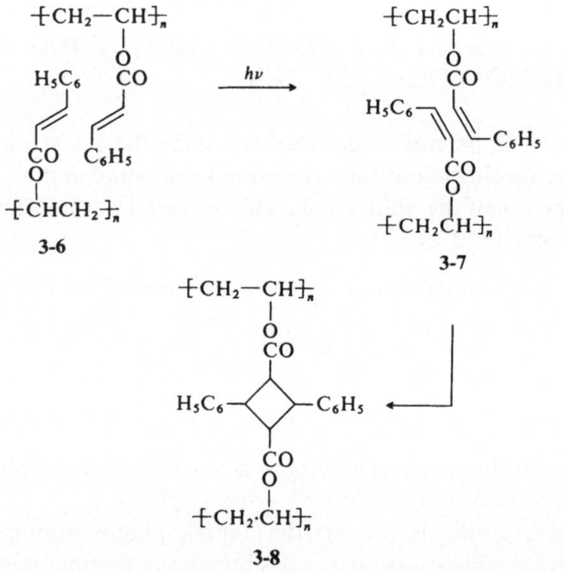

3-6                                                3-7

3-8

The use of the thermal decomposition of endoperoxides to release fluorescent polynuclear aromatic compounds is a known reaction discussed in both Chapter 1 and this chapter. The use of such a reaction in an imaging process was patented[8] by American Cyanamide Company in 1972. The imaging element comprised the endoperoxide adduct (**3-10**) in a polymeric film. The thermal imaging energy triggered the release of the free polynuclear fluorescent form (**3-9**) in an imagewise pattern. A variety of image colors were reported[8] depending on the polynuclear derivative released.

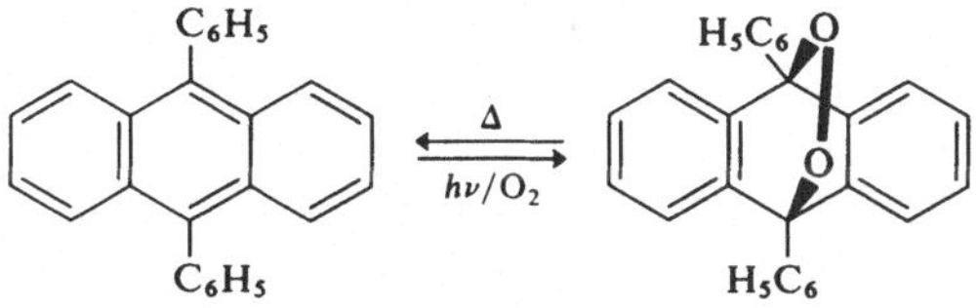

3-9        3-10

The use of photochemical release of surfactants has also been reported[9] as the chemistry for a peel-apart imaging system. Researchers at the 3M Company patented the use of photosensitive blocked surfactants (3-11) in a polymeric film construction used for image reproduction. Upon exposure to the photo-imaging energy, the surfactant (3-13) is released in an imagewise fashion, allowing the delamination of the polymeric film from the base in an imagewise pattern. The photosensitive blocking groups used on the surfactants were derivatives of the 2-nitrobenzyl group.

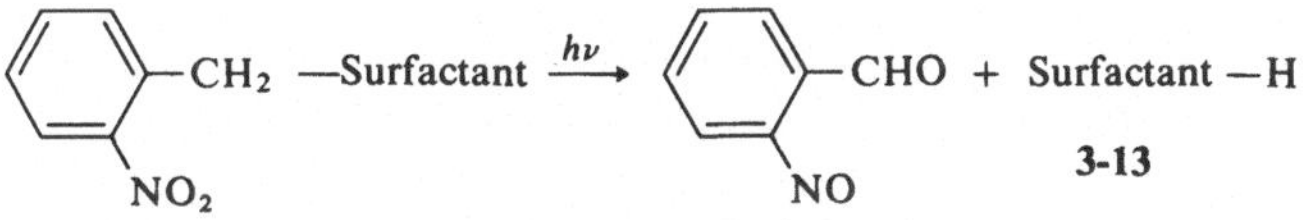

## 3.2. CHEMICAL REACTIONS FOR THE RELEASE OF MONOMERIC SPECIES

### 3.2.1. Thermally Triggered Release

The survey in this chapter will include compounds which upon thermal excitation can fragment or rearrange, giving rise to the release of new molecules.

The released molecule(s) generally is formed from a single parent reactant and not as the result of two reactants combining. Such a process is faster and simpler to conduct, since only a single chemical entity is required at the start.

The uses of such released molecules may vary and it is left to the reader to estimate the potential of the reactions and their possible applications.

Temperatures useful for such release reactions should be of a practical value, and thus the survey will focus on those reactions with release temperatures between 80 and 200°C. Other properties will also be described.

#### 3.2.1.1. Olefins from Esters, Ethers, Lactones, and Amine Oxides

A number of reactions are known to thermally release olefins. These fall within the class of either elimination reactions or ring rupture reactions.

The following are highlights of some of these reactions which provide a variety of olefins derivatives from a number of starting materials and under a range of thermal conditions. One such reaction is the well-known Chugaev reaction,[10] reviewed by Nace[11] with reference to olefinic products, their alcohol precursors, reaction conditions, and yields.

The gas release properties of this reaction have already been reviewed in Chapter 1, part of which is applicable to our discussion here.

The Chugaev reaction can be defined as the dehydration of an alcohol through the thermal decomposition of its xanthate ester (**3-14**), which contains at least one $\beta$-hydrogen, to produce an olefin. Xanthate esters are prepared from the sodium salt of the alcohol with carbon disulfide and methyl iodide.

The reaction was discovered by Chugaev[12] in 1899 and is analogous to the thermal decomposition of carboxylic esters,[13] established by Oppenheim and Precht,[14] carbamates, or carbonates of alcohols to yield olefins.[15]

The reaction proceeds through a cyclic transition state involving a *cis* $\beta$-hydrogen atom of the derivatized alcohol and the thione sulfur atom of the xanthate. Subsequent bond making and bond breaking gives an olefin and an unstable thiocarbonate derivative which decomposes into carbonyl sulfide and a mercaptan. The more electronegative the ester group, the less stable is the xanthate, as shown by the decomposition rate constants of cholesteryl xanthate in Table 3.1.

$$\overset{\overset{\text{S}}{\|}}{RSCOCH_2CH_2R^1} \xrightarrow{\Delta} RSH + COS + CH_2{=}CHR^1$$

**3-14**

In oxygen-containing esters, the stability increases in the order phenyl carbamate < ethyl carbonate < ethyl benzoate < ethyl acetate. Thus, the

*Table 3.1.* First-Order Rate Constants for the Thermal Decomposition of Various Cholesteryl Xanthates at 176°C

| Cholesteryl xanthate | $k \times 10^4$ (min$^{-1}$) |
|---|---|
| $C_6H_5CH_2$ | 214 |
| $4\text{-}ClC_6H_4$ | 295 |
| $4\text{-}NO_2C_6H_4CH_2$ | 623 |
| $(C_6H_5)_3C$ | 1430 |
| $C_2H_5$ | 120 |
| $(C_6H_5)_2CH$ | 495 |
| $2,4\text{-}(NO_2)_2C_6H_3CH_2$ | 706 |
| $CH_3$ | 143 |

Table 3.2. First-Order Rate Constants for the Thermal
Decomposition of Cholesteryl Acetate, Carbonate,
and Carbamate[a]

| Compound | $T(°C)$ | $k \times 10^4 \, (min^{-1})$ |
|---|---|---|
| Cholesteryl ethyl carbonate | 241 | 13.2 |
| Cholesteryl ethyl carbonate | 261 | 59.8 |
| Cholesteryl ethyl carbonate | 281 | 236 |
| Cholesteryl acetate | 281 | 23.5 |
| Cholesteryl acetate | 306 | 128 |
| Cholesteryl acetate | 329 | 557 |
| Cholesteryl choroacetate | 281 | 150 |
| Cholesteryl phenyl carbamate | 241 | 127 |
| Cholesteryl benzoate | 281 | 78.5 |

[a] Ref. 15.

more electronegative the oxygen substituent, the less stable is the corresponding ester toward thermal decomposition. This is reflected in the first-order thermal decomposition rate constants and energies of activation given in Tables 3.2 and 3.3, respectively.

Olefins can also be released from a variety of xanthates of primary, secondary, or tertiary cyclic or acylic alcohols, glycols, or dihaloalcohols. The temperatures used in these reactions are in the 100–250°C range, much milder than in the case of carboxylic esters,[13] which require temperatures of 300–600°C. Carbonate and carbamate esters fall, in the ease of releasing an olefin, between the xanthates and the carboxylic esters. The decomposition of carbamate and carbonate esters is characterized as a clean reaction since the by-products are carbon dioxide gas and an alcohol. On the other hand, xanthates suffer from some difficulties in their preparation and the contamination of the released olefins with sulfur by-products.

Table 3.4 lists a variety of olefins which are released under thermal triggering conditions[16] and their corresponding alcohol precursors.

Table 3.3. Energies of Activation for Thermal
Decomposition of Cholesteryl Acetate and
Ethyl Carbonate[a]

| Compound | $E_a$ (kcal) |
|---|---|
| Cholesteryl acetate | 44.1 |
| Cholesteryl ethyl carbonate | 44.0 |
| Cholesteryl S-methyl xanthate | 32.9 |

[a] Ref. 15.

*Table 3.4.* Thermally Triggered Release of Olefins from Xanthate Esters

| Alcohol precursor of xanthate ester | Released olefin | Percent yield $[T(°C)]$ | Reference |
|---|---|---|---|
| $CH_3CH(OH)C(CH_3)_3$ | $CH_2{:}CHC(CH_3)_3$ | 58 | 16a |
| $CH_3CH(OH)C(CH_3)_2C_2H_5$ | $CH_2{:}CHC(CH_3)_2C_2H_5$ | 67 | 16a |
| $CH_3CH_2CH(OH)C(CH_3)_3$ | $CH_3CH{:}CHC(CH_3)_3$ | 73 | 16a |
| $CH_3(CH_2)_2CH(OH)C(CH_3)_3$ | $CH_3CH_2CH{:}CHC(CH_3)_3$ | 63–82 | 16a, b |
| 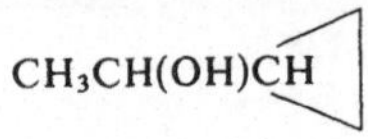 $CH_3CH(OH)\overset{}{CH}$ | 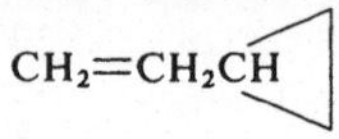 $CH_2{=}CH_2\overset{}{CH}$ | 42 | 16c |
| 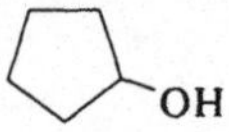 | 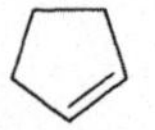 | 70 [255] | 16d |
| 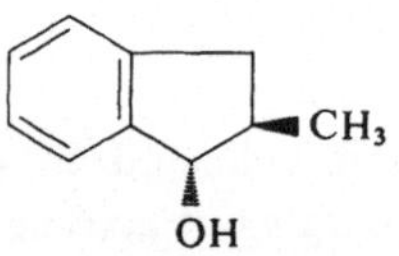 | 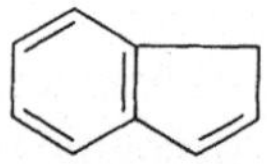 | 98 [100] | 16e, f |
| 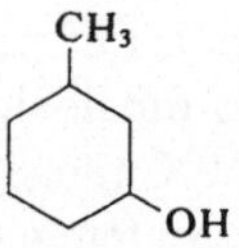 | 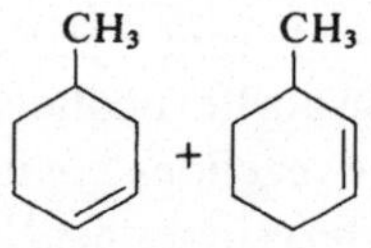 | 70 [100–250] | 16g |
| 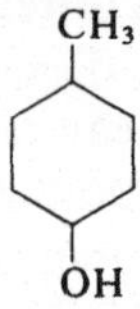 | 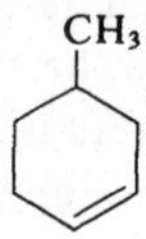 | 27 | 16h |
| 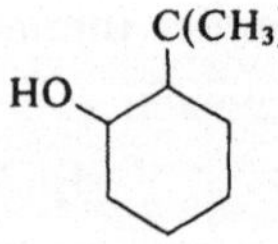 | 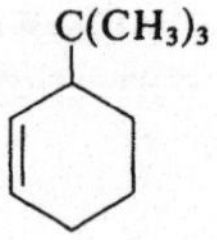 | 60 | 16i |
| 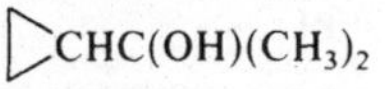 $\triangleright CHC(OH)(CH_3)_2$ | 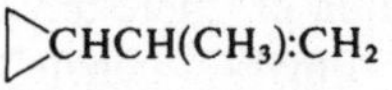 $\triangleright CHCH(CH_3){:}CH_2$ | 24 [130] | 16j |
| 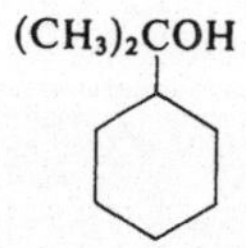 $(CH_3)_2\overset{}{C}OH$ | 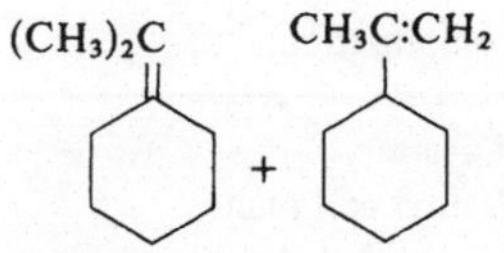 $(CH_3)_2C$ $+$ $CH_3C{:}CH_2$ | 40 [150] | 16k |
| $HOCH_2CH_2OH$ | $HOCH{=}CH_2$ | 26 [200–270] | 16l |

Pyrolysis of *o*-tertiary alkyl *N*-toluene-*p*-sulfonylcarbamates in bulk or in solution yields a mixture of *p*-toluenesulfonamide, carbon dioxide, and the corresponding olefin.[18] These carbamates can be prepared by the reaction of the specific alcohol and toluene-*p*-sulfonyl isocyanate.[19] The release of the olefins occurs slowly in the solid state and the rate increases at the melting point of 100–150°C. The reaction is believed to proceed via an ion pair rather than a concerted mechanism due to the formation of olefinic isomers. The ease of the reaction made it useful for olefin syntheses by the distillation of the parent carbamates.

$$4\text{-}CH_3C_6H_4SO_2NHCOOCR^1R^2CHR^3R^4 \rightarrow 4\text{-}CH_3C_6H_4SO_2NH_2 + CO_2 + R^1R^2C{=}CR^3R^4$$

**3-15**　　　　　　　　　　　　　　　**3-16**

Carboxylic esters seem to be the most difficult to decompose into the corresponding olefins at moderate temperatures. Exceptions are the *t*-butyl esters which were reported by Cain *et al.*[20] to decompose at about 100–150°C, forming carbon dioxide and the corresponding olefins.

**3-17**　　　　　　　　　**3-18**

| **3-18** | $T$(°C) | $t$ (min) | Percent yield |
|---|---|---|---|
| Cyclohexene | 110 | 120 | 38 |
| Norbornene | 135 | 120 | 34 |

Sterically hindered alcohols, when acetylated with *o*-(4-methyl)phenyl chlorothioformate, gave the corresponding thiocarbonates in high yields. These thiocarbonates on heating at temperatures above 160°C decomposed with the release of the corresponding olefins. The reaction is analogous to the Chugaev reaction but proceeds by an ionic mechanism.[21]

The use of *β*-lactones as precursors for the release of olefins has been examined by a number of researchers. The first *β*-lactone was isolated by Einhorn in 1883 during the treatment of *β*-bromo-*o*-nitrohydrocinnamic acid (**3-19**) with sodium carbonate.

**3-19**　　　　　　　　**3-20**

These lactones were later found to decompose thermally at moderate temperatures with the release of olefins and carbon dioxide in quantitative yields.

Adam *et al.*[17] reported the synthesis of a number of $\beta$-lactones [oxetan-2-ones (**3-21**)] from the corresponding $\beta$-hydroxyacids in yields between 40 and 100%. The decomposition temperatures[21,22] fall in the range of 140–160°C. The chemistry of this reaction is also discussed in Chapter 1.

$$
\begin{array}{c}
\text{R}^1 \\
\text{R}^2\!\!-\!\!\!\boxed{\phantom{xx}}\!\!-\!\text{O} \\
\text{R}^3\!\!-\!\!\!\boxed{\phantom{xx}} \\
\text{R}^4
\end{array}
\quad \xrightarrow{\;T\,°C\;}\quad \text{R}^1\text{R}^2\text{C}{=}\text{CR}^3\text{R}^4 + \text{CO}_2
$$

**3-21**                                    **3-22**

| $R^1$ | $R^2$ | $R^3$ | $R^4$ | Percent yield (olefin)[a] |
|-------|-------|-------|-------|----------------------------|
| H | $C_6H_5$ | $C_6H_5$ | $C_6H_5$ | 100 |
| H | $CH_3$ | $C_6H_5$ | $C_6H_5$ | 100 |
| H | $C(CH_3)_3$ | $C_6H_5$ | $C_6H_5$ | 100 |
| $CH_3$ | H | $CH_3$ | $C_6H_5$ | 100 |
| $CH_3$ | $CH_3$ | $C_6H_5$ | $C_6H_5$ | 100 |
| $CH_3$ | $CH_3$ | $CH_3$ | $C_6H_5$ | 100 |
| $CH_3$ | $CH_3$ | $C_6H_5CH_2$ | $C_6H_5$ | 100 |
| $CH_3$ | $C_6H_5CH_2$ | $C_6H_5CH_2$ | $C_6H_5CH_2$ | 100 |
| $CH_3$ | $CH_3$ | H | $C_6H_5$ | 100 |
| $CH_3$ | $CH_3$ | —$(CH_2)$— | | 100 |
| H | H | $C_6H_5$ | $C_6H_5$ | 55 |
| H | H | $C_6H_5CH_2$ | $C_6H_5$ | 64 |
| H | H | $CH_3CH_2$ | $C_6H_5$ | 60 |
| H | $CH_3$ | $C_6H_5CH_2$ | $C_6H_5$ | 99 |
| H | H | $CH(CH_3)_2$ | $C_6H_5$ | 82 |
| H | $CH_3$ | $C_6H_5$ | $C_6H_5CH_2$ | 97 |

[a] Refs. 21, 22.

Ethers have also been reported as precursors of olefins. For example, the reaction of alcohols with a 1-methyl-4-chloropyridinium salt (**3-25**) produces the corresponding 1-methyl-4-alkoxypryidinium salt (**3-23**) which upon heating undergoes decomposition[21] with the release of an olefin. Olefins have been prepared by this reaction in excellent yields at temperatures of 25–185°C. The procedure is a simple one, involving the heating of the alkoxypyridinium salt (**3-23**) until it melts and collection of the volatile olefin. The range of suitable alcohols was reported to include aliphatic cyclic and acyclic alcohols.

Another route to the triggered release of olefins lies in the thermal decompositon of amine oxides (**3-26**). Examples of such a reaction were published[22] in the early literature, but its synthetic applications were not emphasized until 1949 by Cope *et al.*[23] In 1960 an extensive review published by Cope and Trumbull[24] covered the chemistry of the thermal decomposition

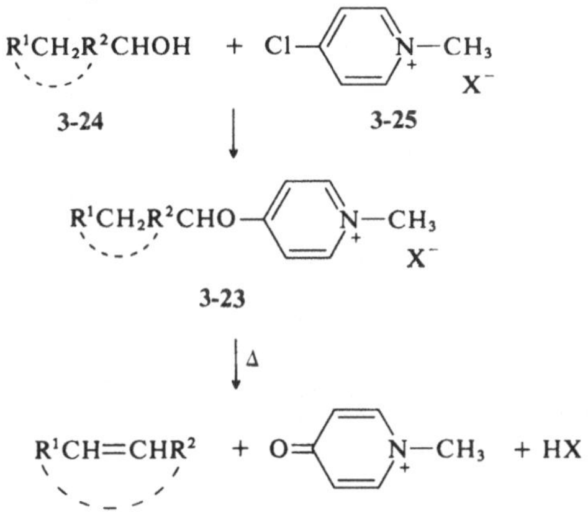

$$R^1CH_2R^2CHOH \quad + \quad Cl-\left\langle\begin{array}{c}\\ \end{array}\right\rangle\!\!\!N\!\!-CH_3$$
$$X^-$$

3-24          3-25

$$R^1CH_2R^2CHO-\left\langle\begin{array}{c}\\ \end{array}\right\rangle\!\!\!N\!\!-CH_3$$
$$X^-$$

3-23

$$\downarrow \Delta$$

$$R^1CH{=}CHR^2 \quad + \quad O{=}\left\langle\begin{array}{c}\\ \end{array}\right\rangle\!\!\!N\!\!-CH_3 \quad + \quad HX$$

| $R^1$—$R^2$ | Percent yield (olefin)[a] | Release temperature (°C) |
|---|---|---|
| Cyclopentyl | 100 | 165 |
| Cyclohexyl | 90 | 165 |
| Cycloheptyl | 87 | 165 |
| Cyclooctyl | 76 | 145–155 |
| $\Delta^2$-Cyclohexenyl | 100 | 25 |
| Menthyl | 100 | 170 |
| 1-Octyl | 100 | 165 |
| 1-Phenyl-1-propyl | 98 (trans) | 150 |
| 1-(4-Methyl)propyl | 99 (trans) | 110 |

[a] Ref. 21.

of amine oxides with tabulation of several reaction results, starting materials, and conditions.

$$R^1R^2CHC(R^3)(R^4)\overset{\overset{\displaystyle O}{\uparrow}}{N}R^5R^6 \overset{\Delta}{\rightarrow} R^1R^2C{=}CR^3R^4 + R^5R^6NOH$$

3-26

The reaction mechanism was determined by Cram *et al.*[25a,b] to be a *cis* elimination. The direction of the elimination parallels the stability of the formed olefin. In general, if exocyclic branching is present, the *endo* olefin (e.g., **3-27** and **3-29**) is preferred.[26] An exception is found in cyclohexyl rings (**3-28**) where the *exo* olefin is preferred.[26]

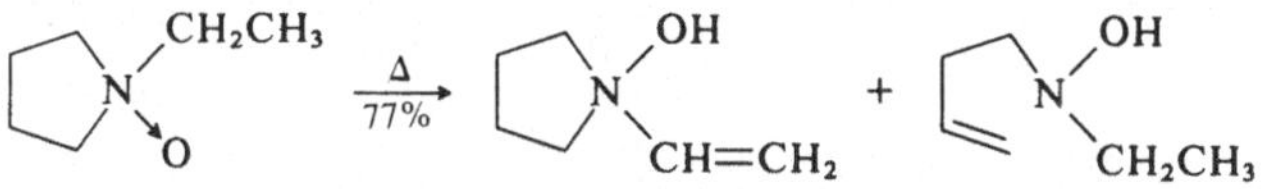

3-27            2.5            97.5

3-28                               97.2                          2.8

3-29                               15.2                          84.8

Table 3.5 contains some highlights reported by Cope et al.[27]

The thermal decomposition of thiiran-1-oxide, thiiran-1,1-dioxide, and thiiren-1,1-dioxide, all of which under mild conditions release olefins and sulfur dioxide, has been reported by Paquette[28] and Carpino et al.[29]; the chemistry is surveyed in Chapter 1.

Alkyl sulfites (3-30) also follow the same reaction pattern as the thiirane oxides and thermally release olefins together with sulfur dioxide. Their chemistry was reported by Price and Berti[30] and is also discussed in Chapter 1.

$$R^1CH_2CH(R^2)OSOOR^3 \rightarrow SO_2 + R^3OH + R^1CH=CHR^2$$
3-30

The release of dienes and sulfur dioxide from 2,5-dihydrothiophene dioxide (3-31) has been reported by Baker and Blass[32] to proceed easily under mild reaction conditions. The reaction is discussed in Chapter 1 in more detail because of its potential use in the release of sulfur dioxide gas.

3-31

Allenes, on the other hand, are released thermally from propargyl vinyl ethers under a Claisen–Cope type of rearrangement. Black and Landor[33] have demonstrated that propargyl vinyl ether (3-32) undergoes a thermally triggered rearrangement with the release of allenic aldehydes (3-33).

$$RR^1C=CHOC(R^2)(R^3)C\equiv CH \xrightarrow{T\,°C} OHCC(R)(R^1)HC=C=CR^2R^3$$
3-32                                        3-33

| R | R$^1$ | R$^2$ | R$^3$ | $T$ (°C) | Percent yield (allene) |
|---|---|---|---|---|---|
| H | H | H | H | 250 | 20–30 |
| H | H | H | CH$_3$ | 200 | 10–20 |
| H | H | CH$_3$ | CH$_3$ | 250 | 10–20 |
| CH$_3$ | CH$_3$ | H | H | 140 | 70 (4 hours) |
| CH$_3$ | CH$_3$ | CH$_3$ | CH$_3$ | 140 | 76 (15 minutes) |

*Table 3.5.* Thermally Triggered Release of Olefins from Amine Oxides

| Amine oxide precursor | Released olefin | Percent yield |
|---|---|---|
| $CH_3CH_2CH(CH_3)NH_2$ | $CH_3CH_2CH:CH_2$ | 67.3 |
| | $CH_3CH:CHCH_3$ | 31.7 |
| $CH_3CH_2CH(C_2H_5)NH_2$ | $CH_3CH_2CH:CHCH_3$ | 100 |
| $CH_3CH_2NH(CH_2)_2CH_3$ | $CH_2:CH_2$ | 62.5 |
| | $CH_3CH:CH_2$ | 37.5 |
| $CH_3CH_2NHCH(CH_3)_2$ | $CH_2:CH_2$ | 27.5 |
| | $CH_3CH:CH_2$ | 72.5 |
| $CH_3CH_2NH(CH_2)_3CH_3$ | $CH_2:CH_2$ | 55.5 |
| | $CH_3CH_2CH:CH_2$ | 44.5 |
| $CH_3CH_2NHCH_2CH(CH_3)_2$ | $CH_2:CH_2$ | 67.5 |
| | $(CH_3)_2CH:CH_2$ | 32.4 |
| $CH_3(CH_2)_2NH(CH_2)_3CH_3$ | $CH_3CH:CH_2$ | 43.1 |
| | $CH_3CH_2CH:CH_2$ | 56.9 |
| $CH_3(CH_2)_2NHCH_2CH(CH_3)_2$ | $CH_3CH:CH_2$ | 58.8 |
| | $(CH_3)_2CH:CH_2$ | 41.2 |
| $CH_3(CH_2)_3NHCH_2CH(CH_3)_2$ | $CH_3CH_2CH:CH_2$ | 64.8 |
| | $(CH_3)_2C:CH_2$ | 35.2 |
| $CH_3(CH_2)_3NH(CH_2)_2CH(CH_3)_2$ | $CH_3CH:CH_2$ | 38.7 |
| | $CH_2:CHCH(CH_3)_2$ | 49.1 |
| | $CH_3CH:C(CH_3)_2$ | 11.2 |
| | $CH_2:C(CH_3)CH_2CH_3$ | 1.0 |

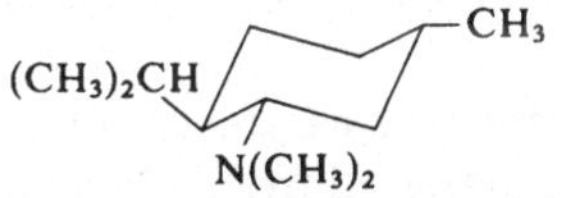

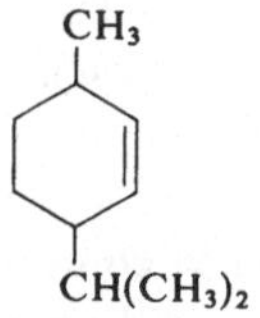

85

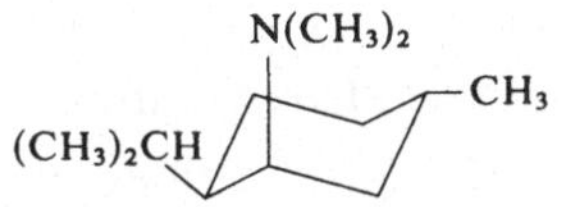

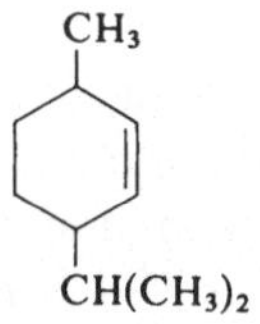

77

| 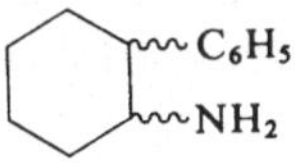 | 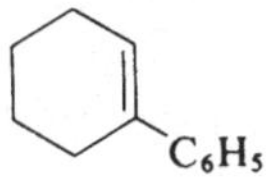 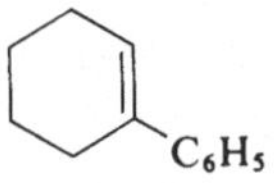 | |
|---|---|---|
| *cis*-Isomer | | 98.0 : 2.0 |
| *trans*-Isomer | | 15.0 : 85.0 |
| $CH_2:CH(CH_2)_3NH_2$ | $CH_2:CHCH_2CH:CH_2$ | 61 |
| $C_6H_5(CH_2)_3NH_2$ | $C_6H_5CH_2CH:CH_2$ | 91 |

The mechanism involves a cyclic, six-membered transition state with C,C-bond formation and C,O-bond breaking simultaneously, analogous to the Claisen rearrangement of vinyl ethers.

In a related reaction, reported by Ziegler,[34] an excellent route to the release of tetraphenylallenes (3-34b) and related compounds is provided by the thermal decomposition of a tertiary alcohol of the type 3-34a at 140–150°C.

$$(C_6H_5)_2CH{=}CHC(OH)(C_6H_4R)_2$$

**3-34a**

$$\Big\downarrow_\Delta \quad (C_6H_5)_2C{=}C{=}C(C_6H_4R)_2$$

**3-34b**

### 3.2.1.2. Aliphatic and Aromatic Hydrocarbons from Diazonium Salts, Aryl Ethers, Heterocycles, Peroxides, and Azo Alkanes

A general route toward the triggered release of aromatic compounds and, in specific, aryl halides is through the thermal decomposition of aryl diazonium salts. The fluoro-, chloro-, and bromoborate salts as well as the fluorophosphates give aryl halides together with nitrogen gas. This reaction, first reported in 1927, is known as the Schiemann reaction[35] and was reviewed by Roe.[36] It is discussed in Chapters 1 and 2 as a route to the release of nitrogen gas, acids, and radicals.

The reaction utilizes diazonium fluoroborate salts as precursors to the corresponding fluoroaryl compounds. Salts are prepared from a large number of amines with overall yields as high as 70% and provide a route to the introduction of a fluorine atom on polynuclear and heterocyclic systems such as naphthalene, phenanthrene, anthracene, biphenyl, fluorene, benzanthrone, pyridine, and quinoline. An extension included the introduction of substituents other than fluorine, for example, an acetoxy group,[37] a nitro group,[38] a nitrile group,[39] a hydrogen atom,[40] and a methoxy group[41] as well as aromatic arsenic,[42] mercury,[43] and copper[44] derivatives. The yields of the thermal release of the fluoroaryl derivatives are a function of the ring substituent, as shown in the following examples.

$$R{-}\underset{\textbf{3-35a}}{\boxed{\bigcirc}}{-}N_2^+BF_4^- \quad \xrightarrow{T\,°C} \quad R{-}\underset{\textbf{3-35b}}{\boxed{\bigcirc}}{-}F \; + \; N_2 \; + \; BF_3$$

| R | Percent yield (3-35b) | $T(°C)$ | Reference |
|---|---|---|---|
| H | 100 | 100 | 45a |
| 2-CH$_3$ | 90 | 106 | 45b |
| 4-CH$_3$ | 99 | 110 | 45c |
| 2-CH$_2$CH$_2$OCO | 87 | 105–118 | 45d |
| 4-CH$_3$CH$_2$OCO | 90 | 93 | 45e |
| 2-Cl | 85 | 171 | 45e |
| 2-Br | 81 | 156 | 45f |
| 4-Br | 75 | 133 | 45g |
| 4-I | 70 | 89–109 | 45h |
| 3-I | 81 | 104–134 | 45h |
| 4-F | 62 | 154 | 45i |
| 2-F | 30 | 154 | 45j |
| 2-CH$_3$CH$_2$O | 36 | 105–135 | 45d |
| 4-CH$_3$CH$_2$O | 53 | 105 | 45d |
| 2-NO$_2$ | 29 | 135 | 45k |
| 4-NO$_2$ | 58 | 156 | 45k |
| 2-COOH | 19 | 125 | 45l |
| 4-COOH | 78–84 | — | 45a |
| 4-(CH$_3$)$_2$N | 17 | 155 | 45m |
| 2-C$_6$H$_5$ | 85 | 81 | 45n |
| 3-C$_6$H$_5$ | 85 | 91 | 45o |
| 4-C$_6$H$_5$ | 88–94 | 116 | 45o |
| 4-(4'-F$_4$BN$_2$)C$_6$H$_4$ | 64–94 | 135–157 | 45i |
| 3-(3'-F$_4$BN$_2$)C$_6$H$_4$ | 98 | 106 | 45o |

Aromatic hydrocarbons can also be released from heterocycles containing sulfur dioxide in their ring structure, such as 1,3-dihydroisothianaphthalene-2,2-dioxide (3-36a), its analogue (3-36b), and thiapin-1,1-dioxides[47] (3-36c) when heated to their melting points. Further discussion of this reaction is included in Chapter 1.

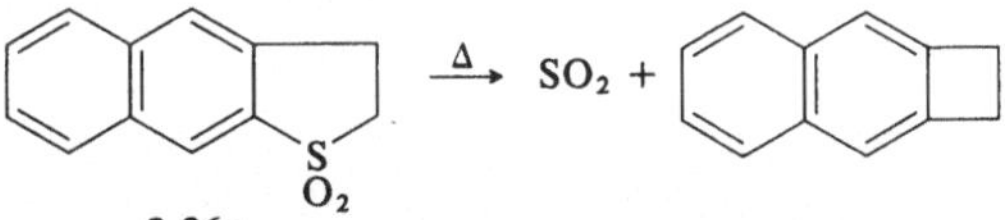

3-36a

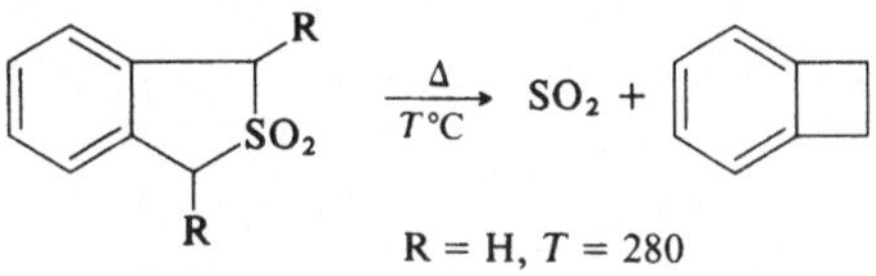

R = H, T = 280
R = C$_6$H$_5$, T = 250

3-36b

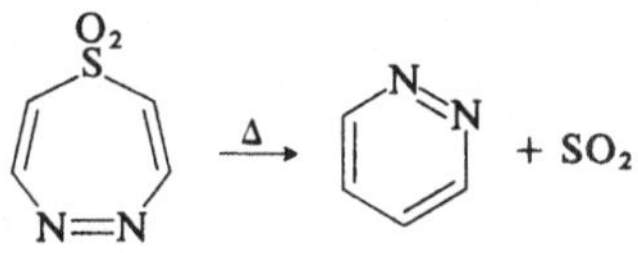

3-36c

Azo compounds (**3-37**) have been reported as precursors for the release of alkanes through their thermal decomposition, which was reviewed by Engel[48a] and is surveyed in depth in Chapters 1 and 2 as a route to the release of nitrogen gas and radicals.

$$R(CH_3)_2CN=NC(CH_3)_2R \xrightarrow{\Delta} [R(CH_3)_2C]_2 + N_2$$

**3-37**

Substituted aryl compounds can also be released through the use of the Smiles type rearrangement which involves the formation of an aromatic ether or amine from the corresponding substituted phenol (**3-38**).

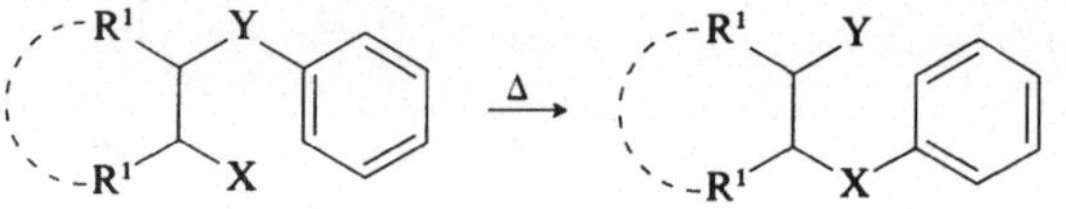

**3-38**     X = N, O; Y = SR, SR$_2$, O

This type of rearrangement has been known since it was reported by Henriques[49] in 1894 and later by Hinsberg,[50] but it was not until the studies of Smiles[51] in the 1930s that it was recognized as an intramolecular nucleophilic rearrangement. The rearrangement involves a six-membered cyclic transition state with bond formation between a nucleophilic oxygen or nitrogen and an adjacent ring and bond breaking between that ring and a sulfur or oxygen atom. The rearrangement requires electronic activation of the electrophilic aromatic ring, commonly by nitro groups.

Early studies by Galbraith and Smiles[52] on the effect of substituents on the reaction rates involved calorimetric measurements of the time taken by derivatives (**3-39**) to complete the rearrangement.

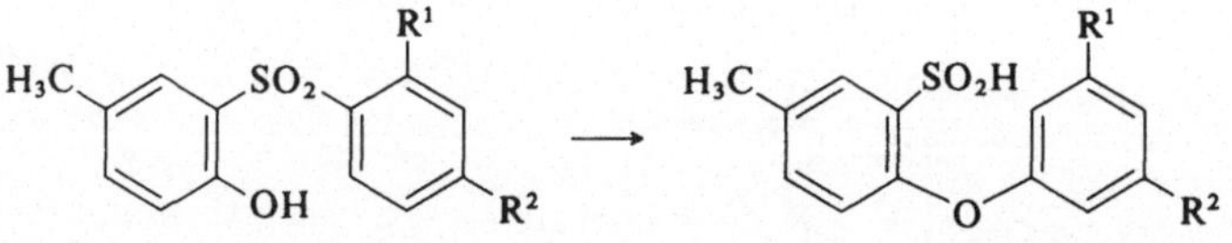

**3-39a**                                                **3-39b**

| R$^1$ | R$^2$ | Release rate of (**3-39b**) (min) |
|-------|-------|-----------------------------------|
| NO$_2$ | NO$_2$ | Very rapid |
| NO$_2$ | C$_6$H$_5$CO | 0.5 |
| NO$_2$ | CO$_2$Na | 50 |
| NO$_2$ | Cl | 70 |
| NO$_2$ | H | 22 |
| H | NO$_2$ | Very slow |
| CO$_2$Na | NO$_2$ | 450 |
| C$_6$H$_5$CO | NO$_2$ | 12 |

The nature of the leaving group X and the nucleophilicity of the group Y are, as expected, essential and interrelated. The chain size and its substituents linking the leaving group X to the attacking group Y also influence the direction of the reaction through electronic or steric factors. For example, steric acceleration of the reaction is observed in derivatives such as (3-40), (3-41), and (3-42), described by McClement and Smiles.[53] The following are a few reported rearrangements. Extensive reviews of the subject are

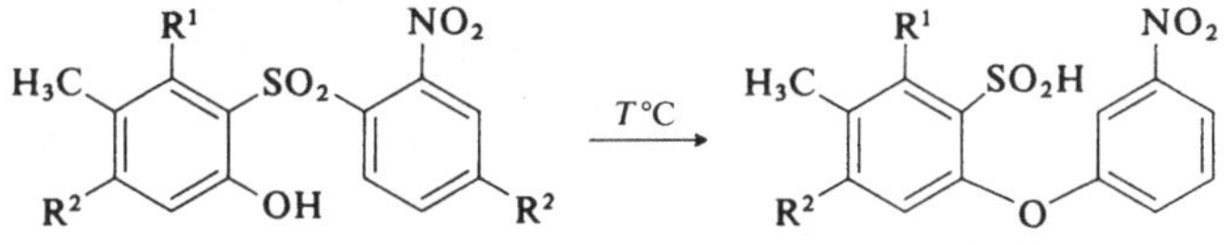

3-40a      $T\,°C$      3-40b

Release rate of (3-40b),
$k \times 10^{-2}\,(\text{min}^{-1})^{a}$

| $R^1$ | $R^2$ | 46°C | 0°C |
|---|---|---|---|
| H | H | 1.94 | — |
| H | CH$_3$ | 1.64 | — |
| H | Cl | 1.44 | — |
| H | Br | 1.23 | — |
| CH$_3$ | H | — | $3.00 \times 10^2$ |
| Cl | H | — | $0.92 \times 10^2$ |
| Br | H | — | $2.1 \ \times 10^2$ |

[a] Refs. 53 and 54.

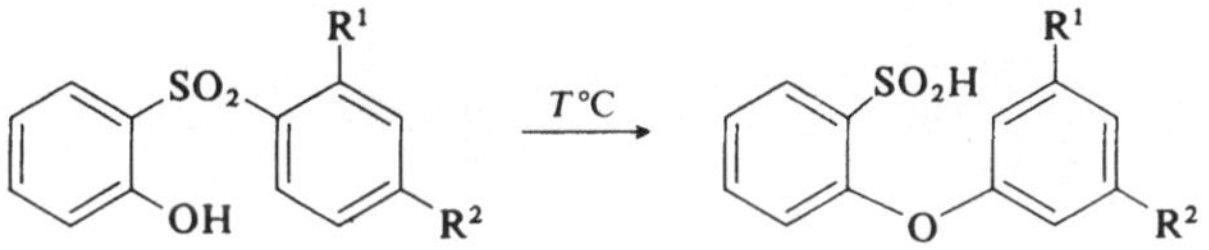

3-41a      $T\,°C$      3-41b

Release time[a] of (3-41b) (min)[b]

| $R^1$ | $R^2$ | 0°C | 50°C |
|---|---|---|---|
| NO$_2$ | NO$_2$ | 6-8 | Very rapid |
| NO$_2$ | C$_6$H$_5$CO | 40 | Less than 0.5 |
| C$_6$H$_5$CO | NO$_2$ | — | 12 |
| NO$_2$ | COONa | — | 50 |
| NO$_2$ | Cl | — | 70 |
| NO$_2$ | H | — | 255 |
| COONa | NO$_2$ | — | 450 |
| H | NO$_2$ | — | Very slow |

[a] The above data show the approximate times required for complete reaction of an equivalent amount of (3-41a) in a methanolic alkaline solution at two temperatures.
[b] Ref. 52.

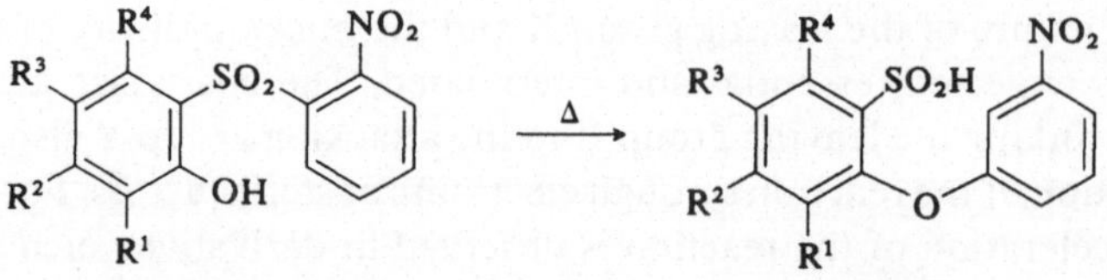

3-42a                                    3-42b

| $R^1$ | $R^2$ | $R^3$ | $R^4$ | Release time[a] of (3-42b) (min) |
|-------|-------|-------|-------|----------------------------------|
| $CH_3$ | H | $CH_3$ | $CH_3$ | Less than 3 |
| $CH_3$ | H | Cl | $CH_3$ | 15 |
| Cl | H | $CH_3$ | $CH_3$ | 12 |
| H | $CH_3$ | Cl | $CH_3$ | 5 |
| H | H | $C_6H_5$ | $C_6H_5$ | 5 |
| $CH_3$ | H | $CH_3$ | H | 95 |
| $CH_3$ | H | Cl | H | More than 150 |
| Cl | H | $CH_3$ | H | More than 150 |
| Cl | $CH_3$ | $CH_3$ | H | More than 150 |

[a] The above times are approximate times to complete the rearrangement at 50°C in an alkaline medium.

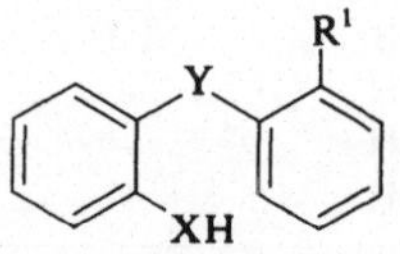

3-42c

| $R^1$ | XH | Y |
|-------|-----|---|
| 2- or 4-$NO_2$ | NHCOR | $SO_2$, SO, S, O |
| | NHR | $SO_2$, O |
| | NHAr, NHR | $SO_2$, O |
| | CONHAr | $SO_2$, O |
| | $CONH_2$ | $SO_2$, O |
| | $CH_2NH_2$, $CH_2NHR$ | O |
| | $SO_2NH_2$ | O |
| | SH | O |
| | OH | $SO_2$, SO |
| | OH | $SO_2$, $CO_2^-$, $SO_3^-$, O, $NHSO_2$ |
| 2-Halogen | NHOH | $SO_2$, SO, S, O |
| | NHR, $NH_2$ | O |
| | $CH_3$, $CH_2R$ | $SO_2$, $P^+$ |
| | $CH_2Ar$, $CHAr_2$ | $SO_2$ |
| | COOH, COOR | $I^+$ |

published by Truce et al.[55] and by Bunnett and Zahler;[54] these contain tabulated examples and results.

The release of polynuclear aromatic hydrocarbons is well exemplified through the thermal reaction of endoperoxides[56] which release molecular oxygen and the parent polynuclear aromatic compound. These endo-peroxides are stable cycloadducts of molecular oxygen and a polynuclear aromatic molecule.

A review of these endoperoxides and their properties has been published by Bergmann and McLean[56] with tabulated results. Chapter 1 also contains a discussion of their reaction because of its potential value as a thermal route to the release of oxygen gas. In this section we will highlight two examples with emphasis on the properties of the released polynuclear molecule.

The thermal decomposition of tetranuclear peroxides was observed by Dufraisse and Priou[57a] when preparing rubene (**3-43**) in the 1920s. Rubene is a red-colored compound which on exposure to light in the presence of oxygen rapidly loses its color with the precipitation of the oxygen adduct (**3-44**).

**3-44**                                    **3-43**

The reaction is thermally reversible with the release of rubene and oxygen from the parent adduct (**3-44**). Analogously, *trans*-peroxide adducts of naphthacene and anthracene have also been formed while those of naphthalene,[57a] acridine,[57b] and phenanthrene[57b] have not.

Substituted phenols can be released thermally from their corresponding allyl ethers. The reaction involves a rearrangement of the allyl group from the oxygen to a carbon atom. The reaction is also common to allyl enols, which in turn release the corresponding ketones. Claisen in 1912 discovered this reaction and since then it has been known as the Claisen rearrangement.[58]

In general, molecular structures suitable for such a rearrangement can be described by (**3-45**), which can represent part of an enol,[59] an allyl vinyl ether, or part of an aromatic phenol ether.[60]

**3-45a**                                    **3-45b**

The first observed example of this type of reaction involved the rearrangement of ethyl $O$-allylacetoacetate under distillation. The triggering temperatures of these reactions tend to be relatively high, between 100 and 200°C, without the presence of a catalyst. The rearrangement of aromatic allyl ethers commonly proceeds from the oxygen atom to the free ortho-carbon atom. If none is available, then the allyl group migrates to the para position. The latter migration is as facile as the ortho migration with yields of about 80%.

Substituents at the aromatic ring seem not to affect the rearrangement; for example, meta-directing groups do not hinder the migration nor do ortho- or para-directing groups enhance it. Some migrations involving allyl ether proceed with displacement of the ortho substituent. For example, $O$-allyl-2,6-diallylsalicylic acid (**3-46**) releases quantitatively at 100°C the corresponding 2,4,6-triallylphenol (**3-47**) and carbon dioxide.[60] Similarly, $O$-allylsalicylic acid (**3-48**) heated to 175–180°C releases 2-allylphenol and 3-allylsalicylic acid (**3-49**) in 23% and 64% yields, respectively.[61]

**3-46**  $R^1 = R^3 = CH_2CH{=}CH_2$
           $R^2 = COOH$

**3-47**  $R^4 = R^5 = R^6 = CH_2CH{=}CH_2$

**3-48**  $R = R^3 = H$
           $R^3 = COOH$

**3-49**  $R^5 = CH_2CH{=}CH_2$
           $R^4 = R^6 = H$

and        $R^4 = CH_2CH{=}CH_2$
           $R^6 = H, R^5 = COOH$

The displacement of a carboxylic function in the para position is also possible, as in the reaction of (**3-46**), which proceeds with 99% yield. Displacement of an aldehyde group as carbon monoxide is less feasible with yields of about 64% and tends to require higher temperatures[62] of 170–180°C.

Tarbell and Wilson[62] also reported that crotyl ethers rearrange more readily than the corresponding allyl ethers. Extension of the Claisen rearrangement to a number of allyloxy derivatives has been reported, among which are anthracene,[63] phenanthrene,[64] hydroindene,[65] fluorene,[66] chromone,[67] and fluorenone.[68]

The generally accepted mechanism is either an intramolecular, cyclic rearrangement to the ortho position or ion or radical dissociation and recombination at the para position.[69] Both types of migration follow first-order kinetics and do not require catalysts.[69a] The first-order rate constants at 200°C for the rearrangement of allyl-*p*-tolyl ether (**3-50a**) showed a 77% conversion in 72 minutes.[70] At longer times the reaction proceeds quantitatively.

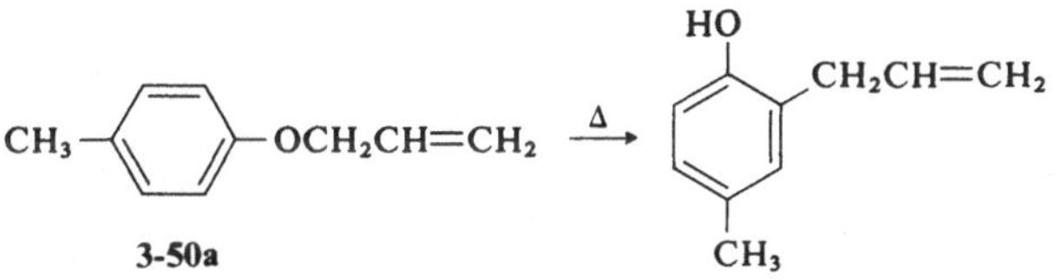

| $t$ (min) | $k \times 10^5 \, s^{-1}$ | Percent yield |
|---|---|---|
| 11.0 | 9.82 | 12.12 |
| 13.00 | 9.65 | 13.81 |
| 24.60 | 10.49 | 26.61 |
| 27.33 | 10.60 | 29.37 |
| 59.33 | 14.81 | 65.16 |
| 69.83 | 16.61 | 75.13 |
| 72.33 | 17.10 | 77.34 |

The para rearrangement[69] at 171°C proceeds with a higher yield than the analogous ortho rearrangement.

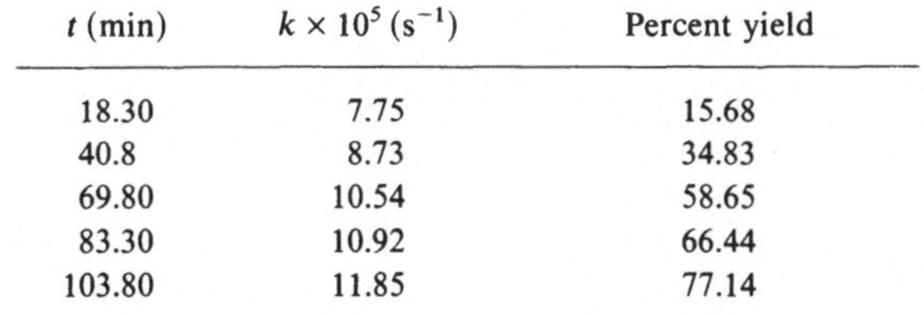

| $t$ (min) | $k \times 10^5 \, (s^{-1})$ | Percent yield |
|---|---|---|
| 18.30 | 7.75 | 15.68 |
| 40.8 | 8.73 | 34.83 |
| 69.80 | 10.54 | 58.65 |
| 83.30 | 10.92 | 66.44 |
| 103.80 | 11.85 | 77.14 |

An extensive review of the Claisen rearrangement by Tarbell[70] provides background, applications, and results. The following are highlights.

$$ArOCH_2CH=CH_2 \xrightarrow{T \, °C} HOArCH_2CH=CH_2$$

**3-51**                              **3-52**

| Ar | $T(°C)$ | Percent yield (3-52) | Reference |
|---|---|---|---|
| $C_6H_5$ | 199–200 | 85 (2-isomer) | 71a |
| $2\text{-}CH_3C_6H_4$ | 207–231 | 70 (6-isomer) | 61 |
| $2\text{-}NO_2C_6H_4$ | 180 | 72 (4-isomer) | 61 |
| $4\text{-}NH_2C_6H_4$ | 185 | 70 (2-isomer) | 71a |
| $4\text{-}CH_3ONHC_6H_4$ | 180 | 85 (2-isomer) | 71a |
| $2\text{-}HOC_6H_4$ | 170–265 | 85 (4-isomer) | 71b |
| $2\text{-}CH_2\text{:}CHCH_2OC_6H_4$ | 180 | — (4-isomer) | 71b |
| $2\text{-}C_6H_5CH_2OC_6H_4$ | 130–280 | 85 (mixture) | 71c |
| $3\text{-}HO\text{-}4\text{-}NO_2C_6H_3$ | 185 | 26 | 71d |
| 2,6-Diallyloxynaphthalene | 190 | 85 | 63 |
| 2-Aliyloxyphenanthrene | 100 | — | 64 |
| 3-Allyloxyphenanthrene | 100 | — | 64 |
| 2,6-Diallyloxyanthracene | 160–180 | 55 | 71e |
| 4-Allyloxy-2-methylquinoline methiodide | 175 | — | 71f |

The release of cyclic alkanes was reported[72] to be possible from the corresponding cyclic di- and triperoxides (3-53). Stalsy and Reichard[72] have shown that a number of diperoxides upon heating release large-membered cyclic alkanes. A review of the chemistry of cyclic peroxides was published by McKay et al.[73]

| Diperoxide | Product | Percent yield |
|---|---|---|
| Cyclohexanone | Cyclodecane | 44 |
| Cycloheptanone | Cyclododecane | 23 |
| Cyclododecanone | Cyclodococane | 20 |
| Cyclopentanone | Cyclododecane | 20 |

### 3.2.1.3. Heterocycles from Other Heterocycles

The release of heteroatom-containing molecules finds use in a variety of applications. A few reactions are known to yield such compounds under triggering conditions. Three-membered heterocycles have been reported[74] to be released on heating large-membered rings. Thiiranes (3-56) are formed in a variety of yields from oxathiolan-5-one (3-55) at temperatures ranging between 150 and 250°C.

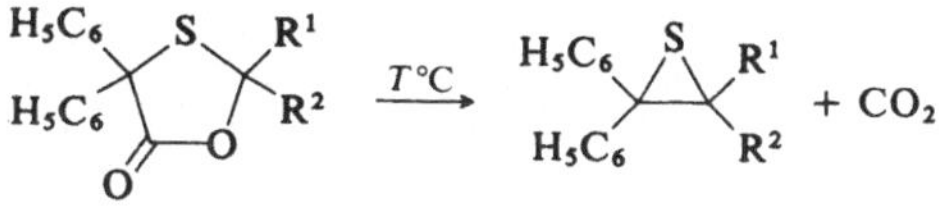


3-55 3-56

| $R^1$ | $R^2$ | $T$ (°C) | $t$ (h) |
|---|---|---|---|
| H | $C_6H_5$ | 150–160 | 2 |
| 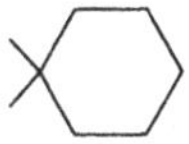 | | 160–200 | 5 |
| 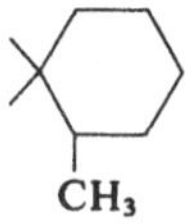  | | 220–240 | 6 |
| 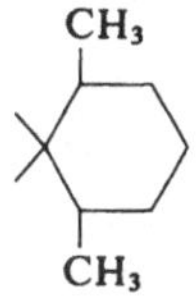  | | 180–320 | 4 |

In an extension of the above reaction, thiiranes are also reported[75] to be easily released from the corresponding ethylene monothiocarbonates (**3-57**).

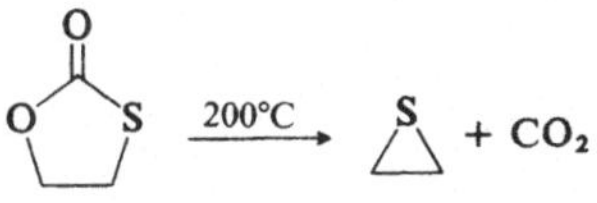


**3-57** **3-58**

Reynolds *et al.*[76] have compared a range of cyclic thiocarbonates to their open-chain analogues such as (**3-59**) and (**3-60**). A more detailed discussion is included in Chapter 1.

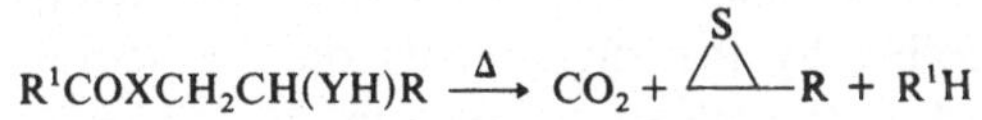


**3-59** $R^1 = C_2H_5O$, X = S, Y = O
**3-60** $R^1 = C_2H_5NH$, X = O, Y = S
$R^1 = C_6H_5NH$, X = O, Y = S

A number of cyclic ethers (**3-62**) of a variety of ring sizes have been reported by Pattison[77] as thermal products from the heating of cyclic carbonates (**3-61**).

$$\text{HO--(CH}_2)_n\text{---O} \quad \xrightarrow{180\text{--}230°C} \quad \text{(CH}_2)_n\text{---O---CH}_2\text{OH} \quad + \quad CO_2$$

$$\textbf{3-61} \qquad\qquad\qquad \textbf{3-62}$$

The release of furans[78] (**3-64**) and pyrroles[79] (**3-66**) as products from the thermal reaction of sultones (**3-63**) and sultames (**3-65**), respectively, has been reported.

$$\textbf{3-63} \quad X = O \qquad\qquad \textbf{3-64} \quad X = O$$
$$\textbf{3-65} \quad X = NH \qquad\qquad \textbf{3-66} \quad X = NH$$

These reactions are valuable due to the large variety of ring substituents which can be included and their high yields. Chapter 1 includes a discussion of both reactions as potential sources of sulfur dioxide. In this chapter interest is focused on the heterocyclic co-products.

Five-membered heterocycles with two or more heteroatoms are also known to be thermally released under mild conditions. The triggered release of 2-oxazolines (**3-68**) from 1-acylaziridines (**3-67**) was first observed by Gabriel and Stelzner[80] and later by Heine.[81] The latter examined the reaction at a range of temperatures and pressures and found that under reduced pressure the aziridine distilled unchanged.

$$RCO\text{---}N \overset{R^1}{\underset{R^2}{\triangleleft}} \quad \xrightarrow{\Delta} \quad R\text{---}\underset{N}{\overset{O}{\diagup}}\underset{R^2}{\overset{R^1}{\diagdown}}$$

$$\textbf{3-67} \qquad\qquad \textbf{3-68}$$

The reaction in the case of unsubstituted aziridines proceeds at temperatures of about 200–300°C. The decomposition temperatures and the yields are controlled by the substituents on the aziridine ring. Certain substituents can drop the release temperature to as low as 5°C from 200°C. The following examples clarify the potential of this reaction.

$$\text{NCOC}_6\text{H}_5 \quad \xrightarrow[320\text{--}330°C]{\Delta} \quad \text{C}_6\text{H}_5$$

3-Benzoyl-3-azatricyclo[3.2.1.0$^{2,4\text{-}exo}$]octane[82a]

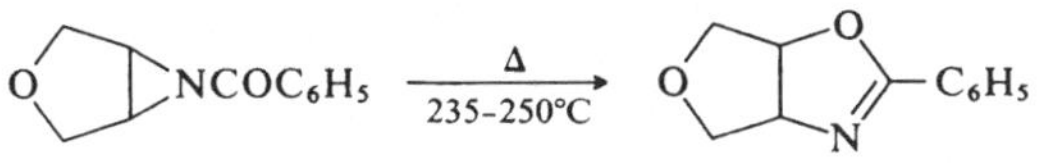

6-Benzoyl-3-oxa-6-azabicyclo[3.1.0]hexane[82b]

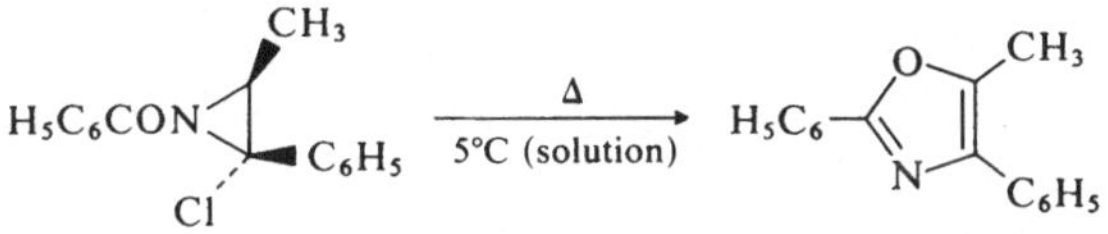

cis- or trans-1-Benzoyl-2-chloro-2-phenyl-3-methylaziridine[82c]

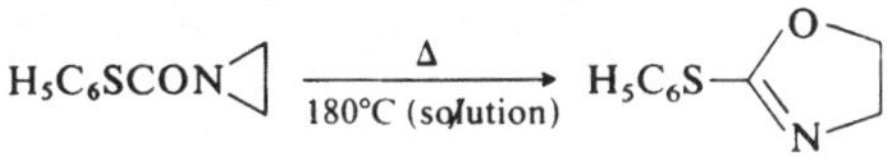

1-Carbothiophenoxyaziridine[82g]

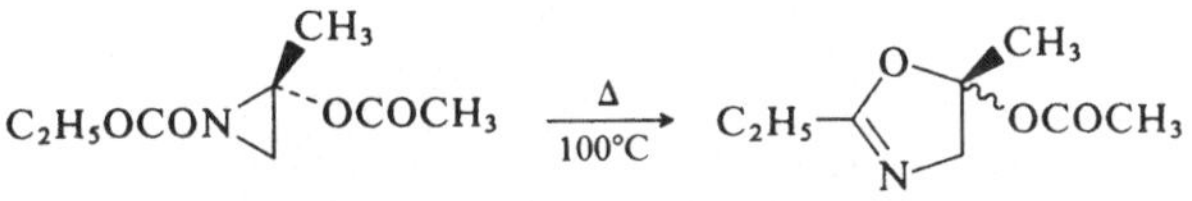

1-Carboethoxy-2-acetoxy-2-methylaziridine[82h]

1-Aziridinethiocarboxyanilides[82i,j]

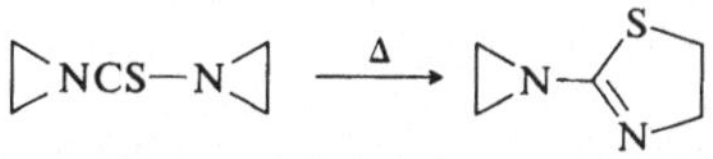

Ref. 82k

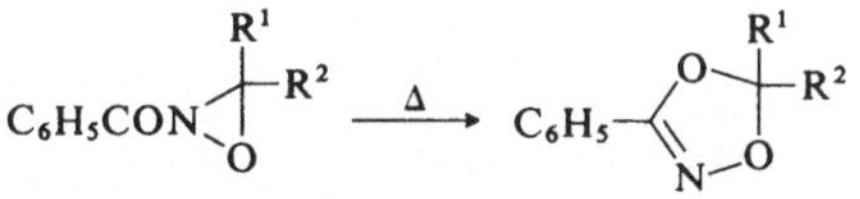

Ref. 82l

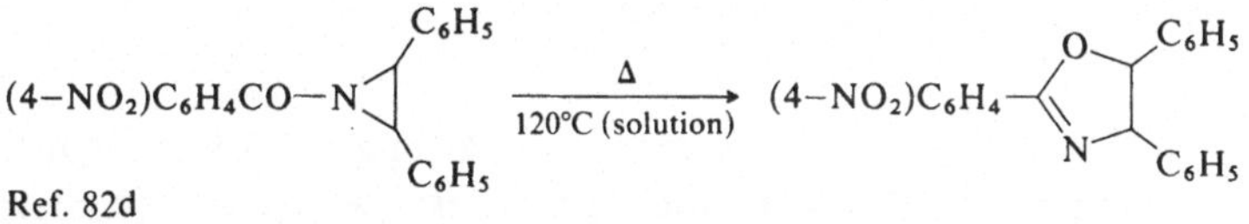

Ref. 82d

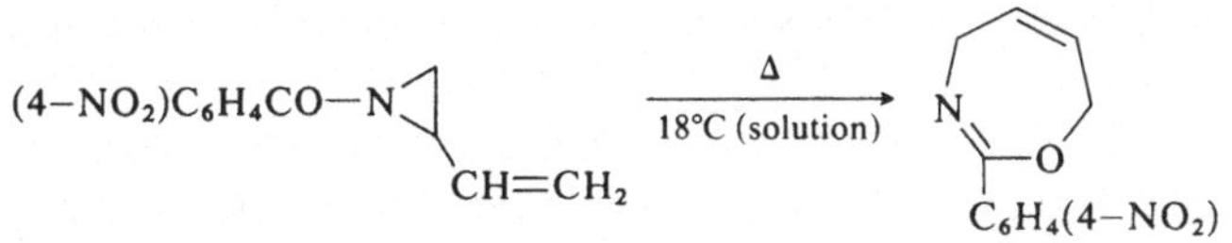

Ref. 82e

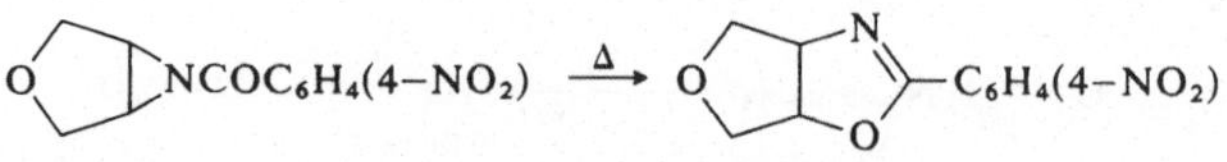

Ref. 82f

Substituted pyrazoles (**3-70**) can be formed from $N$-nitropyrazoles (**3-69**) under thermal triggering.[83] This intramolecular migration of the nitro group can be visualized as a (1,5) sigmatropic shift giving 3-$H$-pyrazole followed by a fast tautomerization.

These $N$-nitropyrazoles can be prepared by the nitration of pyrazole with a preformed mixture of nitric acid and acetic anhydride. The reaction offers a convenient synthesis of 3(5)-nitropyrazole derivatives, which are purple in color.

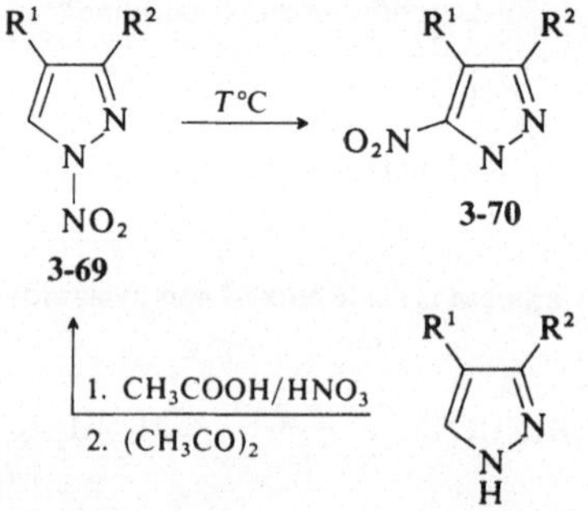

| $R^1$ | $R^2$ | $T(°C)$ | $t$ (h) | Percent yield (**3-70**) |
|-------|-------|---------|---------|--------------------------|
| H | H | 180 | 3 | 80 |
| $CH_3$ | H | 145 | 2 | 80 |
| $C(CH_3)_3$ | H | 130 | 2.5 | 80 |
| $C_6H_5$ | H | 130 | 1.5 | 80 |
| $4\text{-}NO_2C_6H_4$ | H | 140 | 1.5 | 80 |
| $NO_2$ | H | 140 | 10 | 80 |
| H | $CH_3$ | 160 | 20 | 50–80 |
| H | $CH_3CH_2$ | 140 | 50 | 50–80 |
| H | $C_6H_5$ | 120 | 2.5 | 50–80 |
| H | $NO_2$ | 191 | 6 | 50 |
| $C_6H_5$ | $NO_2$ | 140 | 1.5 | 80 |

Recently, other examples of pyrazole rearrangement were reported by Durr and Sergio[85] and Newman and Buchocker,[86] who observed migrations of ester, acyl, and cyano groups which they explained in terms of a (1,5) sigmatropic migration.

Benzo analogues of pyrazoles also undergo thermal release reactions. Fernandes and Habraken[87] have shown that 2-nitroindazole (**3-71**) rearranges on heating to 3-nitroindazole (**3-72**).

R2
R3
R4
R5
N—NO2
Δ
NO2
NH
N

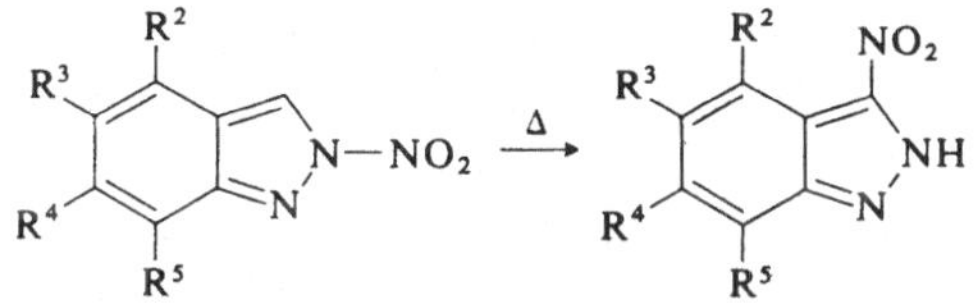

**3-71** **3-72**

$R^2 = R^4 = R^5 = H$, $R^3 = NO_2$
$R^3 = R^4 = R^5 = H$, $R^2 = NO_2$
$R^2 = R^3 = R^4 = H$, $R^5 = NO_2$
$R^2 = R^3 = R^4 = R^5 = NO_2$

The thermal rearrangement of nitroindazoles to the corresponding 3-nitro isomers was triggered in the crystalline state by slowly increasing the temperature to 140°C over about four hours and letting the compound stay at 140°C for a further five hours.

Triazole derivatives have also been reported in the literature to be released from open-chain triazines. Baines *et al.*[88] and Daniels *et al.*[89] reported that 1-aryl-3-cyanomethyltriazine (**3-73**) undergoes cyclization with the release of triazole.

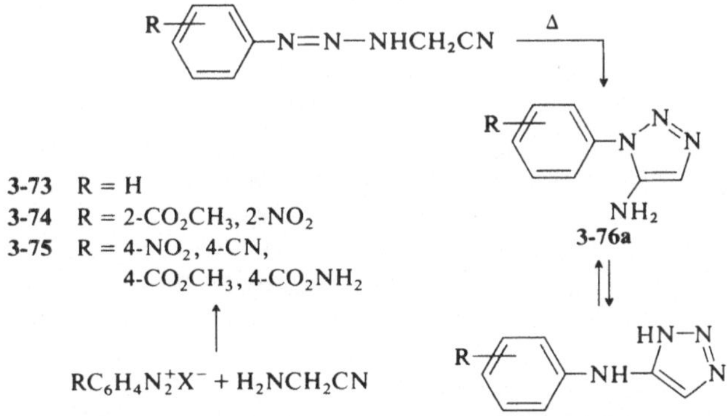

R
N=N—NHCH2CN
Δ
N
NH2
3-76a
HN—N
NH
3-73 R = H
3-74 R = 2-CO2CH3, 2-NO2
3-75 R = 4-NO2, 4-CN,
4-CO2CH3, 4-CO2NH2
RC6H4N2+X− + H2NCH2CN

**3-78** **3-79** **3-76b**

Reaction of the open-chain triazine (**3-75**) proceeded smoothly at about 80–90°C in a matter of minutes. The ring substituents in (**3-73**) affect the distribution of the product among the (**3-76**) isomers. Isomer (**3-76a**) is formed initially as the kinetically preferred isomer, which on further heating is converted into isomer (**3-76b**) in an equilibrium process. For example, 4-nitro- and 4-cyanotriazines (**3-75**) when added quickly to hot ethanol and left to crystallize at room temperature formed isomer (**3-76a**) exclusively in high yields. On further heating for 1–2 hours in ethanol, this product equilibrated with isomer (**3-76b**) in a short time. Triazines with ortho groups such as nitro rapidly cyclized into (**3-76b**) with no trace of (**3-76a**). Triazines with a 2-cyano group gave decomposition products while those with a 2-carbomethoxy group proceeded to form a six-membered ring (3-77). The cyclization yields are high, making this reaction synthetically valuable. The open-chain triazines (3-73) can be prepared through the reaction of a diazonium salt (3-78) with $\alpha$-substituted alkyl amine (3-79) in aqueous media.[89]

3-74                                                    3-77

Analogous reactions studied by Baines et al.[90] involved the cyclization of ethyl $\alpha$-(3-aryl)(triazine) acetate (3-80) into 5-hydroxy-1-aryl-1,2,3-triazole (3-81). The latter triazole upon heating undergoes a ring opening reaction with the release of a dipolar $\alpha$-diazoacetanilide (3-82) in high yields under relatively mild conditions. The interesting feature of this reaction is the color change accompanying the release of the dipolar compound (3-82).

$R$—C$_6$H$_4$—N=NNHCH$_2$CO$_2$CH$_2$CH$_3$

3-80

$(R)C_6H_4$—N        ⇌        $RC_6H_4$—N

3-81a                                  3-81b

$(R)C_6H_4NHCH_2CON=\overset{+}{N}=N^-$

3-82

| R | Color (3-81) | m.p.(°C) | Percent yield (3-82) | Color (3-82) |
|---|---|---|---|---|
| 4-NO$_2$ | Yellow | 125–130 | 95 | Orange |
| 4-CN | Pale yellow | 110–112 | 98 | Yellow |
| 4-Cl | White | 118–122 | 99 | Yellow |
| 4-CO$_2$CH$_2$CH$_3$ | Pale yellow | 181–189 | 92 | Yellow |
| 4-CO$_2$CH$_3$ | Pale yellow | — | 47 | Yellow |

Tetrazines, formed by a number of well-known routes, undergo a cycloaddition reaction with activated olefins to form bicyclic adducts. These adducts decompose with the release of nitrogen gas and pyrazine.[91] Substituents on the tetrazine bicyclic ring play a key role in determining the rate of the reaction. Further discussion can be found in Chapter 1.

The release of substituted phenanthridines (**3-84**) through the thermal extrusion of elemental sulfur from thiazepine derivatives (**3-83**) has been reported to proceed with good yields. Electron-withdrawing groups at the 8-position tend to hinder the reaction while electron-rich groups have the reverse effect. Substituents at the 2-position are less important to the reaction rate.

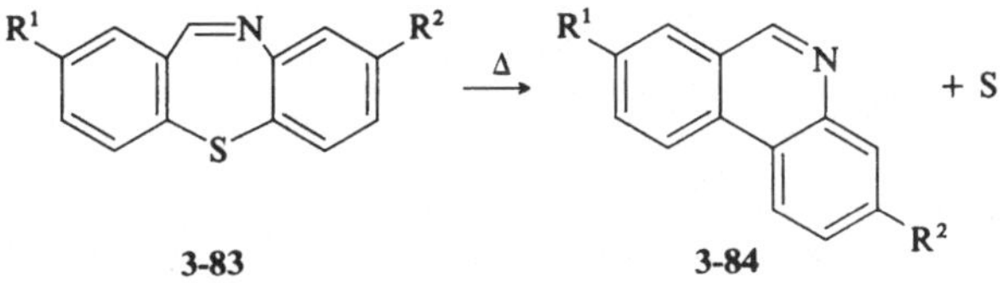

**3-83**                         **3-84**

$R^1$ = H, $CH_3$, $CH_3O$, OH, $NO_2$, Cl, CN, Br
$R^2$ = H, $CH_3$, $CH_3O$, OH, $NO_2$, Cl, CN, Br

The release of indoles (**3-86**) from the appropriate 2-substituted styrenes (**3-85**) is also reported as a valuable synthetic route.[92]

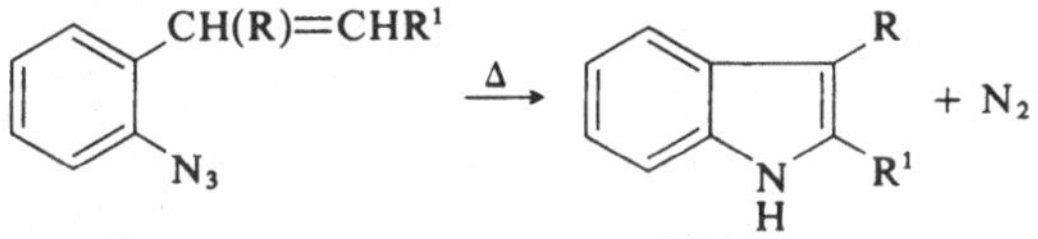

**3-85**                         **3-86**

| R | $R^1$ | Percent yield (**3-86**) |
|---|---|---|
| H | $C_6H_5$ (*trans*) | 88 |
| H | $C_6H_5$ (*cis*) | 18 |
| $CH_3$ | $CH_3$ | 24 |

Smith *et al.* have reported that azidopyrazoles[93] (**3-87**) and azido-triazoles[94] (**3-88**) undergo thermal decomposition above room temperature

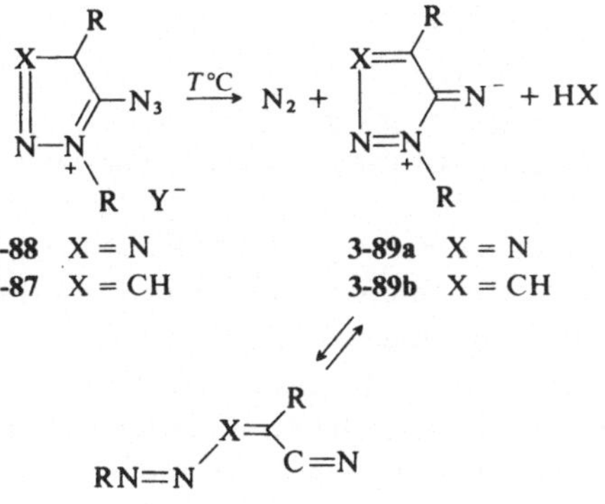

**3-88**  X = N                    **3-89a**  X = N
**3-87**  X = CH                   **3-89b**  X = CH

**3-89c**  X = N
**3-89d**  X = CH

| R$^1$ | R$^2$ | R$^3$ | $T$(°C) | Percent yield (3-89) and color |
|-------|-------|-------|---------|--------------------------------|
| C$_6$H$_5$ | CH$_3$ | H | 100 | 50, deep red |
| 4-CH$_3$C$_6$H$_4$ | C$_6$H$_5$ | C$_6$H$_5$ | 110 | 87, red |
| C$_6$H$_5$ | CH$_3$ | C$_6$H$_5$ | 73–74 | 60, deep red |
| C$_6$H$_5$ | CH$_3$ | 4-NO$_2$C$_6$H$_5$ | 108 | 88, bronze-red |
| C$_6$H$_5$ | CH$_3$ | 4-ClC$_6$H$_5$ | — | —, red |
| C$_6$H$_5$ | CH$_3$ | 4-CH$_3$OC$_6$H$_4$ | — | 80, red |
| C$_6$H$_5$ | H | C$_6$H$_5$ | 68–69 | 90, garnet |

with the release of nitrogen and a red monomeric product which is in equilibrium with 2-azoacrylonitrile, the open-chain isomer. These azido compounds were prepared[93] from the 5-aminopyrazoles by diazotization in the presence of sodium azide.

### 3.2.1.4. Other Small Molecules from Heterocycles, Phenols, Amides, and Ketones

This section will include a brief survey of examples of reactions useful in the release of small molecules not covered by the previous sections.

Isocyanates have been known to be reactive precursors for a number of reactions. A convenient route to them is by the Curtius rearrangement,[95] which involves the thermal decomposition of an acyl azide to release an isocyanate and nitrogen. This reaction has been discussed in Chapter 1 because of its potential use as a route for the release of nitrogen gas.

Isocyanates were also reported[96] to be liberated from 1,3,2,4-dioxathiazole S-oxides (3-90) in quantitative yields at mild temperatures.

**3-90**

$n = 1$; R = H; $T = 80$°C; yield = 90%
$n = 2$; R = 4,4'-C$_6$H$_4$; $T = 130$°C; yield = 99%

The 1,3,2,4-dioxathiazole S-oxides (3-90) can be, in turn, prepared from hydroxamic acids and thionyl chloride or from the cycloaddition of nitrile oxides and sulfur dioxide.

Other routes to the release of isocyanates from a variety of precursors can be found in the early German chemical literature. For example, oxanilic acid melts at 82°C and on further heating above this temperature yields phenyl isocyanate together with the evolution of carbon dioxide and hydrochloric acid gases in high yields.[97]

Phenyl isocyanate is also reported[98] as a by-product of the thermal decomposition of phenylurea. Analogously, phenylthiourea yields phenyl isothiocyanate. Phenylcarbamyl chloride (**3-91**) and methylacetylurea (**3-92**) both thermally release the corresponding phenyl isocyanate and methyl isocyanate, respectively.[99]

$$C_6H_5NHCOCl \xrightarrow{\Delta} HCl + C_6H_5N{=}C{=}O$$

**3-91**

$$CH_3NHCONHCOCH_3 \xrightarrow{\Delta} CH_3N{=}C{=}O + CH_3CONH_2$$

**3-92**

Urethanes sometimes serve as a route to the thermal release of isocyanates. For example, methyl($p$-benzoyl)phenyl urethane (**3-93**) easily changes into $p$-benzoylphenyl isocyanate by fusion at 150°C.[100]

$$C_6H_5CO-\!\!\!\left\langle\!\!\bigcirc\!\!\right\rangle\!\!-NHCO_2CH_3 \xrightarrow{\Delta} C_6H_5CO-\!\!\!\left\langle\!\!\bigcirc\!\!\right\rangle\!\!-N{=}C{=}O + CH_3OH$$

**3-93**

Aminothioformates (**3-94**) decompose between 150 and 200°C into isothiocyanates and mercaptans.[101]

$$RNHCSSR^1 \rightarrow R{-}N{=}C{=}S + R^1SH$$

**3-94**

$$R = H, CH_3, C_6H_5; R^1 = CH_3, CH_2CH{=}CH_2, C_6H_5CH_2, RNCS{-}S{\left(CH_2\right)}_n$$

Iminothiocarbonic esters (**3-95**) also produce thiocyanates and the corresponding mercaptans on gentle heating.[101]

$$(RS)_2C{=}NH \xrightarrow{T°C} RSH + RSC{\equiv}N$$

**3-95**

$$R = CH_3CH_2; T = 100°C$$
$$R = C_6H_5CH_2; T = 160{-}200°C$$

In xylene, 1,3,4-oxathiazol-2-thione (**3-96**) undergoes a thermally triggered decarboxylation with the release of nitrile sulfide.[102] The latter, in the presence of dipolarophiles such as olefins, undergoes cycloaddition reactions. In the absence of dipolarophiles, it decomposes to release a nitrile and sulfur. This reaction found application in the synthetic preparation of thiazoles, through the trapping of the nitrile sulfides with nitrile molecules.

$$\underset{\textbf{3-96}}{R-\overset{O}{\underset{N-S}{\diagdown}}=S} \quad \xrightarrow[-\text{COS}]{\Delta} \quad RC\equiv N^+ - S^- \rightarrow RCN + S$$

| R | m.p. (°C) |
|---|---|
| $C_6H_5$ | — |
| $4\text{-}CH_3OC_6H_4$ | 112 |
| $4\text{-}ClC_6H_4$ | — |
| $CH_3$ | — |
| $CH_3\text{(-}CH_2\text{-)}_{10}$ | 37–38 |

The thermally triggered reverse cycloaddition reaction of tetrazines, which are synthesized through well-known routes, releases the corresponding nitriles and nitrogen gas[103] in good yields.

Nitriles and alcohols are also liberated from the pyrolysis of the corresponding oxime carbonate,[104] which is prepared from an oxime and phenyl chloroformate. Similarly, the decomposition of derivatives of dialkyl hydrogen phosphates,[105] 3-chloro-1,2-benzoisothiazole $S,S$-dioxide,[106] and chlorothionoformate[104] provides a route to nitriles. The thermal decomposition of aliphatic or aromatic oxime carbonates (3-97) produces high yields of the corresponding nitriles.

$$\underset{\textbf{3-97}}{RCH=NOCOOC_6H_5} \xrightarrow[100°C]{} RCN + CO_2 + C_6H_5OH$$

| Aldoxime | Percent yield (nitrile) |
|---|---|
| syn-p-Chlorobenzaldoxime | 90 |
| syn-Benzaldoxime | 85 |
| syn-p-Methylbenzaldoxime | 90 |
| syn-p-Methoxybenzaldoxime | 85 |

Ketones are useful molecules for a number of reactions and their controlled release is valuable. Stevens $et\ al.$[107] have reviewed the thermal release of ketones from $\alpha$-hydroxyimines. The reaction is envisioned to involve an equilibrium between the four isomeric forms (3-98), (3-99), (3-100), and (3-101).

Stevens $et\ al.$[108] also reported the thermal release and interconversion of the various forms. Increase in steric bulk of the $N$-substituents of the $\alpha$-aminoketone (3-98) appears to facilitate rearrangement and the liberation of other isomers. Isopropylamino ketone is found to rearrange to the imine (3-99) at 120°C, a lower temperature than that for the methyl analogue. The

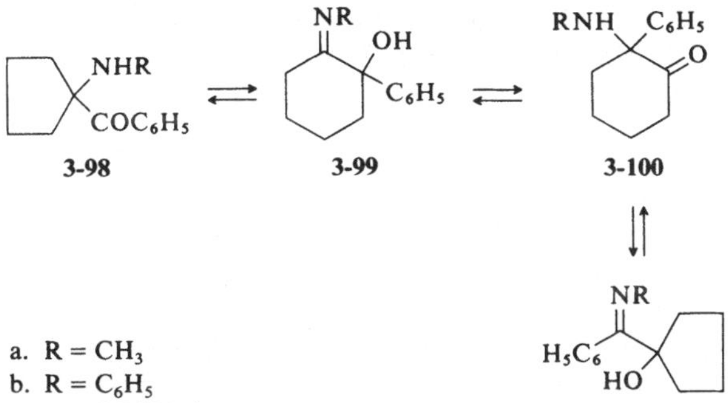

3-98          3-99          3-100

a. R = CH$_3$
b. R = C$_6$H$_5$
c. R = CH(CH$_3$)$_2$

3-101

| Compound | $T$(°C) | $t$ (h) | Percent yield |
|----------|---------|---------|---------------|
| 3-98a    | 200     | 10      | 34 (3-99a)    |
| 3-98a    | 170     | 10      | — (3-99a)     |
| 3-100a   | 190     | 18      | Decomposition |
| 3-101a   | 110     | 42      | 60 (3-100a)   |
| 3-98b    | 184     | 10      | — (3-99b)     |
| 3-99b    | 180     | 45      | 42 (3-98b)    |
| 3-100b   | 180     | 24      | —             |
| 3-101b   | 190     | 4       | 44 (3-100b)   |

effect of the substituents on the conversion of isomers (**3-101**) to (**3-100**) was also examined[109] and the conversion was found to follow first-order kinetics at 208°C. The reaction was also found to be general and a valuable synthetic route to aminoketones and other related compounds which are not easily obtained by other routes.[110]

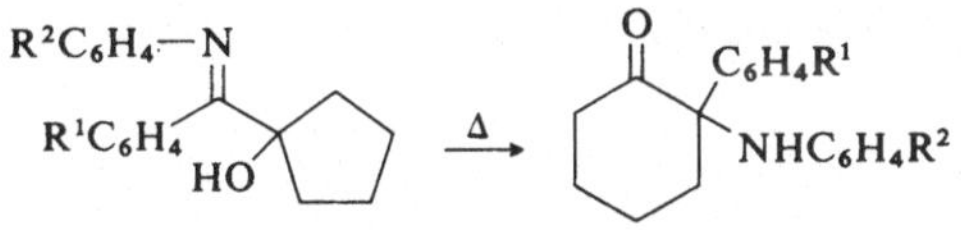

| R$^1$ | R$^2$ | $k_{av.} \times 10^5$ (s$^{-1}$) |
|-------|-------|-----------------------------------|
| H     | 4-CH$_3$O | 6.2 |
| H     | 4-CH$_3$  | 5.7 |
| H     | H         | 4.8 |
| H     | 3-Cl      | 3.9 |
| 4-CH$_3$O | H     | 6.2 |
| 4-CH$_3$  | H     | 5.7 |
| 3-Cl      | H     | 4.0 |

The release of hydroxylamine derivatives (**3-103**) from the corresponding amine oxides (**3-102**) has been reported as the Meisenheimer rearrangement. This reaction was first observed by Pinner[111] in 1895 during the examination of nicotine 1-oxide, but the correct structure of the products was not assigned until 1950 by Rayburn et al.[112]

$$R^1R^2R^3\overset{+}{N}-O^- \overset{\Delta}{\longrightarrow} R^1R^2N-OR^3$$

$$\quad\quad \textbf{3-102} \quad\quad\quad\quad\quad \textbf{3-103}$$

Meisenheimer in 1919 correctly recognized the rearrangement reaction and found that on mild heating of N-methyl-N-allylaniline oxide, O-allyl-N-methyl-N-phenylhydroxylamine was formed in good yield.[113] Cope and Klein[114] later extended the scope of the reaction by successfully preparing the N-oxide compounds which Meisenheimer failed to produce. The reaction bears similarity to the Claisen rearrangement[115] of O-allylphenols.

A number of allylamine oxides (3-104) were converted to the corresponding allylhydroxyl amine derivatives (3-105) by heating at temperatures between 80 and 150°C.[116]

| $R^1$ | $R^2$ | $T$(°C) | $t$ (min) | Percent yield (3-105) |
|---|---|---|---|---|
| $CH_3$ | $CH_3$ | 105–110 | 30 | 51 |
| $CH_2CH_3$ | $CH_2CH_3$ | 125 | 15 | 58 |
| $CH_2CH_2CH_3$ | $CH_2CH_2CH_3$ | 125 | 20 | 80 |
| $CH(CH_3)_2$ | $CH(CH_3)_2$ | 125 | 20 | 67 |
| $CH_2(CH_2)_4CH_3$ | $CH_2(CH_2)_4CH_3$ | — | — | 68 |
| $C_6H_5CH_2$ | $C_6H_5CH_2$ | — | — | 61 |

The reaction also proceeds in competition with the elimination reaction which releases olefins from the amine oxides. Cope et al.[117] found that if hydrogen atoms are present on the carbon beta to the nitrogen, an elimination reaction generally takes place to yield an olefin. This reaction is complex, a number of isomeric olefins being formed together with the released hydroxylamine.

Carruthers and Johnstone[118] extended the ring enlargement aspects of this reaction to the preparation of seven- and eight-membered rings (3-106) and (3-107) in moderate yields. A review of the Meisenheimer rearrangement was published by Johnstone[119] in 1969.

As was the case with isocyanates, the thermal release of water has been reported in a number of dehydration reactions mostly found in the early

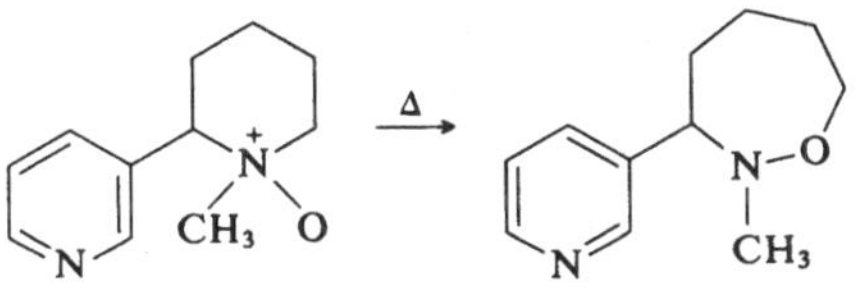

**3-106**

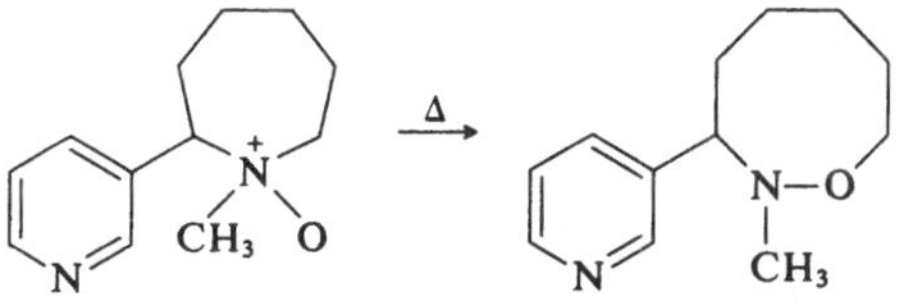

**3-107**

German literature. For example, when 2-hydroxybenzyl alcohol (**3-108**) is heated above 100°C, water and a crystalline resin are observed.[120]

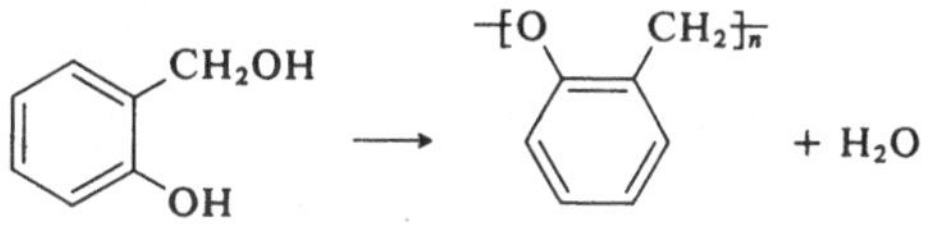

**3-108**

Analogously, 2-hydroxy-$N$-formylaniline (**3-109**) when heated to 160–170°C undergoes a cyclization reaction.[121]

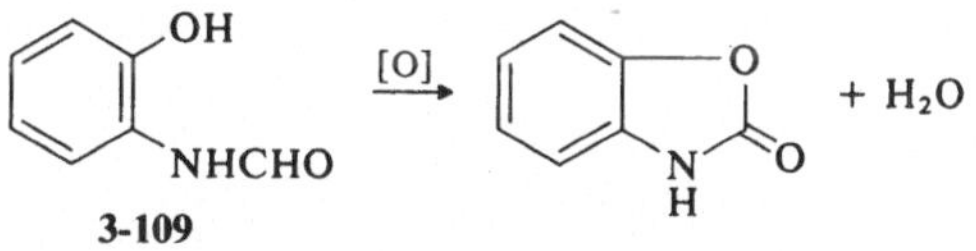

**3-109**

Acetylhydrazide (**3-110**), when heated to 180°C, undergoes a dehydration reaction, forming dimethyldihydrotetrazine (**3-111**) and water.[122]

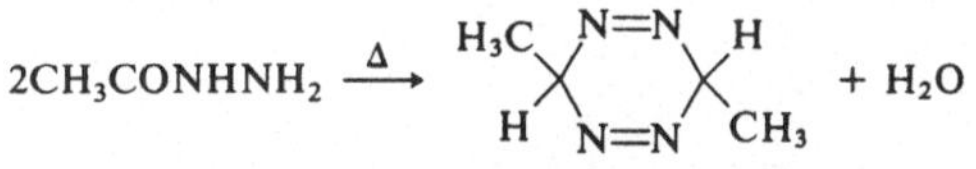

**3-110**          **3-111**

The ring closure of acetylanthranilamide (**3-112**) when heated above its melting point is accompanied by the release of water.[123]

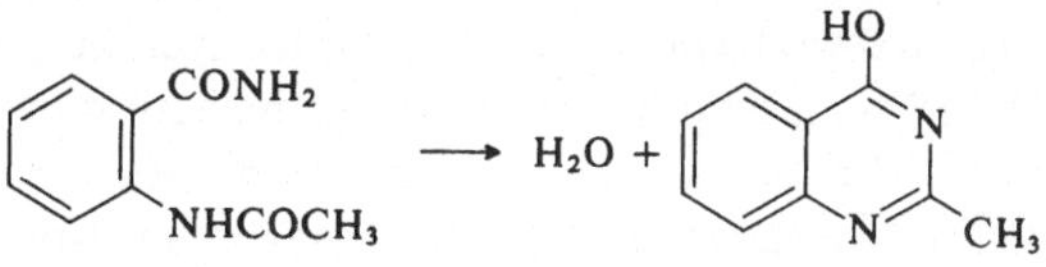

**3-112**

In much the same manner, sodium acetylanthranilate (**3-113**) changes almost quantitatively at 150°C into the sodium salt of 2,4-dihydroxyquinoline (**3-114**), with the release of water.[124]

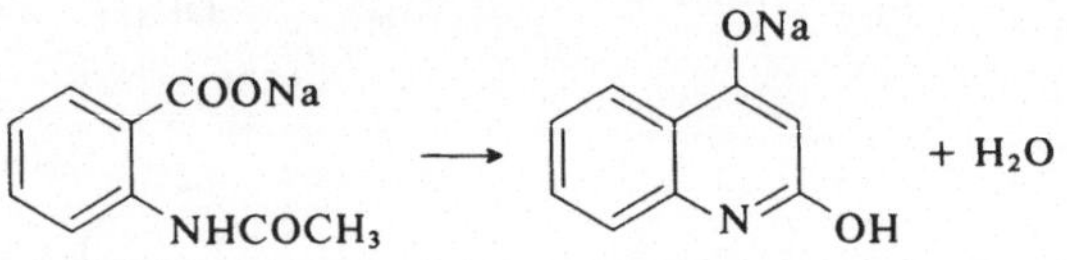

Thiocarbamidoglycolic acid (**3-115**) under heating at 155°C for 30 minutes undergoes a cyclization reaction to form the corresponding imide (**3-116**) with the release of water.[125]

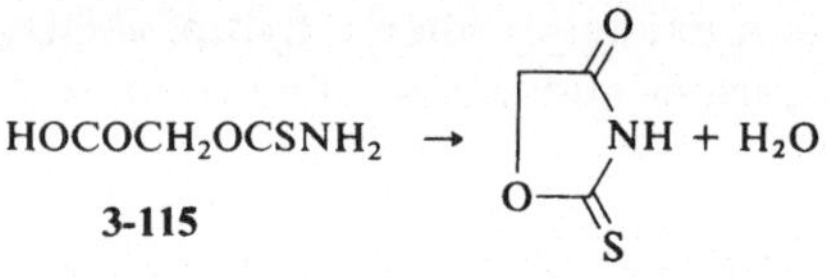

$$HOCOCH_2OCSNH_2 \rightarrow$$

**3-115**

**3-116**

Similarly, imide formation from *N*-substituted phthalamic acids (**3-117**) releases water in high yields.[126]

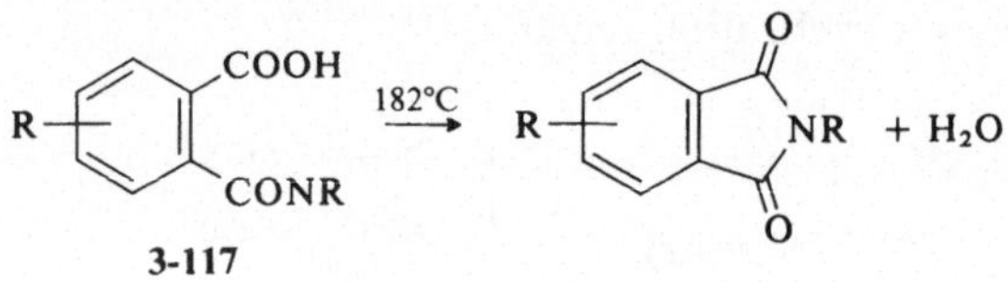

**3-117**

Diphenyl-4,4'-diphthalamic acid is completely converted into diphthalylbenzidine at 130–140°C by heating for about two hours with the release of water.[127]

### 3.2.2. Photochemically Triggered Release

#### 3.2.2.1. Olefins from Electrocyclic Reactions

Electrocyclic rearrangements of cyclic olefins into the corresponding open-chain dienes are well known. For example, cyclobutene (**3-118**) and 1,3-butadiene (**3-119**) can be interconverted by the action of heat and light. Under photochemical triggering conditions, 1,3-dienes are converted into cyclobutenes, mainly because dienes are stronger absorbers of the light energy.

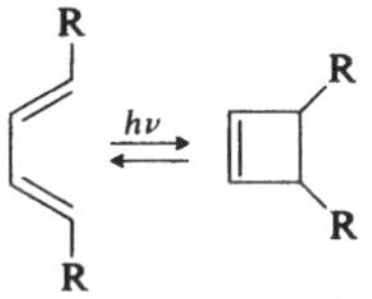

3-119        3-118

A number of examples of this type of reaction have appeared as synthetic transformations. Chapman *et al.*[128] described the photochemical transformation of conjugated cycloheptatrienes (**3-120**) into bicyclo[3.2.0]heptenes (**3-121**) under exposure to light from a mercury arc lamp.

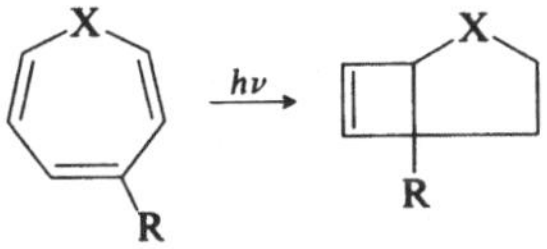

**3-120**  R = H, X = CH$_2$        **3-121**
**3-122**  R = OCH$_3$, X = CO      **3-123**

Analogously, 5-methoxy-2,4-cycloheptadienone (**3-122**) photochemically releases the isomeric bicyclic structure (**3-123**). The reaction can be reversed thermally at 50°C. This reaction is a general type which extends to tropolones,[129] eucarvone,[130] pyrocalciferol,[131] and isopyrocalciferol.[131]

Olefins can also be released under photochemical triggering from four-membered rings containing heteroatoms such as nitrogen, oxygen, or sulfur. Azetidin-2-ones under photochemical excitation undergo ring opening with the release of the corresponding ketene and a Schiff's base.[132] Analogously, oxetanes (**3-124**) undergo ring opening with the release of ethylene and a ketone.[133]

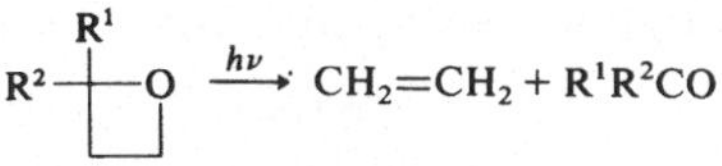

$$R^2 \underset{\qquad}{\overset{R^1}{\boxed{\phantom{x}}}} O \xrightarrow{h\nu} CH_2 = CH_2 + R^1R^2CO$$

**3-124**  R$^1$ = H, CH$_2$, C$_6$H$_5$
          R$^2$ = H, CH$_3$

The photochemical formation of ethylene and formaldehyde from unsubstituted oxetane was shown to proceed via an intramolecular process in 98% yield and with a quantum yield near unity at a wavelength of 200 nm. These oxetanes were studied under exposure from a Hanovia mercury arc lamp at 25°C and were prepared as described by Searles *et al.*[134] and Bennett and Philip.[135]

Six-membered rings and their heterocyclic analogues, such as 2-pyrans, have been shown by Chapman et al.[136] to exhibit a photochemically induced equilibrium between the cyclic structure and an open-chain aldehydoketone which in the presence of an alcoholic solvent forms the corresponding ester.

Cyclic ketones under photochemical triggering conditions have been found to loose carbon monoxide and release olefins. Blacet and Miller[137] have reported a comparative study of cyclic ketones versus their ring sizes. The quantum yields as a function of temperature and the triggering wavelength are discussed in Chapter 1.

Tetrasubstituted 1,3-cyclobutadienones (3-125) were shown by Turro et al.[138] and Heller and Srinivasan[139] to release olefins in 70–80% yields and are discussed in Chapter 1.

$$
\underset{\textbf{3-125}}{\text{[structure]}} \xrightarrow{h\nu} R^1R^2C{=}CR^3R^4 + CO
$$

A Norrish type photochemical rearrangement[140] is another possible route to trigger the release of olefins (3-126) and the corresponding enol (3-127) from a ketone (3-128). These reactions proceed in a facile manner when a $\beta$-hydrogen to the carbonyl group is present and the substituent at the carbonyl group is devoid of either electron donating or withdrawing groups but is commonly a simple aryl ring. These reactions are run using a medium-pressure Hanovia TQ-81 lamp.[140] The reaction was also extended to allow the release of dienes and cyclic olefins. The starting ketonic materials were prepared as reported by Vogel.[141] The following are a few examples to illustrate this type of reaction.

$$
\underset{\textbf{3-128}}{C_6H_5COCH_2{\left(CH_2\right)}_nR} \xrightarrow{h\nu} \underset{\textbf{3-126}}{R{-}CH{=}(CH_2)_{n-1}} + \underset{\textbf{3-127}}{C_6H_5C(OH){=}CH_2}
$$

| R | $n$ | $t$ (h) | Percent yield (3-126) |
|---|---|---|---|
| $CH_3$ | 6 | 9.5 | 74 |
| | 7 | 10 | 84 |
| | 8 | 10 | 75 |
| | 10 | 10 | 35 |
| $C_6H_5CO$ | 6 | 24 | 50 |
| | 7 | 20 | 46 |
| | 8 | 20 | 53 |
| | 5 | 20 | 0 |

### 3.2.2.2. Aromatic and Aliphatic Compounds from Diazonium Salts and Ketones

The release of aryl halides through the photochemically triggered decomposition of aryl diazonium salts in the solid phase is a well-known synthetic route.[142] The nature of the ring halide is related to the anion of the parent diazonium salt. Several salts are known with various counteranions such as fluorophosphate, fluoroborate, chloride, bromide, iodide, oxalate, citrate, and zinc chloride. Salts with a simple anion exhibit the most photosensitivity to the blue and near-ultraviolet regions but are also the most thermally unstable. Salts with complex non-nucleophilic anions have gained the most use because of the increased thermal stability. Hexafluorophosphate (3-129) and tetrafluoroborate (3-130) salts, under photochemical irradiation at 305 nm, release the corresponding fluoroarenes in high yields.

$$R \text{---} \underset{}{\bigcirc} \overset{\overset{+}{N_2}X^-}{} \xrightarrow{h\nu} R \text{---} \underset{}{\bigcirc} F + N_2 + X$$

3-129　$X^- = PF_6^-$; $X = PF_5$
3-130　$X^- = BF_4^-$; $X = BF_3$

The effect of the ring substituents[143] of the diazonium salt on the photosensitivity has also been studied extensively. In general, electron-rich substituents at the ortho positions tend to activate the ring the most. Ando[142a] has reported the relationship between substituents and the counterion in relation to the photochemical reaction. Tsunoda and Yamaoka[144] have correlated the various substituents and their effects on the quantum yields. Since Chapters 1 and 2 contain discussions on the use of diazonium salts as precursors for the release of nitrogen gas and Lewis acids, respectively, the discussion in this chapter will be brief. Other onium salts such as iodonium or sulfonium salts, which bear resemblance to the diazonium salts, have been better known as precursors for the release of Brønsted acids rather than aryl halides and thus are included in Chapter 2.

Polynuclear aromatic compounds are useful in several applications. The irradiation of 2,2'-di(phenylethynyl)biphenyl (3-131) described by

$$C{\equiv}CC_6H_5 \quad \xrightarrow{h\nu} \quad$$

H$_5$C$_6$C$\equiv$C

3-131

H$_5$C$_6$　　C$_6$H$_5$
3-132

White and Seber[145] yields a polynuclear fused aromatic compound (**3-132**) via a diradical intermediate.

Such conversion was accomplished in 80% yield through the irradiation of an ethanolic solution of the biaryl (**3-131**) using a sunlamp with a Pyrex filter for two minutes at 25°C. This reaction was previously observed by Kandil and Dessey[146] but they failed to fully identify the product formed. On the same theme, Stobbe and Lehfeldt[147] and Baddar *et al.*[148] have shown that with catalytic amounts of iodine, photochemical cyclization of certain dibenzylidenesuccinic anhydrides (**3-133**) releases polynuclear anhydrides (**3-134**).

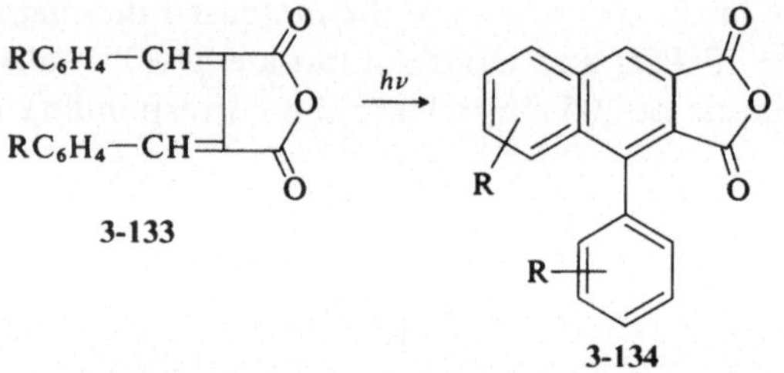

Analogously, the irradiation of stilbene (*cis* or *trans*) solutions using a mercury arc lamp yields phenanthrene. The reaction was first observed by Lewis *et al.*[149] and in later years studied further by Parker and Spoerri[150] and Buckles.[151] Such a reaction involves an isomerization of the thermodynamically more stable *trans*-stilbene into the kinetically favored *cis* isomer, followed by a cyclization reaction.

The photochemical cyclization of *cis*-stilbene derivatives has been extended to cyclic structures (**3-135**), which undergo cyclization to release phenanthrene derivatives (**3-136**), in a facile reaction proceeding in 50–70% yield.[152]

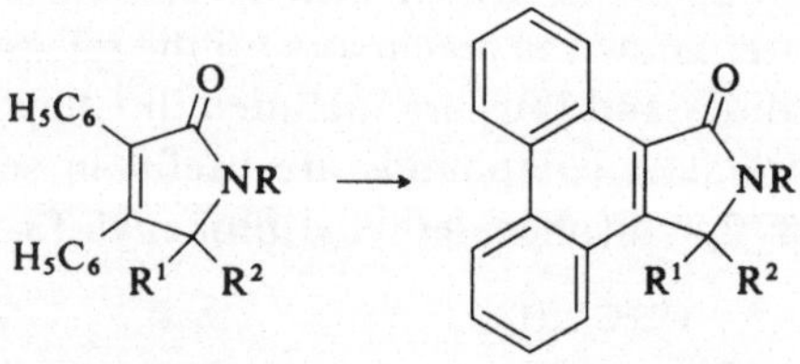

R = H; $R^1 - R^2 = O$
R = $CO_2CH_2CH_3$; $R^1$ = H; $R^2$ = OH

A number of acyclic ketones have been found to lose carbon monoxide[153] with the release of aromatic hydrocarbons. For example, 1,3-

diphenyl-2-propanone (3-137) tends to lose carbon monoxide quantitatively with the release of the corresponding dibenzyl derivative[153a] (3-138). The reaction was run in a solution irradiated with 313-nm light isolated from a medium-pressure mercury lamp using a chromate-carbonate filter.[154]

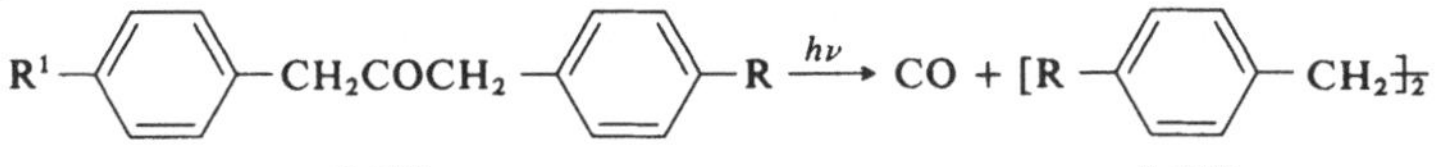

3-137                                    3-138

| $R^1$ | R | $\phi_{313\,nm}$ |
|-------|-----|------|
| H | H | 0.70 |
| $CH_3$ | $CH_3$ | 0.71 |
| H | $CH_3O$ | 0.66 |
| $CH_3O$ | $CH_3O$ | 0.66 |
| CN | CN | 0.02 |

### 3.2.2.3. Phenol and Its Derivatives from Ethers and Esters

The photochemical release of phenols from the corresponding aryl esters of carboxylic acids (3-139) and of sulfonic acids (3-140) was found to be feasible.

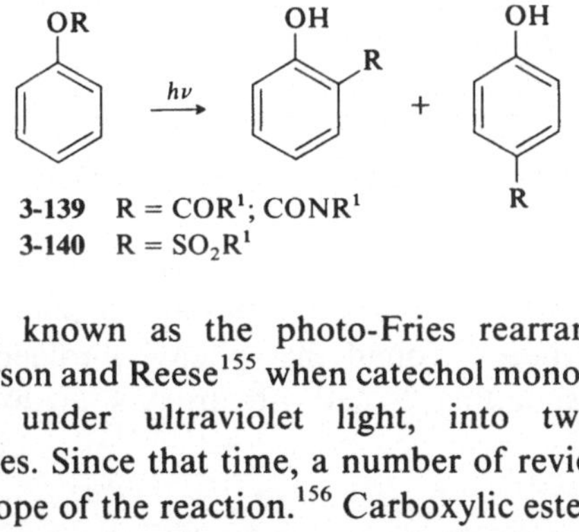

3-139   R = COR¹; CONR¹
3-140   R = SO₂R¹

The reaction, known as the photo-Fries rearrangement, was first observed by Anderson and Reese[155] when catechol monoacetate rearranged, upon irradiation under ultraviolet light, into two isomeric dihydroxyacetophenones. Since that time, a number of reviews have appeared that expand the scope of the reaction.[156] Carboxylic esters that successfully release phenols include phenyl acetate, catechol acetate, p-(t-butyl)phenyl esters, phenyl benzoate, and phenyl ferrocene carboxylate. The reaction mechanism proceeds through photodissociation of the ester and migration of the acyl radical to the ortho or para ring positions. This mechanism accounts for the formation of isomers.

Finnegan and Mattice[157] and Kobsa[156d] have reported that a number of esters (3-141) irradiated with ultraviolet light from a 100-W Hanovia 608A lamp give phenols in about 50% yield. For preparative applications the reaction was found to proceed satisfactorily under irradiation within the 245–330-nm wavelength region.

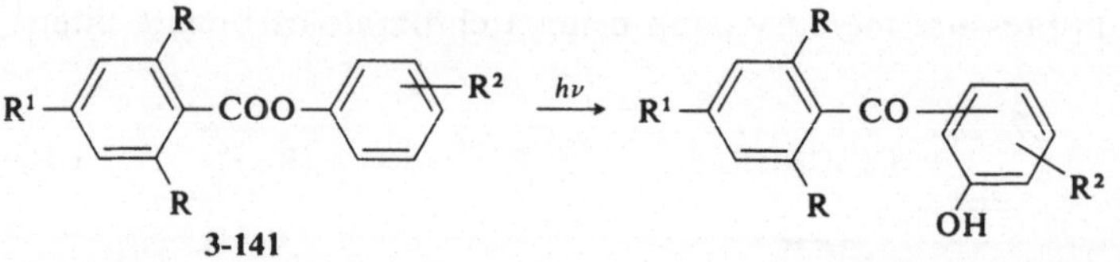

**3-141**

| R | $R^1$ | $R^2$ | $t$ (h) | Percent yield (phenol)[a] | |
| | | | | $o$-Isomer | $p$-Isomer |
|---|---|---|---|---|---|
| H | H | H | 8 | 13 | 20 |
| H | $CH_3$ | H | 19 | 53 | — |
| H | Cl | H | 19 | 51 | — |
| $CH_3$ | H | H | 24 | — | 35 |
| $(CH_3)_2CH$ | H | H | 26 | — | 56 |
| H | H | 2-Cl | 52 | 42 | 14 |
| H | $NO_2$ | 4-$NO_2$ | 35 | 11 | — |
| H | $(CH_3)_3C$ | H | — | 45 | — |
| H | $(CH_3)_3C$ | 4-$(CH_3)_3C$ | — | 48 | — |
| H | $(CH_3)_3C$ | 4-Cl | — | 48 | — |
| H | $(CH_3)_3C$ | 3,4-$Cl_2$ | — | 37 | — |
| H | $(CH_3)_3C$ | 4-CN | — | 48 | — |
| H | $(CH_3)_3C$ | 4-$NH_2$ | — | 12 | — |
| H | $(CH_3)_3C$ | 4-$NO_2$ | — | 10 | — |
| H | $(CH_3)_3C$ | 3,5-$(NO_2)_2$ | — | 9 | — |

[a] Refs. 156d and 157.

Sulfonate salts are less common precursors for the release of phenols, but have been reported to rearrange through a photo-Fries mechanism. Stratenus and Havinga[158] found that phenyl $p$-toluenesulfonate (**3-142**) released the corresponding 2- and 4-hydroxy-4'-methyldiphenyl sulfones (**3-143**).

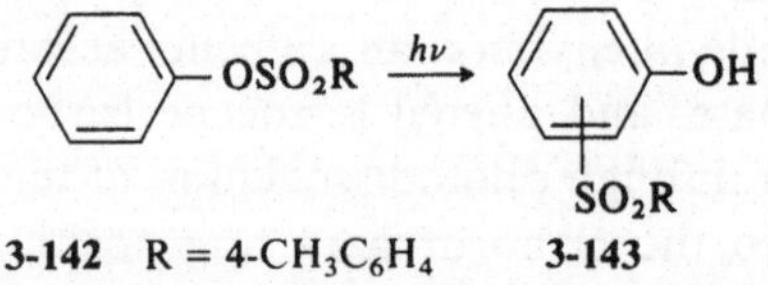

**3-142**   R = 4-$CH_3C_6H_4$   **3-143**

Heterocyclic esters such as 4-pyrimidinyl esters (**3-144**) have shown analogous rearrangement with the release of the ring hydroxy group in 60% yield.

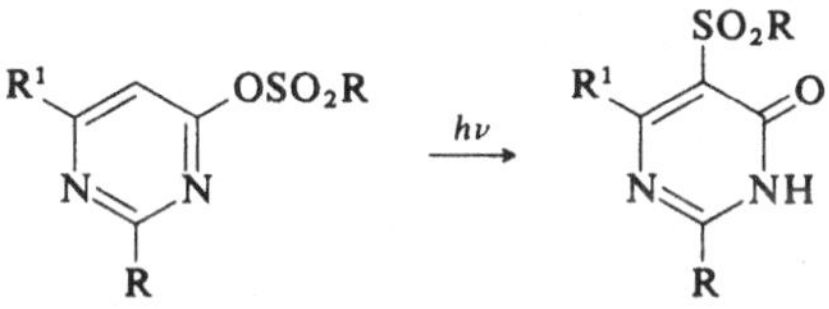

3-144                          3-145

The 2,4-dinitrobenzenesulfenyl esters (**3-146**) of phenols (and carboxylic acids) are also good precursors for the photochemical release of phenols.[159] Irradiation is generally carried out using 125-W high-pressure mercury lamp.

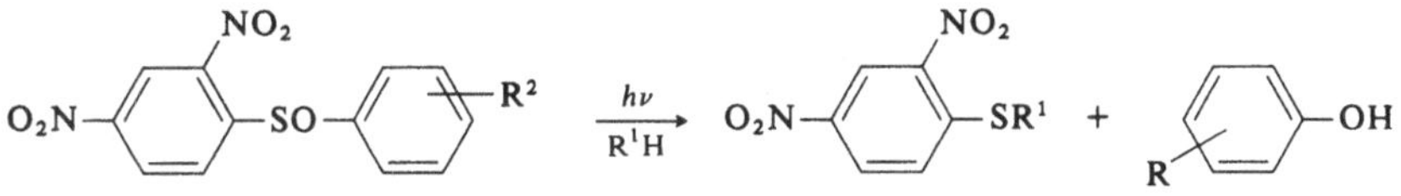

3-146                                          3-147

R = COCH$_3$

| R$^1$ | $t$ (min) | Percent yield (3-147)$^a$ |
|---|---|---|
| C$_6$H$_5$ | 60 | 73 |
| 4-(1-CH$_3$CO)C$_6$H$_4$ | 60 | 74 |
| 4-(2-Cl)C$_6$H$_4$ | 60 | 14 |
| 2-(1,4-C$_2$H$_5$O)C$_6$H$_3$ | 20 | 55 |
| 4-[1-(CH$_3$)$_2$N]C$_6$H$_4$ | 70 | 27 |
| 2-[4-CH$_3$-1-(CH$_3$)$_2$N]C$_6$H$_3$ | 40 | 27 |
| 1-C$_{10}$H$_8$ | 15 | 41 |
| 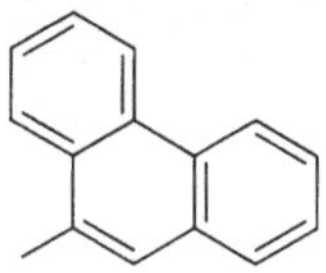 | 10 | 82 |
| 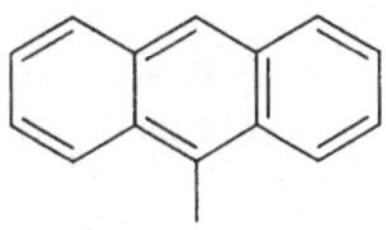 | 15 | 74 |
| 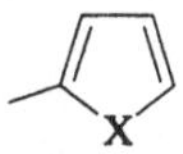 X = S | 45 | 13 |
| X = O | 60 | 28 |

$^a$ Ref. 160.

The nitrophenyl phosphate and sulfate esters have also been reported by Havinga *et al.*[161] to undergo photochemically triggered release of the corresponding 3-nitro or 3,5-dinitro derivatives together with phosphoric or sulfuric acids. This reaction gained more applications as a source for the release of mineral and carboxylic acids, as discussed in detail in Chapter 2. Also, as a result of the wide variety of solvents which can be used and the reaction of the released radicals with these solvents, a secondary use has emerged: the release of aryl thioethers in yields[160] between 10 and 80%.

The photochemical release of phenols from aminoaryl ethers via a photo-Smiles type reaction was reported by Matsui et al.[162] in the early 1970s. The reported conversion of aminoaryl ethers (**3-148**) into $N$-substituted aminophenols was triggered by radiation from a high- or low-pressure mercury lamp at wavelengths of 253 and 300 nm.

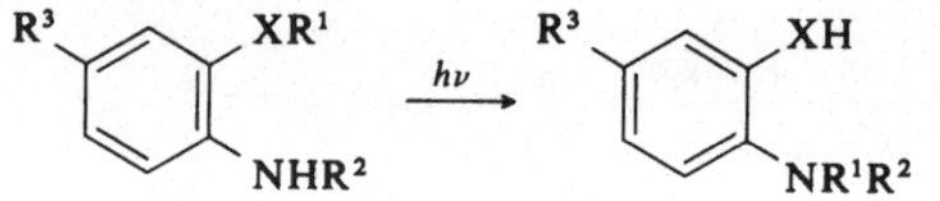

| $R^1$ | $R^2$ | $R^3$ | X | Irradiation time (h) |
|---|---|---|---|---|
| 2,6-$(NO_2)_2C_6H_3$ | H | H | O | 24 (300 nm) |
| 3,5-$(CH_3O)_2$-$s$-triazine | H | H | O | 50 (300 nm) |
| 3,5-$(CH_3)_2$-$s$-triazine | H | H | S | 30 (253 nm) |

The formation of free phenols from several phenyl and $t$-butylphenyl esters (**3-149**) under irradiation has been reported by Horspool and Pauson.[224] Better yields were also reported using oxalate and formate esters.

$$2ArOCOR \longrightarrow ArOH + 2\text{-}RCOArOH + RCHO$$

$$\textbf{3-149} \qquad\qquad \textbf{3-150a} \qquad \textbf{3-150b}$$

| Ar | R | Percent yield (phenol derivative) |
|---|---|---|
| $C_6H_5$ | $C_6H_5$ | 4.5 |
| $C_6H_5$ | 2-$HOC_6H_4$ | 19 |
| 4-$(CH_3)_3CC_6H_4$ | 4-$(CH_3)_3CC_6H_4O$ | 22 |
| $C_6H_5$ | $C_6H_4COO$ | 23 |
| 4-$(CH_3)_3CC_6H_4$ | 4-$(CH_3)_3CC_6H_4COO$ | 40 |
| $C_6H_5$ | H | 49 |
| 4-$(CH_3)_3CC_6H_4$ | H | 80 |
| 4-$(CH_3)_3CC_6H_4$ | $CH_3$ | 34 |

### 3.2.2.4. Alcohols from Amides, Esters, Ketones, and Azide Derivatives

Alcohols, because of their synthetic value, have been examined in various release reactions. Mutai et al.[163] have reported that 1-(p-nitrophenyl)-$\omega$-anilinoalkanes (3-151) on irradiation yield N-(p-nitrophenyl)-$\omega$-anilino-1-alkanols (3-152). The reaction proceeded by a mechanism analogous to the known photo-Smiles rearrangement. The reaction rate in acetonitrile was dependent on the length of the carbon chain between the amino nitrogen and the ether oxygen in (3-151). Such a reaction is accompanied by a color change. The colorless amines were converted into yellow-colored alcohols with a reaction rate dependent on the chain length under irradiation, with a 100-W high-pressure mercury lamp, in acetonitrile or methanol.

3-151   n = 2-5                          3-152

Other derivatives[163] also showed a similar reaction, for example, the release of (3-154) from (3-153).

3-153                          3-154

The conversion of the colorless amines (3-151) into colored alcohols was accelerated by the addition of amines such as triethylamine or pyridine, as shown in Table 3.6.

Tosylates (3-155) of alcohols have been shown to be photosensitive precursors of alcohols.[165] The high yields made this a valuable synthetic route, especially in natural product syntheses. These tosyl derivatives were prepared from p-toluenesulfonyl chloride and the corresponding alcohol.

Table 3.6. Relative Rates of Conversion of (3-151) in Acetonitrile[a]

|            | n = 2 | n = 3 | n = 4 | n = 5 |
|------------|-------|-------|-------|-------|
| k(rel)     | 1.0   | 2.0   | 420.0 | 120.0 |
| k*(rel)[b] | 710.0 | 400.0 | 110.0 | 8.6   |

[a] Ref. 164.
[b] k* is rate in the presence of triethylamine.

Irradiation at about 253 nm was accomplished using the RPR 2537 Rayonet and the Philips TUV 15 units.

$$R^1 \text{—} \bigcirc \text{—} SO_2OR \xrightarrow{h\nu} ROH$$

3-155

| ROH | Reaction solvent | Percent yield (ROH) |
|---|---|---|
| $3\beta(5\alpha)$-Cholestane | $(C_2H_5)_2O$ | 67 |
| $3\beta$-Dimethyl-4,4($5\alpha$)-cholestane | $(C_2H_5)_2O/CH_3OH$ | 92 |
| $3\beta$-Dimethyl-4,4-cholest-5-ene | $(C_2H_5)_2O/CH_3OH$ | 97 |
| $2\beta$-Dimethyl-4,4($5\alpha$)-cholestane | $(C_2H_5)_2O$ | 48 |
| $3\beta$-Cholest-5-ene | $(C_2H_5)_2O$ | 96 |
| Cyclohexanol | $(C_2H_5)_2O$ | 95 |

Cyclic alcohols can also be released via the photoreduction of the corresponding acyclic ketones.[165] These ketones (3-156) need to contain a $\gamma$-hydrogen for the reaction to proceed. Applications find their place in the synthesis of cyclobutanols[166] (3-157) and in natural product[167] chemistry. A stepwise mechanism has been proposed[166a] involving a $\gamma$-hydrogen abstraction followed by cyclization of the diradical intermediate.

$$RCOCH_2CH_2CHR^1R^2 \xrightarrow{h\nu} R^2 \underset{\underset{\text{3-157}}{}}{\overset{R^1 \ OH}{\square}} R$$

3-156

$R^1 = H; R^2 = H, C_3H_7, C_4H_9; R = CH_3$

Barton et al.[168] have demonstrated the production of a variety of alcohols (and carboxylic acids) from the $\beta$-($o$-azidophenyl)ethyl derivatives (3-158). On irradiation at wavelengths longer than 350 nm, a nitrene intermediate was formed along with nitrogen. The nitrene then inserted itself into the alkyl chain at the ortho position, releasing an indole and the alcohol molecule. The reaction has found more synthetic success and use in the release of carboxylic acids because of its quantitative synthetic yields, as discussed in Chapter 2. Few results were published to exploit its full potential as an alcohol-producing reaction.

3-158   R = H, CH$_3$

### 3.2.2.5. Ketones from Silyl Ethers, Ketals, and Thioketal Derivatives

Examples of the photochemical release of ketones and aldehydes have been reported by Pinnick and Lajis.[169] Trimethylsilyl ethers (**3-159**), which are prepared in high yields from alcohols and chlorotrimethylsilane, when irradiated in the presence of N-bromosuccinimide, give high yields of the corresponding ketones. These ketones are the oxidation products of the starting alcohols.[170]

$$R^1R^2CHOH + (CH_3)_3SiCl \rightarrow R^1R^2CHOSi(CH_3)_3$$

**3-159**

$$R^1R^2C{=}O \xleftarrow[\text{NBS}]{h\nu}$$

| R$^1$ | R$^2$ | Irradiation[a] | | Percent yield (ketone) |
|-------|-------|-------|-------|-------|
| | | T (°C) | t (h) | |
| C$_6$H$_5$ | H | 0 | 5 | 48 |
| C$_6$H$_5$ | CH$_3$ | 25 | 3.5 | 76 |
| CH$_3$(CH$_2$)$_5$ | CH$_3$ | 25 | 3.5 | 55 |

[a] Sunlamp exposure unit.

Diketones (**3-161**) can be released[171] through the irradiation of their 2-arenesulfonyloxy derivatives (**3-160**). The 2-arenesulfonyloxy-2-cyclo-hexanone (**3-160**) is prepared from cyclohexan-1,2-dione and arylsulfonyl chloride.

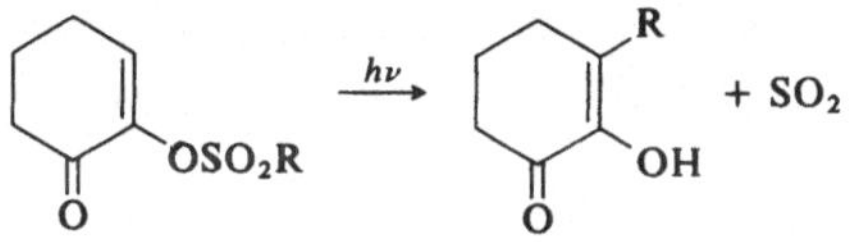

**3-160**          **3-161**

| R | Percent yield (**3-161**)[a] |
|---|---|
| C$_6$H$_5$ | 40 |
| 4-CH$_3$C$_6$H$_4$ | 50 |
| 4-CH$_3$OC$_6$H$_4$ | 50 |

[a] Irradiation at 366 nm.

Ketones can also be released from the corresponding thioketals upon irradiation. Takahashi et al.[172] have reported that the dethioketalization of thioketal-containing steroid (**3-162**) under photochemical irradiation gave the parent ketone (**3-163**).

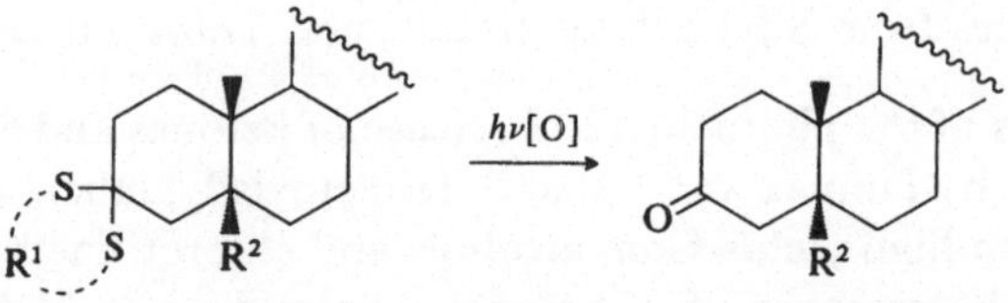

3-162                                            3-163

| $R^1$ | $R^2$ | Irradiation time (h) | Percent yield (3-163) |
|---|---|---|---|
| $CH_2CH_2$ | $\alpha$-H | 3.5 | 77 |
| $CH_2CH_2$ | $\Delta^2$-compound | 2.0 | 57 |
| $CH_2CH_2$ | $\beta$-H | 4.0 | 75 |
| 2-($C_6H_5CH_2$) | $\alpha$-H | 3.0 | 80 |

The procedure is simple and generally gives rise to high yields of the released ketones. In general, a solution of the thioketal and a sensitizer, such as benzophenone, is irradiated with a high-pressure mercury lamp mounted in a Pyrex immersion tube. A number of thioketals (3-164a) were studied and the effect of adjacent substituents was found to be minor. These thioketals are usually prepared in quantitative yields from the dithiol and ketone in the presence of a Lewis acid at room temperature.[173]

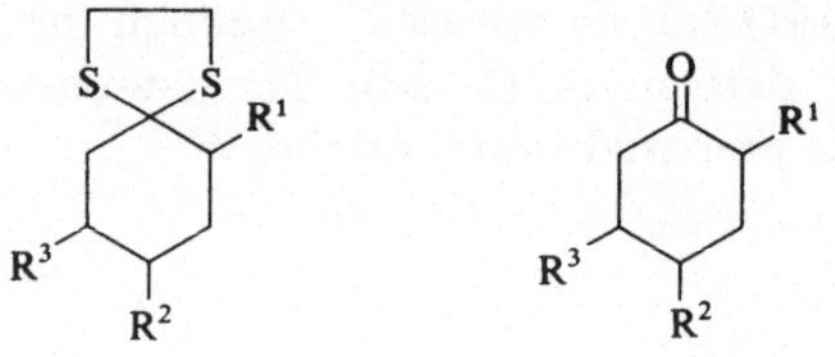

3-164a                                            3-164b

| $R^1$ | $R^2$ | $R^3$ | Irradiation time (h) | Percent yield (3-164b) |
|---|---|---|---|---|
| H | H | H | 4.5 | 65 |
| $CH_3$ | H | H | 2.5 | 70 |
| H | $(CH_3)_3C$ | H | 5.5 | 60 |
| $(CH_3)_2CH$ | H | $CH_3$ | 4.5 | 87 |

Heterocycles can also be used as precursors of aldehydes: for example, 1-benzyloxy-1,2,3-benzotriazole (3-165) and 1-methoxy-1,2,3-benzotriazole (3-166), described by Feld et al.[174] and Savre,[175] respectively.

Irradiation of (3-165) at 300 nm for about eight hours gave benzaldehyde in yields between 38 and 69%, depending on the nature of the solvent used. The reaction was also found to be sensitive to spectral sensitization using 3-methoxyacetophenone. Benzotriazoles (3-165) and (3-166) are

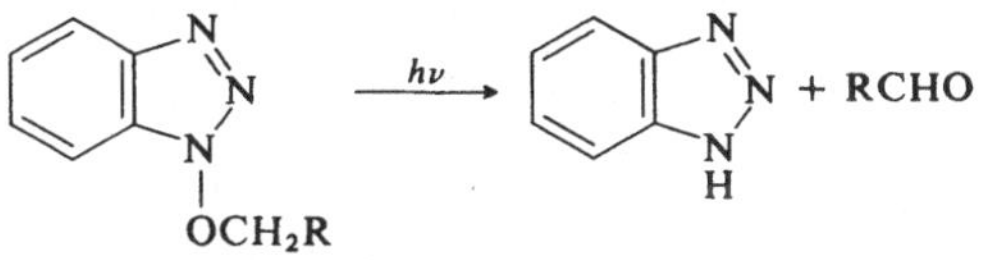

3-165   R = C$_6$H$_5$
3-166   R = H

easily prepared from the reaction of benzotriazole and methyl iodide or benzyl bromide in the presence of a base.[175]

Derivatives of diphenylmethane, such as benzoyldiphenylmethane (3-167), were first reported by Schonberg et al.,[176] and later by others, to be photosensitive. Under the action of sunlight, these compounds decompose to release benzaldehyde and 1,1,2,2-tetraphenylethane.

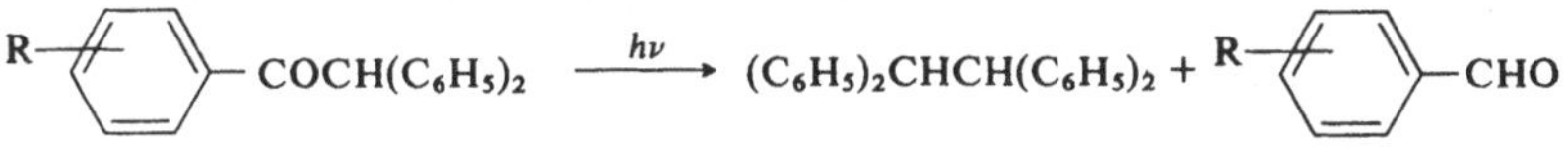

3-167

An analogous reaction is reported[177] with the irradiation of desoxybenzoin derivatives (3-168). Desoxybenzoin in the absence of air and under photochemical triggering conditions releases benzaldehyde, dibenzyl, and didesyl via a free radical mechanism. In the presence of air, desoxybenzoin and its derivatives are photo-oxidized to the corresponding aldehydes and acids.

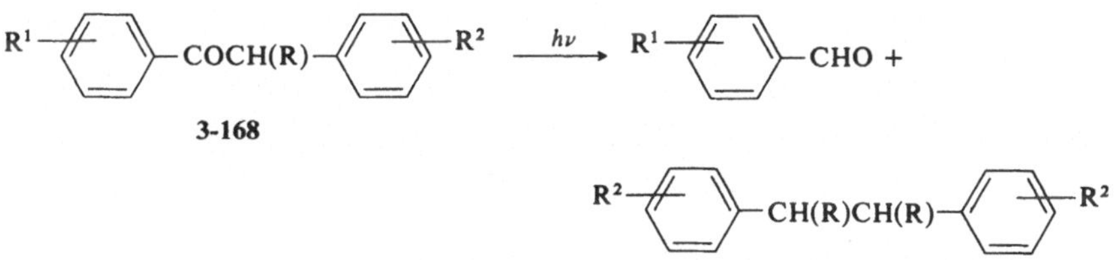

3-168

$R^1 = CH_3, R^2 = H; R^1 = R^2 = CH_3O; R^1 = NO_2, R^2 = H$

Other light-induced oxidation–reduction reactions are those of aromatic nitro compounds (3-169) containing a carbon–hydrogen bond ortho to the nitro group. This type of reaction has been much investigated[178] and used in a number of applications. The reaction mechanism involves the migration of an oxygen atom from the nitro group onto the carbon atom of the ortho substituent, thus forming nitroso aldehydes or ketones. The reaction is characterized by high yields of the carbonyl compounds together with acids[179] or amines[179] from the corresponding 2-nitrobenzyl esters (3-169) or amides (3-170). The high yields make it synthetically valuable.

Irradiation, generally, utilizes a medium-pressure mercury lamp with a Pyrex filter for 2–4 hours to yield 70–80% products.

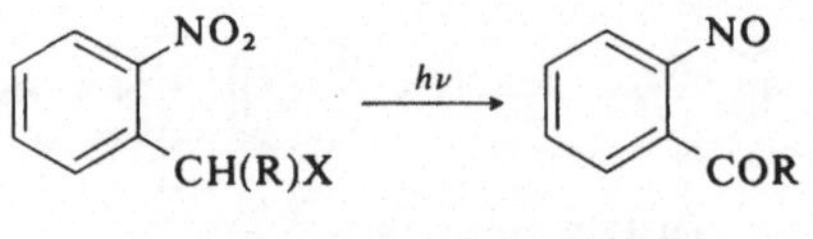

3-169   X = R¹COO
3-170   X = R¹NH

When the released carbonyl compound is a nitroso aldehyde, a further dehydrative coupling reaction is observed, yielding an azobenzene.[179,180] This is discussed in Chapter 2 as a source of acids and amines.

Similarly, ketones have been formed[181] using 2-nitrophenylethylene glycol acetals (3-171). Irradiation of these acetals with light of a wavelength longer than 320 nm, using a Rayonet 3500A lamp in a Pyrex vessel, for 1–2 hours releases the desired ketone in 30–80% yield. The acetals can be prepared in high yields from 2-nitrostyrene glycol derivatives[184] (3-172) and the desired ketone.

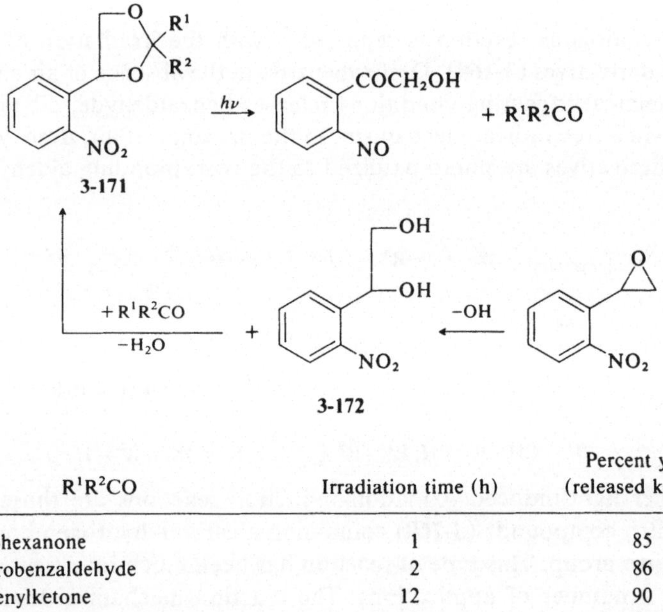

| R¹R²CO | Irradiation time (h) | Percent yield (released ketone) |
|---|---|---|
| Cyclohexanone | 1 | 85 |
| 4-Nitrobenzaldehyde | 2 | 86 |
| Diphenylketone | 12 | 90 |
| 20-Oxo-5-pregnen-3$\beta$-ol acetate | 7 | 83 |
| 17-Oxo-5-androsten-3$\beta$-ol acetate | 2 | 74 |
| Testosterone | 20 | 31 |

Phenacyl ester derivatives, and in particular, 4-methoxy and 2-methyl derivatives, have also been used to provide ketones and acids on irradiation.[183] Yields are usually between 70 and 80% with irradiation at wavelengths longer than 313 nm. The reaction is discussed in Chapter 2 because of its synthetic value as a latent source for carboxylic acids and amines.

A slightly different route to carbonyl-containing molecules was reported by Pedersen. He extended[184] the use of the nitrone group in $N,N'$-disubstituted $p$-quinonediimine $N,N'$-dioxides (3-173) to release $p$-quinones.

| R$^1$ | R$^2$ | $\lambda_{max}$ (nm) ($\varepsilon \times 10^3$) | Percent yield ($p$-quinone) |
|---|---|---|---|
| CH$_3$ | CH$_3$ | 385 (17)$^a$ | 17 |
| CH$_3$(CH$_2$)$_2$ | CH$_3$(CH$_2$)$_2$ | 405 (66)$^b$ | 11 |
| (CH$_3$)$_2$C(CN) | (CH$_3$)$_2$C(CN) | 406 (85)$^b$ | 52 |
| $\alpha$-Carbamoylisopropyl | $\alpha$-Carbamoylisopropyl | 410 (45)$^a$ | 57 |
| $\alpha$-Carboethoxyisopropyl | $\alpha$-Carboethoxyisopropyl | 406 (78)$^b$ | 69 |
| Cyclohexyl | Cyclohexyl | 404 (65) | 77 |
| 1-Cyanocyclohexyl | 1-Cyanocyclohexyl | 415 (63)$^c$ | 79 |
| Cyclohexyl | C$_6$H$_5$ | 415 (47)$^c$ | 30 |
| C$_6$H$_5$ | C$_6$H$_5$ | 417 (47)$^a$ | 83 |
| 4-Dodecylphenyl | 4-Dodecylphenyl | — — | 36 |
| C$_6$H$_5$ | $\beta$-Naphthyl | 432 (45)$^c$ | 60–80 |
| $\beta$-Naphthyl | $\beta$-Naphthyl | 435 (35)$^c$ | 62 |
| 2-CH$_3$-3-ClC$_6$H$_3$ | 2-CH$_3$-3-ClC$_6$H$_3$ | 414 (62)$^c$ | 60–80 |
| 4-CH$_3$OC$_6$H$_4$ | 4-CH$_3$OC$_6$H$_4$ | 431 (40)$^c$ | 60–80 |
| C$_6$H$_5$ | 2,4-(NO$_2$)$_2$C$_6$H$_3$ | 408 (34)$^c$ | 59 |

$^a$ In CH$_3$OH.
$^b$ In CH$_3$CH$_2$OH.
$^c$ In CHCl$_3$.

Upon irradiation of (3-173) with light of 300–450-nm wavelength, it decomposed rapidly and quantitatively into an $N$-substituted $p$-quinonimine $N$-oxide and an azo molecule. Such a reaction is accompanied by a color change: the yellow-colored $N$-dioxide converted to the colorless quinone under irradiation from a mercury vapor lamp with a filter allowing only light of 390-nm wavelength to pass. These dioxides were prepared from the corresponding $p$-quinonimines and peracid.[185]

### 3.2.2.6. Heterocycles from Other Heterocycles, Aryl Ketones, and Nitrone Derivatives

The photochemistry of heterocycles has recently gained much interest and provides several routes for the release of heterocyclic rings in a variety of reactions. Some highlights are included in this section with reference to two excellent reviews on the subject of the photochemistry of heterocycles by Reid.[225]

Three-membered heterocycles such as oxiranes can be released from larger rings conaining a photosensitive keto group. For example, tetramethyloxetan-3-one (**3-174**) under photochemical triggering releases oxirane (**3-175**) and carbon monoxide.[186]

**3-174**                                              **3-175**

Oxaziridines (**3-177**), however, can be photochemically liberated from the corresponding nitrones[187] (**3-176**). The substituents control the light absorption characteristics of the nitrone and the stability of the released oxaziridine. On heating, oxaziridines may either revert to the parent nitrone or rearrange to release the isomeric amide. The chemistry of nitrones was reviewed by Smith[188] in 1938 and later by Hamer and Macaluso[189] and DelPierre and Lamchen[190] in 1964 and 1965, respectively. Nitrones can be prepared by oxidation of nitrogen-containing molecules with peroxides or peracids. A large selection of nitrones have been reported in the literature and the following are a few examples.

**3-176**                                              **3-177**

| $R^1$ | $R^3$ | $\lambda_{max}$ (nm) ($\varepsilon \times 10^3$) |
|---|---|---|
| $C_6H_5$ | $C_6H_5$ | 227 (9.8), 236 (9.0), 315 (14.0) |
| 4-$CH_3OC_6H_4$ | $C_6H_5$ | 227 (9.8), 280 (7.95), 237 (11.8), 329 (26.9), 240 (16.7), 330 (11.8) |
| 4-$NO_2C_6H_4$ | $C_6H_5$ | 228 (8.5), 266 (10.4), 237 (11.9), 247 (8.6), 352 (20.0), 265 (11.8), 350 (10.4) |
| 4-$HOC_6H_4$ | $C_6H_5$ | 228 (16.3), 280 (8.0), 310 (9.7), 330 (7.8), 228 (8.4), 237 (8.4) |
| $C_6H_5$ | 4-$ClC_6H_4$ | 228 (8.4), 237 (8.4), 254 (7.6), 318 (20.9) |
| $C_6H_5$ | 4-$CH_3C_6H_4$ | 236 (10.3), 251 (5.7), 251 (5.1), 316 (15.8), 229 (9.0) |

Analogously, an azoxy group can be photochemically triggered to release an oxadiziridine ring (**3-179**), as exemplified by the irradiation of azoxy-*t*-butane (**3-178**) at 212 nm in pentane at 10°C with a Hanovia type L lamp.[191] The reaction is also thermally reversible in quantitative yields.

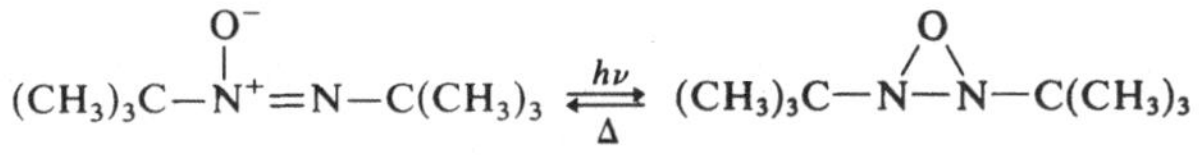

$$(CH_3)_3C-\overset{\overset{O^-}{|}}{N^+}=N-C(CH_3)_3 \underset{\Delta}{\overset{h\nu}{\rightleftharpoons}} (CH_3)_3C-\overset{\overset{O}{\wedge}}{N}-N-C(CH_3)_3$$

**3-178**                                          **3-179**

These azoxy compounds can be prepared[192] by the condensation of N-alkylhydroxylamines and nitrosobenzene or through peracid N-oxidation of a hydrazine.

The triggered release of larger heterocycles can be exemplified by the photochemically induced rearrangement of nitrofurans (**3-180**) and nitropyrroles (**3-181**) to release their corresponding keto derivatives.[193] The reaction yields are between 15 and 75% with irradiation by a 100-W Hanovia medium-pressure mercury lamp. These nitro compounds can be prepared as reported in the literature.[194]

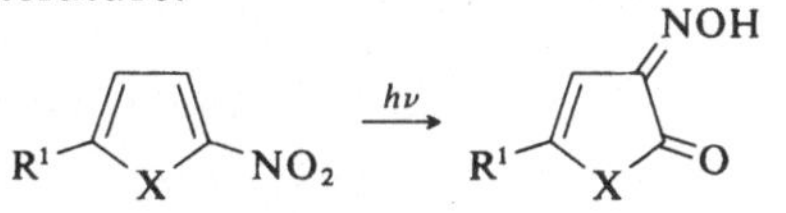

**3-180a**   R$^1$ = H, X = O, yield = 74%
             R$^1$ = CH$_3$, X = O, yield = 32%
**3-181**    R$^1$ = H, X = NH, yield = 15%

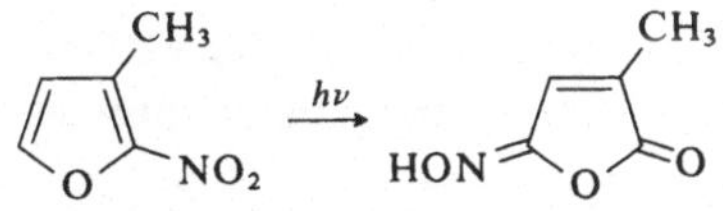

**3-180b**

In contrast to these reactions, furan (**3-182**), benzofuran (**3-183**), and indole (**3-184**) derivatives give rise to oxazine derivatives on irradiation[195] using a medium-pressure mercury lamp and a Pyrex filter.

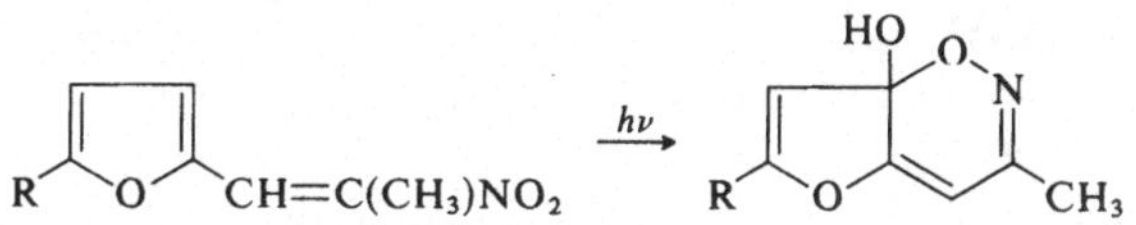

**3-182**   R = H, CH$_3$

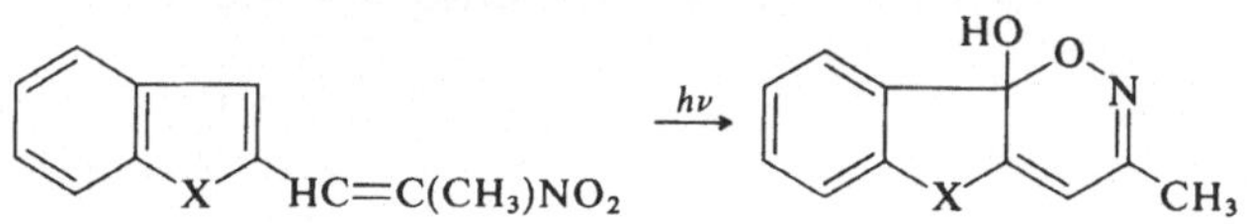

**3-183**   X = O
**3-184**   X = NH

Benzo derivatives of furan (**3-186a**) and thiophene (**3-186b**) have been reported[196] to be formed in high yields from the corresponding benzylphenyl ketones (**3-185a**) and thioketones (**3-191b**), respectively, by a photochemically induced cyclization.

| 3-185a  X = O | 3-186a  X = O |
| 3-185b  X = S | 3-186b  X = S |

When the R substituent is a good leaving group, other than a thio group, the reaction proceeds in high yields with the release of benzofuran; for example, when R group is a trialkylammonium group the yields are over 70%. The reaction is also sensitive to the ring substituents. When (**3-185a**) contains 4,4'-dimethoxy groups the yield of the released benzofuran is about 1.0%, while the 3,3'-dimethoxy analogue gives the corresponding 5- and 7-methoxybenzofuran in 48% yields. Sheehan *et al.*[196a] have described a number of derivatives (**3-187**) and their photochemical reaction under irradiation with a high-pressure mercury lamp. The quantum yield was determined to be 0.644, using a 550-W Hanovia 673-36 medium-pressure mercury lamp emitting at 336 nm through a filter solution. Sensitivity of this reaction was demonstrated with other light sources such as nitrogen lasers emitting at 337.1 nm and argon-ion lasers emitting at 351.1 and 362.8 nm. The syntheses of these benzoin derivatives from the cyanohydrin of benzaldehyde and a Grignard reagent of interest are well documented by Sheehan *et al.*[196a]

3-187a   $R^1-R^4 = R^{10} = H$, $R = OCOCH_3$
3-187b   $R^1-R^4 = R^{10} = H$, $R = Cl$
3-187c   $R^1-R^4 = R^{10} = H$, $R = OSO_2C_6H_4(4\text{-}CH_3)$
3-187d   $R^1-R^4 = R^{10} = H$, $R = (CH_3)_3N^+Cl^-$
3-187e   $R^1 = R^3 = R^{10} = H$, $R^2 = R^4 = CH_3O$, $R = OCOCH_3$
3-187f   $R^{10} = R^1 = CH_3O$, $R^2-R^4 = H$, $R = OCOCH_3$

3-187g   $R^{10} = R^1 = CH_3O$, $R^2-R^4 = H$, $R = COCH-N$

3-187h  $R^2 = CH_3O$, $R^{10} = R^1 = R^3 = R^4 = H$, $R = OCOCH_3$
3-187i  $R^1 = CH_3O$, $R^{10} = R^2 \text{-} R^4 = H$, $R = OCOCH_3$
3-188a  $R^5 \text{-} R^9 = H$
3-188b  $R^5 = R^7 = R^8 = H$, $R^6 = R^9 = CH_3O$
3-188c  $R^5 = R^6 = R^9 = H$, $R^7 = R^8 = CH_3O$
3-188d  $R^6 = R^7 = R^9 = H$, $R^5 = R^8 = CH_3O$
3-188e  $R^5 = R^7 = CH_3O$, $R^6 = R^8 = R^9 = H$
3-188f  $R^6 = CH_3O$, $R^5 = R^7 \text{-} R^9 = H$
3-188g  $R^5 = CH_3O$, $R^6 \text{-} R^9 = H$
3-188h  $R^7 = CH_3O$, $R^5 \text{-} R^8 = H$

| 3-187 | Irradiation time (h) | Percent yield (3-188) |
|-------|----------------------|------------------------|
| a | 17 | 15 (a) |
| b | 18 | 1 (a) |
| c | 17 | 3 (a) |
| e | 17 | 1 (b) |
| f | — | 94 (e) |
| g | — | 87 (e) |
| h | — | 10 (f) |
| i | — | 21 (g) |
|   |   | 67 (h) |

Benzoxazole and its iso-, nitrogen, and sulfur analogues have been reported by Grellmann and Tauer[197] to be released from ortho-substituted aryl Schiff's bases (3-189), thioamides (3-190), and salicylaldehyde oxime (3-191) via a photochemically triggered oxidative cyclization.

In a similar reaction, 2′-hydroxychalcones[198] (3-192), prepared by a Fries rearrangement of phenyl cinnamates,[199] undergo a photochemically triggered cyclization reaction with the release of flavanone (3-193).

Multi-ring heterocycles, for example, carbazoles, have been reported to be formed in high yields from 2-azidobiphenyls by a photochemical cyclization.[200] This reaction is discussed in Chapter 1 as a source for the release of nitrogen.

Heterocycles such as azepines (3-195) have been reported to be liberated from bicyclic adducts such as 7-azanorbornadiene derivatives (3-194) on photolysis. Prinzbach et al.[201] have described the synthesis of a number of bicyclic adducts and their corresponding photochemical reactions. These bicyclic adducts were prepared by the thermal cycloaddition reaction of pyrrole derivatives with activated acetylenes. Irradiation to drive the reaction was in the region above 290 nm and gave azepines in high yields.

The release of carbazole (3-197) together with a nitrile was also observed on irradiation of benzaldehyde diphenylhydrazone (3-196) at room temperature through a Vycor filter with a 450-W Hanovia high-pressure mercury lamp.[202] Other aryl diphenylhydrazones were also found to yield analogous

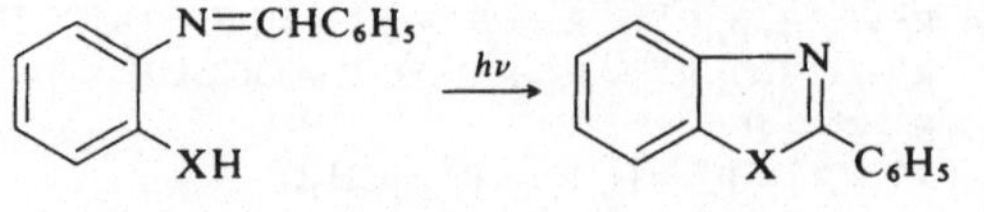

**3-189**   X = O, S, NH

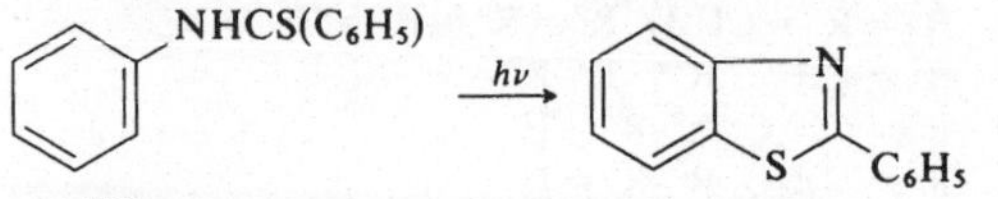

**3-190**

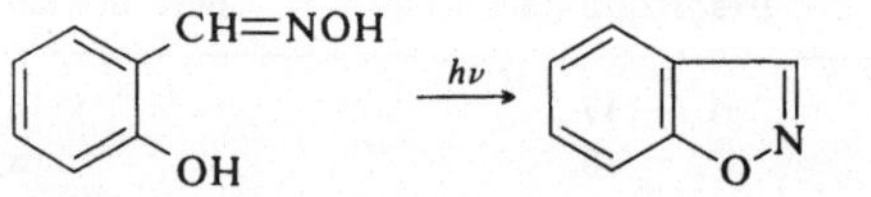

**3-191**

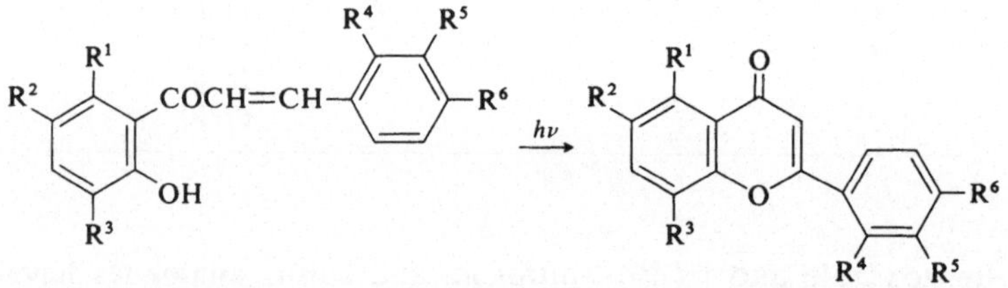

**3-192**                                                      **3-193**

**a** $R^1$-$R^6$ = H                                    **f** $R^2$ = $R^3$ = $R^4$ = OH, $R^1$ = $R^5$ = $R^6$ = H

**b** $R^1$ = $CH_3O$, $R^2$-$R^6$ = H         **g** $R^2$ = $R^3$ = OH, $R^4$ = $CH_3O$, $R^1$ = $R^5$ = $R^6$ = H

**c** $R^2$ = OH, $R^1$ = $R^3$-$R^6$ = H       **h** $R^2$ = $R^3$ = $R^5$ = $R^6$ = OH, $R^1$ = $R^4$ = H

**d** $R^3$ = OH, $R^1$ = $R^2$ = $R^4$-$R^6$ = H   **i** $R^2$ = $R^3$ = $R^6$ = OH, $R^5$ = $CH_3$, $R^1$ = $R^4$ = H

**e** $R^2$ = $R^3$ = OH, $R^1$ = $R^4$-$R^6$ = H   **j** $R^3$ = $R^5$ = OH, $R^5$ = $R^6$ = $CH_3O$, $R^1$ = $R^4$ = H

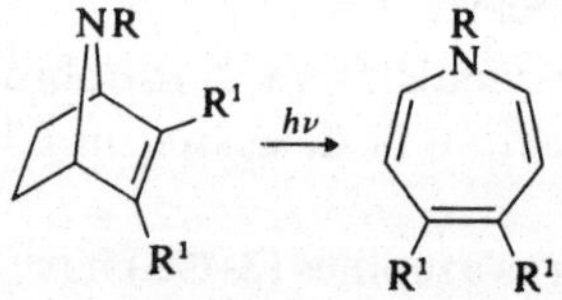

**3-194**   $R^1$ = $COOCH_3$           **3-195**

| R | Percent yield (3-195) |
|---|---|
| $C_6H_5$ | 25 |
| 4-$ClC_6H_4$ | 25 |
| 4-$BrC_6H_4$ | 25 |
| 4-$NO_2C_6H_4$ | 65 |
| 4-$CH_3C_6H_4$ | 94 |
| $COOCH_3$ | 42 |
| $CONH_2$ | 36 |
| $COCH_3$ | 45 |

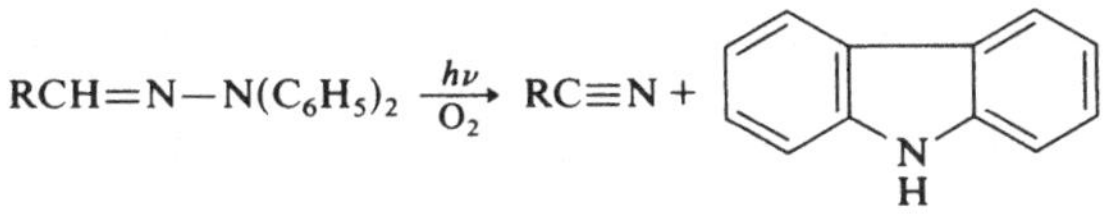

$$RCH=N-N(C_6H_5)_2 \xrightarrow[O_2]{h\nu} RC\equiv N +$$

**3-196**                                                **3-197**

| Aryl diphenylhydrazone | Percent yield (nitrile) |
|---|---|
| Cinnamaldehyde | 43 |
| Phenylacetaldehyde | 41 |
| 1-Naphthaldehyde | 55 |
| Benzaldehyde | 75 |

products. These hydrazones were prepared from diphenylhydrazine and the corresponding aldehyde.

Coumarins (**3-199**) were made by the irradiation of $\gamma$-lactone derivatives.[203] For example, irradiation of 3-oxabicyclo[3.1.0]hexan-2-one or 3-(2-hydroxybenzylidene)-4,5-dihydrofuran-2-(3$H$)-one (**3-198**) in methanol triggers a *trans-cis* isomerization followed by transesterification to yield the corresponding coumarin.[204]

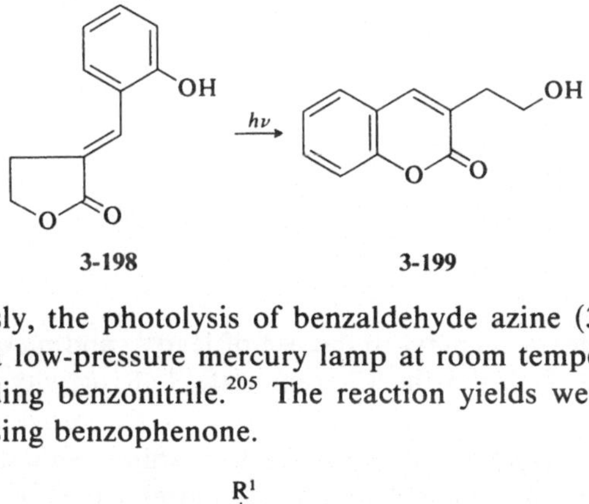

**3-198**                                                **3-199**

Analogously, the photolysis of benzaldehyde azine (**3-200**) in cyclohexane using a low-pressure mercury lamp at room temperature released the corresponding benzonitrile.[205] The reaction yields were enhanced by sensitization using benzophenone.

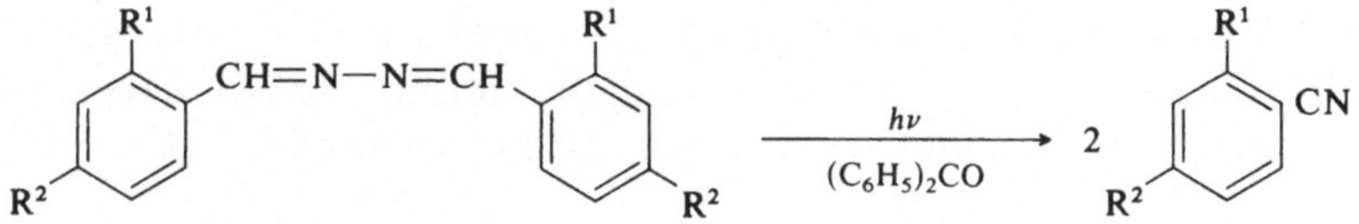

**3-200**                                                **3-201**

| $R^1$ | $R^2$ | Percent yield (3-201)[a] |
|---|---|---|
| H | H | 85 |
| Cl | H | 88 |
| H | $CH_3O$ | 82 |
| H | $NO_2$ | 95 |
| H | $(CH_3)_2N$ | 80 |

[a] Ref. 205.

### 3.2.3. Electrochemically Triggered Release

As discussed in Chapter 2, the use of an electric current in a chemical reaction has been known since the Kolbe reaction, but it has been less investigated than thermal and photochemical energies.

In this section examples of electrochemical release reactions will be discussed only briefly, since most of the reported reactions involve the release of acids and bases, which is considered in Chapter 2.

The release of small molecules such as alcohols and thioalcohols via the electrochemical reduction of hydroxyketones[206] and thiones[207] has been examined. Reviews by Maironovsky,[208] Lund,[209] Allen,[210] Popp and Schultz,[211] and Perrin[212] include extensive discussions of electrochemical reactions and their synthetic applications.

Precursors useful for the release of alcohols and thioalcohols usually incorporate benzoyl, tosyl, trimethyl,[213] diphenylmethyl, benzyl, cinnamyl,[214] or phenyl groups covalently bonded to the hydroxyl or thiol group. Electroreduction of that bond forms alcohols in 70–90% yields. For example, the electrochemical release of alcohols from tosyl esters (**3-202**) and benzoyl esters (**3-203**), was first reported by Horner and Newman[215] and Wawzonek et al.,[216] gave 60–70% yields.

$$4\text{-}CH_3C_6H_4SO_2OR \xrightarrow{2e} 4\text{-}CH_3C_6H_4SO_2H + ROH$$

**3-202**

$$C_6H_5COOR \xrightarrow{2e^-} C_6H_5COOH + ROH$$

**3-203**

The reaction was general to the use of lead[217] and mercury electrodes and showed high yields at the 10-gram scale.[217] Reactivity varies with the substituents and the solvent. Substituents on the aroyl portion of the molecule (**3-204**) which increase the electron affinity enhance the reaction rate; for example, a cinnamyl group is more activating than a benzyl group. Halogens on the alcohol portion of the molecule (**3-205**) have similar activating power, as shown by Semmelhack and Heinsohn.[218] A variety of

$$RCOOR^1 \xrightarrow{2e^-} RCOOH + R^1OH$$

**3-204**

| R | $E_{1/2}$ (volts) |
|---|---|
| $C_6H_5CH_2$ | −2.75 |
| $C_6H_5CH{=}CH_2$ | −2.50 |
| $4\text{-}NO_2C_6H_4$ | −1.30 |

$$C_6H_5COOCH_2R \xrightarrow{2e^-} RCH_2OH + C_6H_5COOH$$

**3-205**

| R | $E_{1/2}$ (volts) |
|---|---|
| CHCl$_2$ | −1.85 |
| CCl$_3$ | −1.65 |
| CBr$_3$ | −0.70 |

solvents, such as dimethylformamide, dimethylsulfoxide, acetonitrile,[219] pyridine, benzonitrile,[220] and dimethylformamide with aqueous alcohol,[221] are useful.

The release of thiols by similar routes has also been studied but to a lesser extent. Semmelhack and Heinsohn[222] have shown an analogous reaction with phenyl thioethers (**3-206**). The experimental set up generally is analogous to that reported in chapter 2 for the release of acids and bases.

$$RSR^1 \xrightarrow[NH_3]{e^-} RSH$$

**3-206**

Heterocycles have been reported as products in electrochemically triggered release reactions. A review of these has been published by Lund[209] in 1970 and updated in 1984.

Electrochemical cyclization[223] reactions such as those that aromatic compounds (**3-209**) undergo are common. For example, 2-nitroformanilide (**3-207**) releases the corresponding benzimidazole (**3-208**) upon electrochemical triggering.

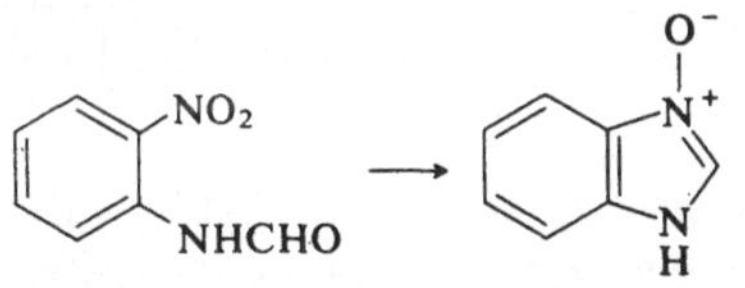

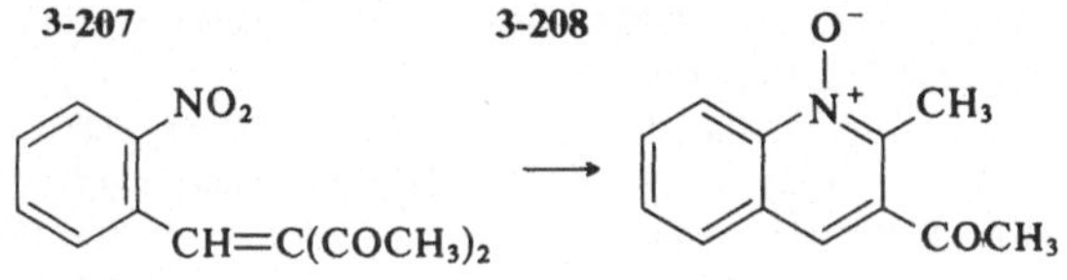

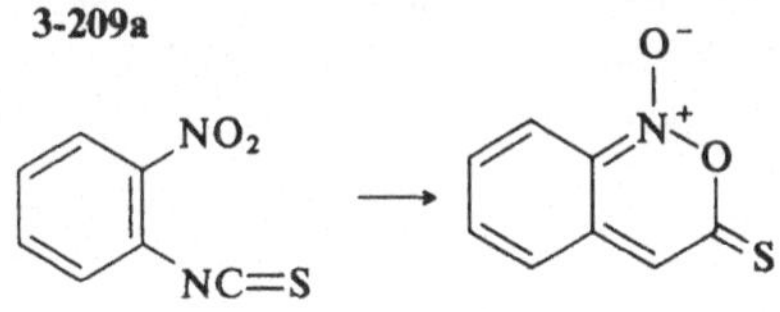

3-209c

# REFERENCES

1. W. Neugebauer and M. Tomanek, U.S. Pat. 3,070,443 (1962).
2. M. Morrison, U.S. Pat. 2,732,299 (1956).
3. Buchet & Cie, British Pat. 909,491 (1962).
4. (a) J. Hammer and A. Macaluso, *Chem. Rev.* *64*, 472 (1964); (b) G. R. DePierre and M. Lamchen, *Quart. Rev. (London)* *19*, 329 (1965).
5. S. I. Schlesinger, *SPSE Seminar Novel Imaging Systems*, 109 (1969).
6. H. Stobbe, *Ber.* *52B*, 670 (1919).
7. (a) L. M. Minsk, *J. Appl. Polym. Sci.* *2*, 302 (1959); (b) L. M. Minsk, J. G. Smith, W. P. Van Deusen, and J. R. Wright, *J. Appl. Polym. Sci.* *2*, 308 (1959); (c) J. R. Williams, Preprints of Society of Plastics Engineers, Regional Technical Conference, Ellenville, New York, Oct. 1973, p. 43.
8. T. H. Bronwlee and K. Matsuda, U.S. Pat. 3,683,336 (1972).
9. G. L. Eian and J. E. Trend, U.S. Pat. 4,369,244 (1983).
10. C. H. DePuy and R. W. King, *Chem. Rev.* *60*, 431 (1960).
11. H. R. Nace, *Org. Reactions* *12*, 57 (1962).
12. L. Chugaev, *Ber.* *32*, 3332 (1899).
13. C. H. Hurd and F. H. Blunch, *J. Am. Chem. Soc.* *60*, 2419 (1938).
14. A. Oppenheim and Precht, *Ber.* *9*, 325 (1876).
15. G. L. O'Connor and H. R. Nace, *J. Am. Chem. Soc.* *75*, 2118 (1953).
16. (a) I. Schurmann and C. E. Boord, *J. Am. Chem. Soc.* *55*, 4930 (1933); (b) R. A. Benkeser, J. J. Hazdra, and A. E. Burrows, *J. Am. Chem. Soc.* *81*, 5374 (1959); (c) C. G. Overberger and A. E. Borchert, *J. Am. Chem. Soc.* *82*, 4896 (1960); (d) J. D. Roberts and C. W. Sauer, *J. Am. Chem. Soc.* *17*, 3925 (1949); (e) E. R. Alexander and A. Mudrak, *J. Am. Chem. Soc.* *73*, 59 (1951); (f) E. R. Alexander and A. Mudrak, *J. Am. Chem. Soc.* *72*, 3194 (1950); (g) T. Markowrnikov and N. Stadnikov, *J. Russian Phys. Chem. Soc.* *35*, 392 (1903); (h) S. Nametkin and L. Brusson, *Ber.* *56*, 1807 (1923); (i) F. G. Bordwell and P. S. Landis, *J. Am. Chem. Soc.* *80*, 6379 (1958); (j) R. Van Volkenburgh, K. W. Greenlee, J. M. Derfer, and C. F. Boord, *J. Am. Chem. Soc.* *71*, 172 (1949); (k) R. A. Benkeser and J. J. Hazdra, *J. Am. Chem. Soc.* *81*, 228 (1959); (l) V. E. Tischenko and A. F. Kosternaya, *J. Gen. Chem. USSR* *7*, 1366 (1937).
17. W. Adam, J. Baeza, and J-C Liu, *J. Am. Chem. Soc.* *94*, 2000 (1972).
18. L. C. Roach and W. H. Daly, *J. Chem. Soc., Chem. Commun.*, 606 (1970).
19. H. Ulrich and A. A. R. Sayigh, *Angew. Chem. Intl. Ed. Engl.* *5*, 704 (1966).
20. E. N. Cain, R. Vukov, and S. Masamue, *J. Chem. Soc., Chem. Commun.*, 98 (1969).
21. G. H. Schmid and A. W. Wolkoff, *Can. J. Chem.* *80*, 1181 (1972).
22. (a) W. Wernick and R. Wolffenstein, *Ber.* *31*, 1553 (1898); (b) L. Mamlock and R. Wolffenstein, *Ber.* *33*, 159 (1900).
23. A. C. Cope, T. T. Foster, and P. H. Towle, *J. Am. Chem. Soc.* *71*, 3929 (1949).
24. A. C. Cope and E. R. Trumbull, *Org. Reactions* *11*, 317 (1960).
25. (a) D. J. Cram and J. E. McCarty, *J. Am. Chem. Soc.* *76*, 5740 (1954); (b) D. J. Cram and M. R. V. Sahyun, *J. Am. Chem. Soc.* *85*, 1263 (1963).

26. A. C. Cope, C. L. Bumgardner, and E. E. Schweizer, *J. Am. Chem. Soc. 79*, 4729 (1957).
27. (a) A. C. Cope, N. A. LeBel, H-H. Lee, and W. R. Moore, *J. Am. Chem. Soc. 79*, 4720 (1957); (b) A. C. Cope and C. L. Bumgardner, *J. Am. Chem. Soc. 79*, 960 (1957).
28. L. A. Paquette, *Acc. Chem. Res. 1*, 209 (1968).
29. L. A. Carpino, L. V. McAdams III, R. H. Rynbrandt, and J. W. Spiewak, *J. Am. Chem. Soc. 93*, 476 (1971).
30. C. C. Price and G. Berti, *J. Am. Chem. Soc. 76*, 1212 (1954).
31. H. Gerlach, T. T. Huong, and W. Muller, *J. Chem. Soc., Chem. Commun.*, 1215 (1972).
32. H. J. Baker and A. H. Blass, *Recl. Trav. Chim. Pays-Bas 61*, 785 (1942).
33. (a) D. K. Black and S. R. Landor, *J. Chem. Soc.*, 6784 (1965); (b) D. K. Black and S. R. Landor, *J. Chem. Soc.*, 5225 (1965).
34. (a) K. Ziegler, *Ann. 434*, 74 (1923); (b) K. Ziegler, *Ber. 55*, 2274 (1922).
35. G. Balz and G. Schiemann, *Ber. 60B*, 1186 (1927).
36. A. Roe, *Org. Reactions 5*, 193 (1949).
37. (a) L. E. Smith and H. L. Haller, *J. Am. Chem. Soc. 61*, 143 (1939); (b) H. L. Haller and P. S. Schaffer, *J. Am. Chem. Soc. 55*, 4954 (1933).
38. E. B. Starkey, *Org. Syntheses, Coll.*, Vol. 2, 225 (1943).
39. P. Ruggli and E. Casper, *Helv. Chim. Acta 18*, 1414 (1935).
40. (a) M. S. Lesslie and E. E. Turner, *J. Chem. Soc.*, 1590 (1933); (b) F. C. Schmelkes and M. Rubin, *J. Am. Chem. Soc. 66*, 1631 (1944).
41. M. E. Smith, E. Elisberg, and M. L. Sherrill, *J. Am. Chem. Soc. 68*, 1301 (1946).
42. A. W. Ruddy, E. B. Starkey, and W. H. Hartung, *J. Am. Chem. Soc. 64*, 828 (1942).
43. (a) M. F. W. Dunker, F. B. Starkey, and G. L. Jenkins, *J. Am. Chem. Soc. 58*, 2308 (1936); (b) M. F. W. Dunker and F. B. Starkey, *J. Am. Chem. Soc. 61*, 3005 (1939).
44. (a) F. A. Bolth, W. M. Whaley, and E. B. Starkey, *J. Am. Chem. Soc. 65*, 1456 (1943); (b) W. M. Whaley and E. B. Starkey, *J. Am. Chem. Soc. 68*, 793 (1946).
45. (a) M. F. W. Dunker, E. B. Starkey, and G. L. Jenkins, *J. Am. Chem. Soc. 58*, 2308 (1936); (b) G. Balz and G. Schiemann, *Ber. 60*, 1186 (1927); (c) G. Schiemann, *Ber. 62*, 1794 (1929); (d) G. Schiemann, *Z. Physik. Chem. A156*, 397 (1931); (e) E. B. Starkey, *J. Am. Chem. Soc. 57*, 1479 (1937); (f) E. Bergmann, L. Engel, and S. Sander, *Z. Physik. Chem. 10B*, 106 (1930); (g) L. A. Bigelow, J. H. Person, L. B. Cook, and W. T. Miller, Jr., *J. Am. Chem. Soc. 55*, 4614 (1934); (h) G. Schiemann, *J. Prakt. Chem. 140*, 97 (1929); (i) F. M. Meigs, U.S. Pat. 1,916,327 (1933); (j) G. Schiemann and J. Pillarsky, *Ber. 62*, 3035 (1929); (k) M. F. W. Dunker and E. B. Starkey, *J. Am. Chem. Soc. 61*, 3005 (1939); (l) J. F. J. Dippy and F. R. Williams, *J. Chem. Soc.*, 1466 (1934); (m) R. Stolle and K. Th. Gunzert, *J. Prakt. Chem. 139*, 141 (1934); (n) G. Schiemann and W. Roselius, *Ber. 62*, 1805 (1929); (o) G. Schiemann and W. Roselius, *Ber. 65*, 737 (1932).
46. (a) M. P. Cava and A. A. Deana, *J. Am. Chem. Soc. 81*, 4266 (1959); (b) M. P. Cava and R. L. Shirley, *J. Am. Chem. Soc. 82*, 654 (1960).
47. (a) W. L. Mock, *J. Am. Chem. Soc. 89*, 1281 (1967); (b) J. D. Loudon and L. B. Young, *J. Chem. Soc.*, 5496 (1963); (c) J. D. Loudon and L. B. Young, *J. Chem. Soc.*, 3262 (1962).
48. (a) P. S. Engel, *Chem. Rev. 80*, 99 (1980); (b) S. Patai, ed., *Chemistry of Hydrazo, Azo & Azoxy Groups*, J. Wiley and Sons, New York (1975).
49. R. Henriques, *Ber. 27*, 2993 (1894).
50. (a) O. Hinsberg, *J. Prakt. Chem. 90*, 345 (1914); (b) O. Hinsberg, *J. Prakt. Chem. 91*, 307 (1915).
51. L. A. Warren and S. Smiles, *J. Chem. Soc.*, 956, 1327 (1930).
52. F. Galbraith and S. Smiles, *J. Chem. Soc.*, 1234 (1935).
53. C. S. McClement and S. Smiles, *J. Chem. Soc.*, 1016 (1937).
54. J. F. Bunnet and R. E. Zahler, *Chem. Rev. 49*, 362 (1951).
55. W. E. Truce, E. M. Kreider, and W. W. Brand, *Org. Reactions 18*, 99 (1970).
56. W. Bergmann and M. J. McLean, *Chem. Rev. 28*, 367 (1941).

57. (a) C. Dufraisse and R. Priou, *Bull. Soc. Chim. Fr. 6,* 1649 (1939); (b) C. Dufraisse and J. Houpillart, *Bull. Soc. Chim. Fr. 5,* 626 (1938).

58. L. Claisen, *Ber. 45,* 3157 (1912).

59. W. M. Lauer and E. I. Kilburn, *J. Am. Chem. Soc. 59,* 2586 (1937).

60. C. D. Hurd and M. A. Pollack, *J. Am. Chem. Soc. 60,* 1905 (1938).

61. L. Claisen and O. Eisleb, *Ann. 401,* 21 (1913).

62. D. S. Tarbell and J. W. Wilson, *J. Am. Chem. Soc. 64,* 607 (1942).

63. L. F. Fieser and W. C. Lothrop, *J. Am. Chem. Soc. 58,* 749 (1936).

64. L. F. Fieser and M. N. Young, *J. Am. Chem. Soc. 53,* 4120 (1931).

65. W. C. Lothrop, *J. Am. Chem. Soc. 62,* 132 (1940).

66. W. C. Lothrop, *J. Am. Chem. Soc. 61,* 2115 (1939).

67. S. Rangaswami and T. R. Seshadri, *Proc. Indian Acad. Sci. 9A,* 1 (1939); *Chem. Abstr. 33,* 4244 (1939).

68. E. Bergmann and T. Berlin, *J. Am. Chem. Soc. 62,* 316 (1940).

69. (a) J. F. Kinacid and D. S. Tarbell, *J. Am. Chem. Soc. 61,* 3085 (1939); (b) D. S. Tarbell and J. F. Kinacid, *J. Am. Chem. Soc. 62,* 728 (1940).

70. D. S. Tarbell, *Org. Reactions 2,* 1 (1957).

71. (a) L. Claisen, *Ann. 418,* 97 (1926); (b) S. Kawai, *Sci. Papers Inst. Phys. Chem. Research, Tokyo, 3,* 263 (1925); (c) G. Hahn and W. Stenner, *Z. Physiol. Chem. 181,* 88 (1929); (d) W. Baker and O. M. Lothian, *J. Chem. Soc.,* 274 (1939); (e) L. F. Fieser and W. C. Lothrop, *J. Am. Chem. Soc. 57,* 1459 (1935); (f) B. Mander-Jones and V. M. Trikojius, *J. Proc. Roy. Soc., N.S. Wales 66,* 300 (1932).

72. S. W. Stalsy and D. W. Reichard, *J. Am. Chem. Soc. 90,* 817 (1968).

73. A. F. McKay, I. M. Billy, and E. J. Tarlton, *J. Org. Chem. 29,* 291 (1964).

74. (a) D. H. R. Barton and B. J. Willis, *J. Chem. Soc., Chem. Commun.,* 1225 (1970); (b) C. T. Peterson, *Acta Chim. Scand. 22,* 247 (1968).

75. (a) D. D. Reynolds, D. L. Fields, and D. L. Johnson, *J. Org. Chem. 26,* 5130 (1960); (b) D. D. Reynolds, *J. Am. Chem. Soc. 79,* 4951 (1957).

76. (a) D. D. Reynolds, M. K. Massad, D. F. Fields, and D. L. Johnson, *J. Org. Chem. 26,* 5109 (1961); (b) D. D. Reynolds, D. L. Johnson, and D. L. Fields, *J. Org. Chem. 26,* 5125 (1961).

77. D. B. Pattison, *J. Am. Chem. Soc. 79,* 3455 (1957).

78. T. Mortel and P. E. Verkade, *Recl. Trav. Chim. Pays-Bas 70,* 35 (1951).

79. (a) B. Helferich, *Ann. 646,* 45 (1961); (b) B. Helferich and W. Klebert, *Ann. 657,* 79 (1962).

80. S. Gabriel and R. Stelzner, *Chem. Ber. 28,* 2929 (1895).

81. H. W. Heine, *Mech. Mol. Migr. 3,* 145 (1971).

82. (a) C. W. Woods, A. B. Borkovec, and F. M. Hart, *J. Med. Chem. 7,* 371 (1964); (b) P. E. F. Fanta and E. N. Walsh, *J. Org. Chem. 31,* 59 (1966); (c) F. W. Fowler and A. Hassner, *J. Am. Chem. Soc. 90,* 2875 (1968); (d) H. W. Heine and M. S. Kaplan, *J. Org. Chem. 32,* 3069 (1967); (e) P. G. Mentz, H. W. Heine, and G. R. Scharoubim, *J. Org. Chem. 33,* 4547 (1968); (f) P. E. Fanta, R. J. Smat, and J. R. Krikau, *J. Heterocycl. Chem. 5,* 419 (1968); (g) H. W. Heine, *J. Am. Chem. Soc. 85,* 2143 (1963); (h) J. F. W. Keana, S. B. Keana, and D. Beetham, *J. Chem. Soc. 32,* 3057 (1967); (i) Y. Iwakura and A. Nabeya, *Bull. Tokyo Inst. Technol. 42,* 69 (1961); (j) Y. Iwakura and A. Nabeya, *J. Chem. Soc., Japan, Pure Chem. Soc. 77,* 773 (1956); (k) D. A. Tomalia, *J. Heterocycl. Chem. 4,* 419 (1967); (l) E. Schmitz and S. Schramm, *Chem. Ber. 100,* 2593 (1967).

83. J. W. A. M. Janssen, H. J. Koeners, C. G. Kruse, and C. L. Habraken, *J. Org. Chem. 38,* 1777 (1973).

84. C. L. Habraken, P. C. Fernandes, S. Balian, and K. C. van Erk, *Tetrahedron Lett.,* 479 (1970).

85. H. Durr and R. Sergio, *Tetrahedron Lett.,* 3479 (1972).

86. M. F. Newmann and C. Buchecker, *Tetrahedron Lett.*, 937 (1972).

87. P. C. Fernandes and C. L. Habraken, *J. Org. Chem. 36*, 3084 (1971).

88. K. M. Baines, T. W. Rourke, and K. Vaughan, *J. Org. Chem. 46*, 856 (1981).

89. T. A. Daniels, S. Sidi, and K. Vaughan, *Can. J. Chem. 55*, 3751 (1977).

90. K. M. Baines, K. Vaughan, D. L. Hooper, and L. F. Lever, *Can. J. Chem. 6*, 1549 (1983).

91. (a) R. A. Carboni, R. V. Lindsey, and C. D. Nenitzech, *Chem. Ber. 81*, 4342 (1959); (b) R. H. B. Galtz and J. D. Loudon, *J. Chem. Soc.*, 885 (1959).

92. P. A. S. Smith, in: *Nitrenes* (W. Lwowski, ed.), p. 136, Interscience Pub. Inc., New York (1970).

93. P. A. Smith, G. J. W. Breen, M. K. Hajek, and D. V. C. Awang, *J. Am. Chem. Soc. 35*, 2215 (1970).

94. P. A. S. Smith, L. O. Krebechhek, and W. Resemann, *J. Am. Chem. Soc. 86*, 2025 (1964).

95. P. A. S. Smith, *Org. Reactions 3*, 337 (1946).

96. E. H. Burke and D. D. Carlos, *J. Heterocycl. Chem. 7*, 177 (1970).

97. O. Aschan, *Ber. 23*, 1823 (1890).

98. T. L. Davis and H. W. Underwood, *J. Am. Chem. Soc. 44*, 2597 (1922).

99. A. W. Hofmann, *Ber. 14*, 2727 (1881).

100. P. Dinglinger, *Ann. 311*, 130 (1900).

101. (a) J. von Braun, *Ber. 35*, 3372 (1902); (b) J. von Braun, *Ber. 42*, 4571 (1909).

102. (a) A. M. Damas, R. O. Gould, M. M. Harding, P. M. Paton, J. F. Ross, and J. Crosby, *J. Chem. Soc., Perkin Trans. 2*, 2991 (1981); (b) R. K. Howe and J. E. Franz, *J. Org. Chem. 39*, 962 (1974).

103. J. G. Erickson, P. E. Wiley, and V. P. Wystrach, *The Chemistry of Heterocyclic Compounds*, Vol. 5, Chapter 5, Interscience Pub. Inc., New York (1956).

104. D. L. Clive, *J. Chem. Soc., Chem. Commun.*, 1014 (1970).

105. P. J. Folley, Jr., *J. Org. Chem. 34*, 2805 (1969).

106. H. Hettler and H. Neygenfind, *Chem. Ber. 103*, 1397 (1970).

107. C. L. Stevens, P. M. Pillai, M. E. Munk, and K. G. Taylor, *Mech. Mol. Migr. 3*, 271 (1963).

108. C. L. Stevens, I. L. Klundt, M. E. Munk, and M. D. Pillai, *J. Org. Chem. 30*, 2967 (1965).

109. C. L. Stevens, A. Thuillier, and F. A. Daniher, *J. Org. Chem. 30*, 2962 (1965).

110. K. L. Nelson, J. C. Robertson, and J. J. Duvall, *J. Am. Chem. Soc. 86*, 684 (1964).

111. A. Pinner, *Chem. Ber. 28*, 456 (1895).

112. C. H. Rayburn, W. R. Harlan, and H. R. Hanmer, *J. Am. Chem. Soc. 72*, 1721 (1950).

113. J. Meisenheimer, *Chem. Ber. 52*, 1667 (1919).

114. A. C. Cope and R. F. Klein, *J. Am. Chem. Soc. 66*, 1929 (1944).

115. D. S. Tarbell, *Org. Reactions 2*, 1 (1944).

116. A. C. Cope and P. H. Towle, *J. Am. Chem. Soc. 71*, 3423 (1949).

117. A. C. Cope, T. T. Foster, and P. H. Towle, *J. Am. Chem. Soc. 71*, 3929 (1949).

118. W. Carruthers and R. A. W. Johnstone, *J. Chem. Soc.*, 1653 (1965).

119. R. A. W. Johnstone, *Mech. Mol. Migr. 2*, 249 (1969).

120. R. Piria, *Ann. 56*, 41 (1845).

121. (a) E. Bamberger, *Ber. 36*, 2051 (1903); (b) E. Bamberger, *Ber. 30*, 1264 (1897).

122. R. Stolle, *J. Prakt. Chem. 68*, 464 (1908).

123. R. Weddige, *J. Prakt. Chem. 36*, 143 (1887).

124. *Chem. Zentr. 1*, 236 (1901).

125. A. Ahlquist, *J. Prakt. Chem. 99*, 45 (1919).

126. (a) G. Schroeter, *Ann. 426*, 1 (1922); (b) R. H. F. Manske, N. H. Perkin, Jr., and R. Robinson, *J. Chem. Soc.*, 1 (1927).

127. A. Shimomura, *Mem. Coll. Sci. Kyoto Imp. Univ. A8*, 19 (1925); *Chem. Abstr. 19*, 2196 (1925).

128. O. L. Chapman, D. J. Pasto, G. W. Borden, and A. A. Griswold, *J. Am. Chem. Soc. 84*, 1220 (1962).

129. O. L. Chapman and D. J. Pasto, *J. Am. Chem. Soc. 82*, 3642 (1960).
130. G. Buchi and E. M. Burgess, *J. Am. Chem. Soc. 82*, 4333 (1960).
131. W. G. Dauben and G. J. Fonken, *J. Am. Chem. Soc. 81*, 4060 (1959).
132. M. Fisher, *Chem. Ber. 101*, 2669 (1968).
133. J. D. Margerum, J. N. Pitts, J. G. Rutgers, and S. Searles, *J. Am. Chem. Soc. 81*, 1549 (1959).
134. (a) S. Searles, *J. Am. Chem. Soc. 73*, 124 (1951); (b) S. Searles, K. A. Pollart, and E. F. Lutz, *J. Am. Chem. Soc. 79*, 948 (1957).
135. G. M. Bennett and W. G. Philip, *J. Chem. Soc.*, 1937 (1928).
136. O. L. Chapman, C. L. McIntoch and J. Pacansky, *J. Am. Chem. Soc. 95*, 614 (1973).
137. F. E. Blacet and A. Miller, *J. Am. Chem. Soc. 79*, 4327 (1957).
138. N. J. Turro, P. A. Leermakers, H. R. Wilson, D. C. Weckers, and G. F. Vesley, *J. Am. Chem. Soc. 87*, 2613 (1965).
139. I. Heller and R. Srinivasan, *J. Am. Chem. Soc. 87*, 1144 (1965).
140. R. M. Kellog, W. L. Prins, and B. Schoustra, *J. Org. Chem. 36*, 1838 (1971).
141. A. I. Vogel, *Practical Organic Chemistry*, p. 732, Longmans, Green and Co., London (1962).
142. (a) W. Ando, in: *The Chemistry of Diazonium and Diazo Groups*, Part I (S. Patai, ed.), Chapter 4, p. 341, J. Wiley and Sons, New York (1978); (b) M. Andressen, *Chem. Zentr. 66*, 530 (1895).
143. (a) M. S. Dinaburg, *Photosensitive Diazo Compounds*, Focal Press, New York (1964); (b) H. Zollinger, *Azo and Diazo Chemistry*, Interscience, London (1961); (c) D. J. Brown, *Chem. Ind. 22*, 146 (1944).
144. T. Tsunoda and T. Yamaoka, *J. Photogr. Sci. Japan 29*, 197 (1966).
145. E. H. White and A. A. F. Seber, *Tetrahedron Lett.*, 2713 (1967).
146. S. A. Kandil and R. E. Dessey, *J. Am. Chem. Soc. 88*, 3027 (1960).
147. H. Stobbe and A. Lehfeldt, *Ber. 58*, 2415 (1925).
148. F. G. Baddar, L. S. El-Assal, and M. Gindy, *J. Chem. Soc.*, 1270 (1948).
149. G. N. Lewis, T. T. Magel, and D. Lipkin, *J. Am. Chem. Soc. 62*, 2973 (1940).
150. C. O. Parker and P. E. Spoerri, *Nature 166*, 603 (1950).
151. R. E. Buckles, *J. Am. Chem. Soc. 77*, 1040 (1955).
152. (a) I. Izawa, K. Yokoi, and H. Tomioka, *Chem. Lett.*, 1473 (1983); (b) K. Yakushijin, M. Kozuka, and H. Furukawa, *Chem. Pharm. Bull. 28*, 2178 (1980).
153. (a) W. K. Robbins and R. H. Eastman, *J. Am. Chem. Soc. 92*, 6076 (1970); (b) J. E. Starr and R. H. Eastman, *J. Org. Chem. 31*, 1393 (1966); (c) O. L. Chapman, D. J. Pasto, G. W. Borden, and A. A. Griswold, *J. Am. Chem. Soc. 84*, 1220 (1962); (d) D. I. Schuster, R. R. Scholnick, and F. T. H. Lee, *J. Am. Chem. Soc. 90*, 1300 (1968).
154. P. J. Wagner, *J. Am. Chem. Soc. 98*, 5898 (1967).
155. J. C. Anderson and C. B. Reese, *Proc. Chem. Soc.*, 217 (1960).
156. (a) D. Bellus and P. Hrdlovic, *Chem. Rev. 67*, 599 (1967); (b) V. T. Stenberg, in: *Organic Photochemistry*, Vol. 1 (O. L. Chapman, ed.), p. 127, Dekker, New York (1967); (c) D. Bellus, in: *Advances in Photochemistry*, Vol. 8 (J. N. Pitts, G. S. Hammond, and W. A. Noyes, Jr., eds.), p. 109, Wiley-Interscience, New York, (1971); (d) H. Kobsa *J. Org. Chem. 27*, 2293 (1962).
157. R. A. Finnegan and J. J. Mattice, *Tetrahedron 21*, 1015 (1965).
158. J. L. Stratenus and E. Havinga, *Recl. Trav. Chim. Pays-Bas 85*, 434 (1966).
159. D. H. R. Barton, Y. L. Chow, A. Cox, and G. W. Kirby, *J. Chem. Soc.*, 3571 (1965).
160. D. H. R. Barton, T. Nakano, and P. G. Sammes, *J. Chem. Soc.*, 322 (1968).
161. E. Havinga, R. O. DeJong, and W. Dorst, *Recl. Trav. Chim. Pays-Bas 75*, 378 (1956).
162. K. Matsui, N. Maeno, S. Susuki, H. Shizuka, and T. Morita, *Tetrahedron Lett.*, 1467 (1970).
163. K. Mutai, S-I. Kanno, and K. Kobayashi, *Tetrahedron Lett.*, 1273 (1978).
164. A. Abad, D. Mellier, J. P. Pete, and C. Portella, *Tetrahedron Lett.*, 4555 (1971).
165. I. Orban, K. Schaffner, and O. Jeger, *J. Am. Chem. Soc. 85*, 3033 (1963).

166. (a) N. C. Yang and D. H. Yang, *J. Am. Chem. Soc. 80*, 2913 (1958); (b) N. C. Yang and D. H. Yang, *Tetrahedron Lett.*, 10 (1960).

167. (a) P. Buchschacher, M. Cereghetti, H. Wehrli, K. Schaffner, and O. Jeger, *Helv. Chim. Acta 42*, 2122 (1959); (b) M. Ceregetti, H. Wehrili, K. Schaffner, and O. Jeger, *Helv. Chim. Acta 43*, 354 (1960).

168. D. H. R. Barton, P. G. Sammes, and G. G. Weingarten, *J. Chem. Soc., Sect. C,* 721 (1971).

169. H. W. Pinnick and W. H. Lajis, *J. Org. Chem. 43*, 371 (1978).

170. (a) M. E. Jung *J. Org. Chem. 41*, 1479 (1976); (b) E. J. Corey and B. B. Snider, *J. Am. Chem. Soc. 94*, 2549 (1972).

171. A. Feigenbaum, J. P. Pete, and D. Scholler, *Tetrahedron Lett.*, 537 (1979).

172. T. T. Takahashi, C. Y. Nakamura, and J. Y. Sato, *J. Chem. Soc., Chem. Commun.*, 680 (1977).

173. E. Fujita, Y. Nago, and K. Kaneko, *Chem. Pharm. Bull. 26*, 3743 (1978).

174. W. A. Feld, R. Paessun, and M. P. Serve, *J. Heterocycl. Chem. 7*, 1309 (1980).

175. M. P. Savre, *J. Org. Chem. 39*, 3788 (1974).

176. A. Schonberg, A. K. Fateen, and S. M. A. R. Omran, *J. Am. Chem. Soc. 18*, 1224 (1956).

177. J. Kenyon, A. R. A. A. Rasoul, and G. Soliman, *J. Chem. Soc.*, 1774 (1956).

178. (a) P. DeMayo, *Adv. Org. Chem. 2*, 367 (1960); (b) H. A. Morrison, in: *The Chemistry of the Nitro and Nitroso Groups* (H. Fever, ed.), Part I, p. 185, J. Wiley and Sons, New York (1970).

179. J. A. Baltrop, P. J. Plant, and P. Schofield, *J. Chem. Soc., Chem. Commun.*, 822 (1966).

180. A. Patchornik, B. Amit, and R. B. Woodward, *J. Am. Chem. Soc. 92*, 6333 (1970).

181. J. Herbert and D. Gravel, *Can. J. Chem. 52*, 187 (1974).

182. C. O. Guss, *J. Org. Chem. 17*, 678 (1952).

183. J. C. Sheehan and K. Umezawa, *J. Org. Chem. 38*, 3771 (1973).

184. C. J. Pedersen, *J. Am. Chem. Soc. 79*, 5014 (1957).

185. C. J. Pedersen, *J. Am. Chem. Soc. 79*, 2295 (1957).

186. D. J. Wagner, C. A. Stout, S. Searles, and G. S. Hammond, *J. Am. Chem. Soc. 88*, 1242 (1966).

187. (a) M. J. Kamlet and L. A. Kaplan, *J. Org. Chem. 22*, 576 (1957); (b) J. S. Splitter and M. Calvin, *J. Org. Chem. 23*, 651 (1958); (c) R. Bonnett, V. M. Clark, and A. Todd, *J. Chem. Soc.*, 2102 (1959).

188. L. I. Smith, *Chem. Rev. 23*, 193 (1938).

189. J. Hamer and A. Macaluso, *Chem. Rev. 64*, 474 (1964).

190. G. R. DelPierre and M. Lamchen, *Quart. Rev. 19*, 329 (1965).

191. S. S. Hecht and F. D. Greene, *J. Am. Chem. Soc. 89*, 6761 (1967).

192. J. P. Freeman, *J. Org. Chem. 28*, 2508 (1963).

193. (a) R. Hunt and S. T. Reid, *J. Chem. Soc., Perkin Trans. 1*, 2527 (1972); (b) R. Hunt and S. T. Reid, *J. Chem. Soc., Chem. Commun.*, 1576 (1970).

194. (a) M. R. Marquis, *Ann. Chim. Phys. 4*, 196 (1905); (b) I. J. Rinkes, *Recl. Trav. Chim. Pays-Bas 49*, 1120, 1225 (1930); (c) K. J. Morgan and D. P. Morrey, *Tetrahedron 22*, 57 (1966).

195. (a) R. Hunt, S. T. Reid, and K. T. Taylor, *Tetrahedron Lett.*, 2861 (1974); (b) J. S. Cridland and S. T. Reid, *J. Chem. Soc., Chem. Commun.*, 125 (1969).

196. (a) J. C. Sheehan, R. M. Wilson, and A. W. Oxford, *J. Am. Chem. Soc. 93*, 7222 (1971); (b) J. C. Sheehan and R. M. Wilson, *J. Am. Chem. Soc. 86*, 5277 (1964); (c) J. R. Collier and J. Hill, *J. Chem. Soc., Chem. Commun.*, 640 (1969); (d) A. Schonberg, A. K. Fateen, and S. M. A. R. Omran, *J. Am. Chem. Soc. 78*, 1224 (1956).

197. K. H. Grellmann and E. Tauer, *Tetrahedron Lett.*, 1909 (1967).

198. (a) F. R. Sternmitz, J. A. Adamovics, and J. Geigert, *Tetrahedron 31*, 1593 (1975); (b) R. Matsushima and I. Hirao, *Bull. Chem. Soc. Japan 53*, 518 (1980).

199. V. T. Ramakrishnan and J. Kagan, *J. Org. Chem. 35*, 2901 (1970).
200. P. A. Smith and B. B. Brown, *J. Am. Chem. Soc. 73*, 2435 (1951).
201. (a) H. Prinzbach, R. Fuchus, and R. Kitzing, *Angew. Chem. Intl. Ed. Engl. 7*, 67 (1968);
      (b) R. Kitzing, R. Fuchus, M. Joyeux, and H. Prinzbach, *Helv. Chim. Acta 51*, 888 (1968).
202. R. W. Binkley, *Tetrahedron Lett.*, 2085 (1970).
203. P. C. M. Van Noort and H. Cerfontain, *J. Chem. Soc., Perkin Trans. 2*, 757 (1978); *ibid.*, 249 (1979).
204. I. R. Bellobono, L. Zanderighi, S. Omarini, B. Marcandalli, and C. Parini, *J. Chem. Soc., Perkin Trans. 2*, 1529 (1975).
205. J. H. Hodgkins and J. A. King, *J. Am. Chem. Soc. 85*, 2679 (1963).
206. M. Fedoronko, *Chem. Zvesti 12*, 17 (1958).
207. P. Zuman, O. Manousek, and V. Horak, *Collect. Chech. Chem. Commun. 29*, 2906 (1964).
208. V. G. Maironovsky, *Angew. Chem. Intl. Ed. Engl. 15*, 281 (1976).
209. H. Lund, in: *Advances in Heterocyclic Chemistry* (A. R. Katritzky and A. J. Boulton, eds.), Vol. 12, p. 213 and Vol. 36, p. 237, Academic Press, London (1970 and 1984).
210. M. J. Allen, *Organic Electrode Processes*, Chapman and Hall, London (1958).
211. F. D. Popp and H. P. Schultz, *Chem. Rev. 62*, 19 (1962).
212. C. L. Perrin, *Prog. Phys. Org. Chem. 3*, 165 (1965).
213. V. G. Mairanovsky, A. Ja. Veinberg, and G. I. Samokhvalov, *Avt. Svid. N222*, 353 (1966).
214. V. G. Mairanovsky, L. A. Vakulova, and G. I. Samokhvalov, *Electrokhimiya 3*, 23 (1967).
215. L. Horner and H. Newman, *Chem. Ber. 98*, 1715, 3462 (1965).
216. S. Wawzonek, H. Laitinen, and S. Kwiatkowski, *J. Am. Chem. Soc. 66*, 830 (1944).
217. K. Okumura, T. Iwasaki, M. Matsuoka, and K. Matsumoto, *Chem. Ind. (London)*, 929 (1971).
218. M. Semmelhack and G. Heinsohn, *J. Am. Chem. Soc. 94*, 5139 (1972).
219. V. M. Maremjae, *Org. Reactions, USSR 4*, 573 (1967).
220. V. M. Maremjae, *Org. Reactions, USSR 5*, 943, 953 (1968).
221. R. V. Vizgeri, M. A. Kovbur, and A. V. Senko, *Org. Reactions, USSR 9*, 763 (1972).
222. M. F. Semmelhack and G. E. Heinsohn, *J. Am. Chem. Soc. 94*, 5139 (1972).
223. H. Lund and L. G. Feokitskov, *Acta Chem. Scand. 23*, 3482 (1969).
224. W. M. Horspool and P. L. Pauson, *J. Chem. Soc.*, 5162 (1965).
225. (a) S. T. Reid, in: *Advances in Heterocyclic Chemistry* (A. R. Katritzky and A. J. Boulton, eds.), Vol. 11, p. 1, Academic Press, London (1970); (b) S. T. Reid, in: *Advances in Heterocyclic Chemistry* (A. R. Katritzky, ed.), Vol. 30, p. 1, Academic Press, London (1982).

# 4

# Triggered Reactions of Polymers

## 4.1. PROCESSES UTILIZING THE TRIGGERED REACTIONS OF POLYMERS

Polymers with photochemically or thermally labile groups have found numerous applications in applied chemistry. The following examples set the scene for the contents of this chapter.

Within the area of image reproduction, polymeric azides (**4-1**) have found a number of applications.[1-4] For example, Photozid[R], a commercially available azide polymer from the Upjohn Company, is used to form dyed images.[5] It is a copolymer (**4-1**) of maleic anhydride and vinyl ether with light-sensitive arene sulfonylazide groups. The acidic groups in the polymers act as anchoring sites for basic dyes which give the image color.

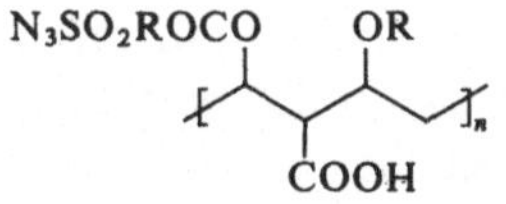

$$N_3SO_2ROCO \qquad OR$$

**4-1**

The light exposure triggers the decomposition of the azide groups with the release of nitrenes. The latter add across adjacent polymer chains causing cross-linking. During the development wash-off step, the areas which have not been exposed tend to dissolve more easily and wash off quicker than the exposed cross-linked polymer areas. This results in the generation of a relief image made up from the cross-linked polymer.

Polymers from 4-diazodiphenylamine and formaldehyde have found use as sensitized resins for printing plate applications.[6,7]

Poly(vinyl cinnamates) (**4-2**) are classic examples where the triggered photocycloaddition between the C,C-double bonds of the cinnamyl groups has been utilized in negative-acting photoresists. Several commercial resists are based on such photocycloaddition chemistry.[7-9]

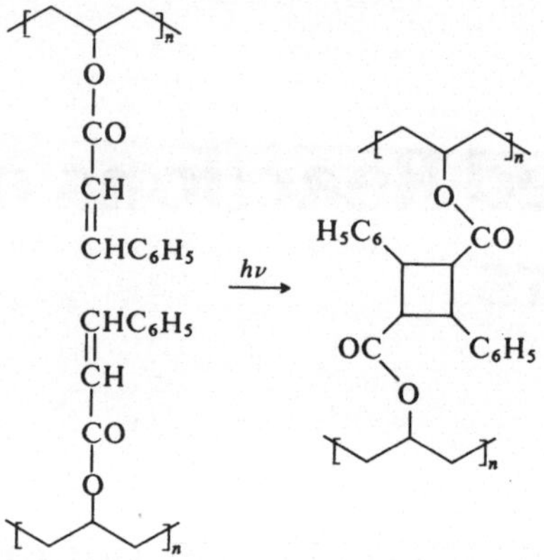

**4-2**

Poly(alkyl sulfones) (**4-3**) have recently gained much interest as photo-resists.[10,11] Under photochemical triggering conditions, the polymer decomposes with the release of sulfur dioxide and the corresponding alkene. This increases the volatility of the exposed over the unexposed areas under thermal development which results in a positive polymeric image.

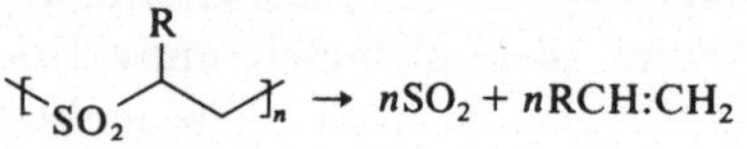

**4-3**

Thermally triggered release of radicals from polymeric azo photo-initiators (**4-4**) has found use in the syntheses of block polymers, such as in the preparation of Telechelics[12-14] and polymeric emulsifiers.[15]

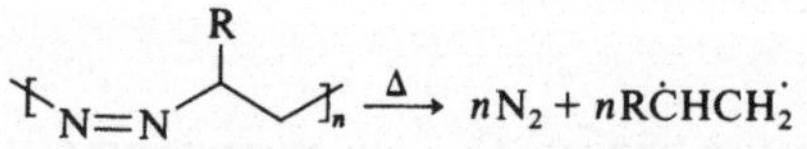

**4-4**

The polymeric azo initiators thermally decompose to release high-molecular-weight monomeric radicals. These can recombine to form block homopolymers[15] or copolymers with other monomers.

Unsymmetrical block polymers are synthesized using an unsymmetrical polymeric azo initiator which decomposes into two different radical blocks and recombines to form an unsymmetrical block polymer. For example, polymeric azo initiators containing blocks derived from polydiols (**4-5**), such as poly(ethylene oxide) and poly(propylene oxide), have been used[16] to prepare block copolymers with styrene (**4-6**) at 60–80°C.

$$\text{+}(OR)_N\text{—OCO—C}(CH_3)_2N\text{=}N\text{—C}(CH_3)_2OCO\text{+}_n$$

**4-5** $\quad\downarrow C_6H_5CH\text{=}CH_2$

$$\text{+}(OR)_N\text{—OCO—C}(CH_3)_2\text{—+}CH_2CH(C_6H_5)\text{+}_x\text{+}(C_6H_5)CHCH_2\text{+}_y C(CH_3)_2OCO\text{+}_n$$

**4-6**

Thermally triggered intramolecular cyclization within a polymeric chain has been reported to be useful in the formation of heat-stable polyimides (**4-7**) by DuPont[17,18] and of copolymers of amidoimidazolines (**4-10**). The latter were used in antistatic finishes for textiles, as reported by Valko *et al.*[19] The open-chain polyamide precursor (**4-9**) is heated on the fiber surface at 160–180°C to release the insoluble and substantive polymer (**4-10**).

$$\text{+}CONHCH_2CH_2NHCH_2CH_2NHCOR\text{+}_n$$

**4-9**

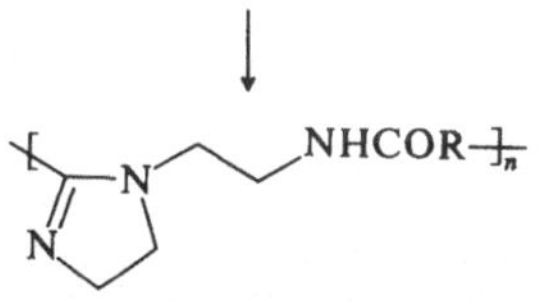

**4-10**

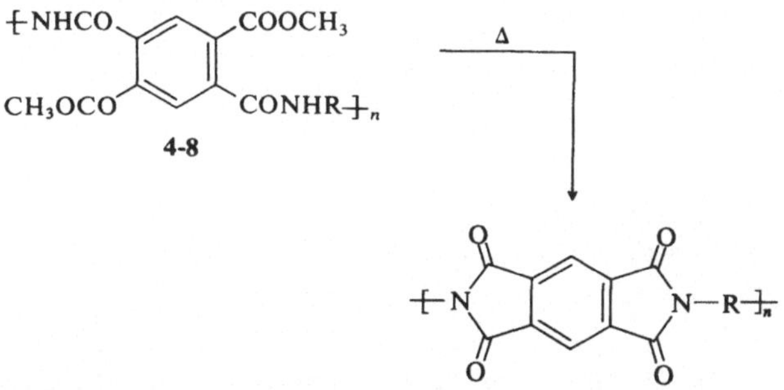

**4-8**

**4-7**

# 4.2. CHEMICAL REACTIONS FOR THE RELEASE OF MODIFIED POLYMERS AND MONOMERS

## 4.2.1. Thermally Triggered Release

This section is a survey of polymers that are labile and sensitive to a thermally triggered release of their derivatives. Such derivatives may be polymeric or monomeric in nature but with distinctly different properties from their precursors in terms of color, solubility, or reactivity.

### 4.2.1.1. Polyimides from Polyamides

The formation of the imide ring is achieved by an intramolecular cyclization of an open-chain polyamide whose carbonyl groups are structurally situated for easy cyclization. Polyimides as a chemical class of compounds have been available[20,21] since 1926 and a number of polyimide structures (**4-11**) have been reported.[22]

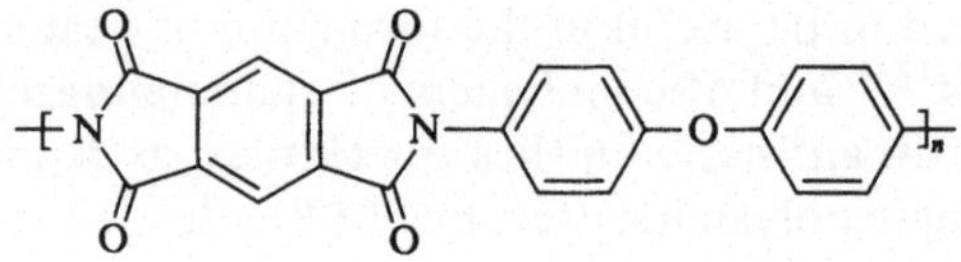

**4-11a**

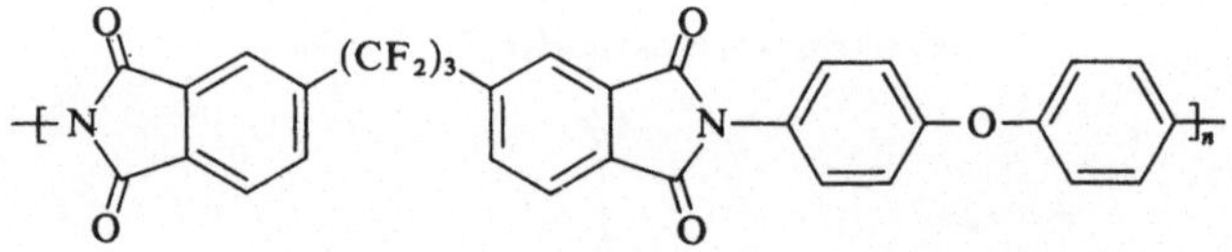

**4-11b**

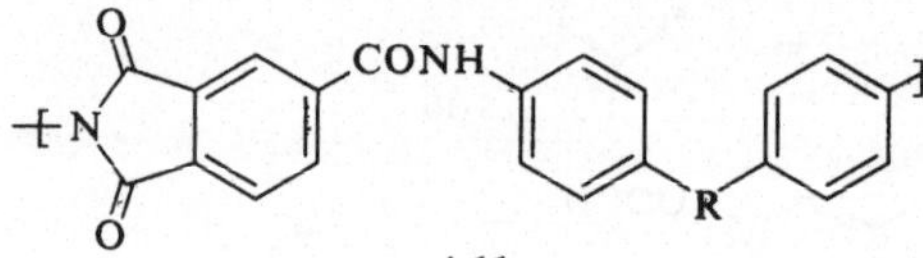

**4-11c**

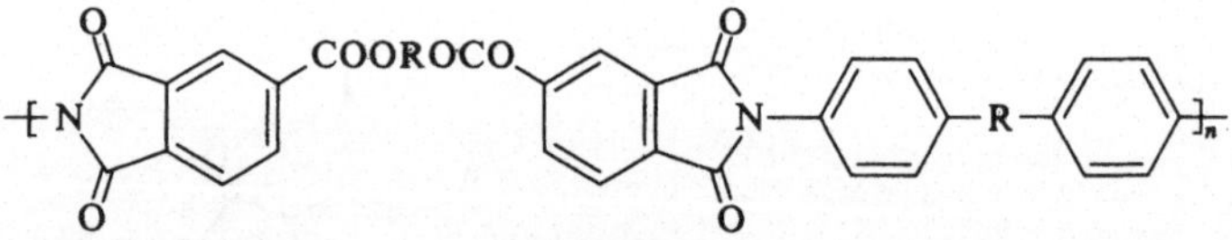

**4-11d**

Aromatic poly(amic acids) (**4-12**) have gained much interest as precursors of polyimides. These are generally derived from dianhydrides or acids and the aryl diamine. Heating the poly(amic acid) triggers a cyclization reaction which gives the corresponding polyimide.[23]

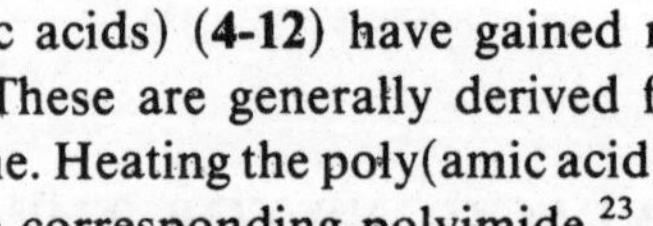

**4-12**

Aliphatic polyimides (**4-13**) have also been reported to be formed from the reaction of polymeric amides. The latter are formed from tetrasubstituted carboxylic acids and aliphatic diamines.[17,18] Upon heating, they undergo cyclization to release the corresponding polyimide.

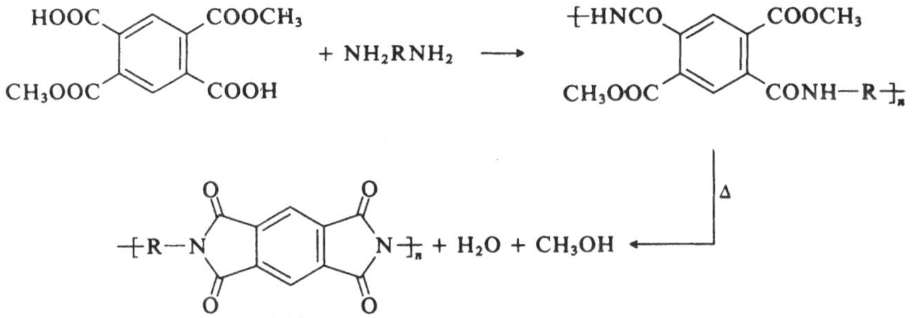

The reaction of aromatic tetramines such as 1,2,4,5-tetraaminobenzene (**4-14**) with pyromellitic dianhydride (**4-15**) forms an addition polyamide (**4-16**) in an exothermic reaction.[24-30] Such a polymer contains free reactive amino and carboxylic acid groups and upon thermal triggering releases the corresponding polyimide (**4-17**). Further application of heat leads to additional condensation and rearrangement to give poly(benzoylene-benzimidazole) derivatives (**4-18**).

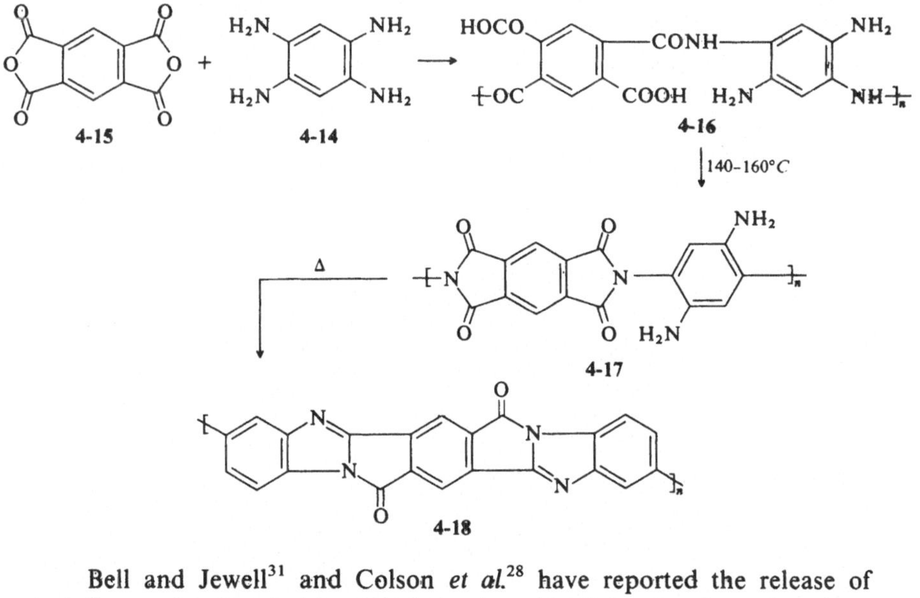

Bell and Jewell[31] and Colson et al.[28] have reported the release of poly(benzoylenebenzimidazole) derivatives (**4-19**) by the thermal cyclization of polyamide intermediates (**4-20**).

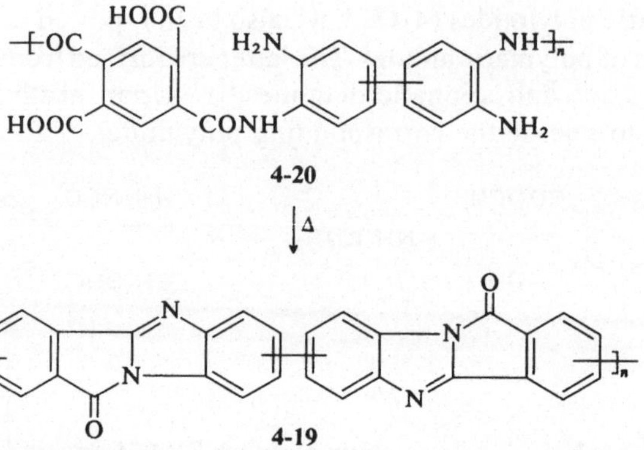

4-20

↓Δ

4-19

Dawans and Marvel[32] reported the formation of a number of other heterocyclic polymers (**4-21–4-24**) related to the poly(benzoylene-benzimidazoles) from the corresponding polyamides. The latter are prepared from phthalic or pyromellitic anhydride and 1,3-dianilino-4,6-diaminobenzene, 3,3'-diaminobenzidine, or 3,3',4,4'-tetraaminophenyl ether. Poly(benzimidazobenzophenanthroline) derivatives (**4-25**) were also reported by Van Deusen.[33]

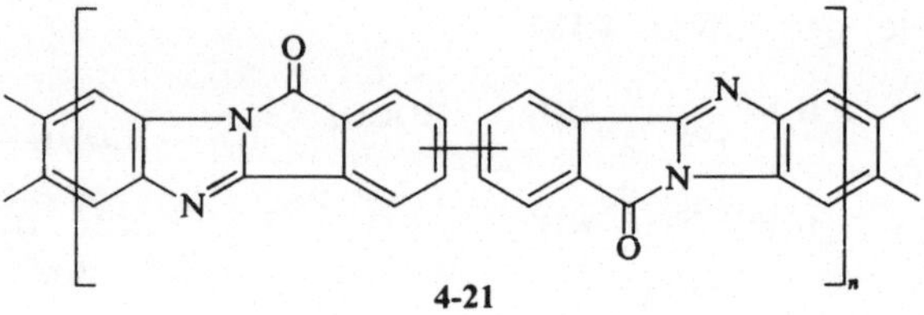

4-21

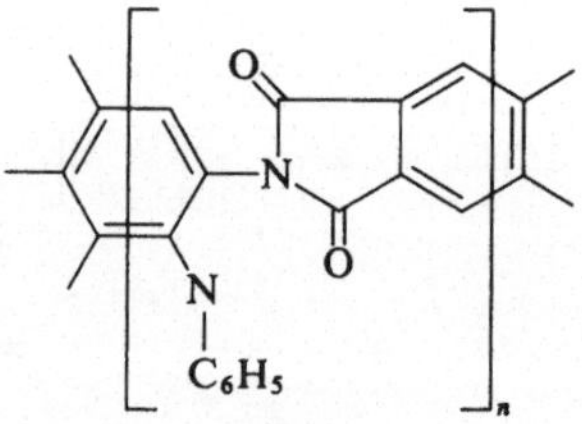

4-22

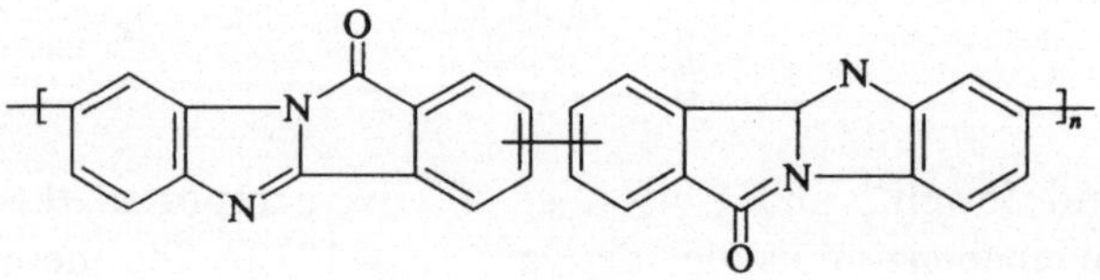

4-23

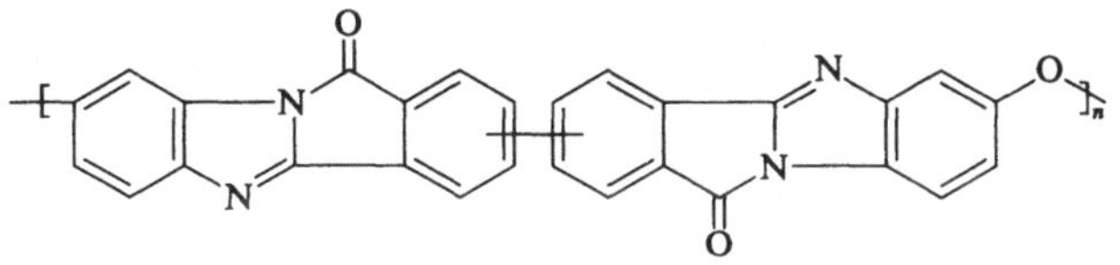

4-24a

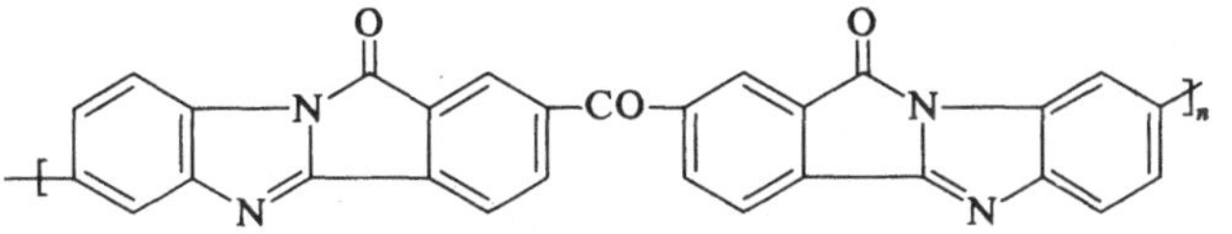

4-24b

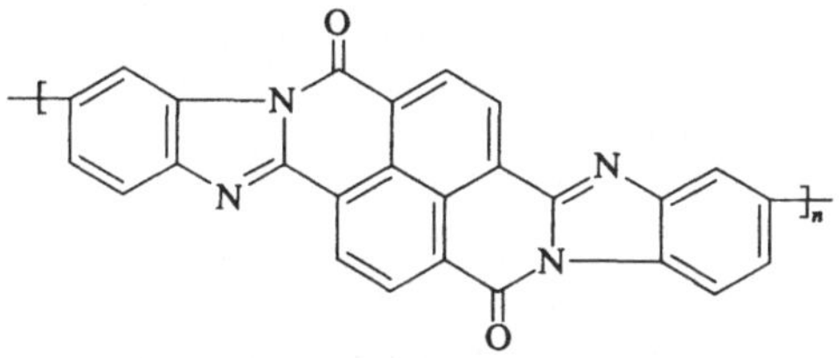

4-25

Bower and Frost[34] published the preparation of a number of poly-pyromellitamides from *m*-phenylenediamine, benzidine, 4,4′-diamino-diphenyl ether, 3,4′-diaminodiphenyl ether, 4,4′-diamonodiphenyl sulfide, methylene dianiline, and 4,4′-diaminophenyl benzoate. These are good film-forming polymers which upon heating undergo an intramolecular cyc-lization with the release of the corresponding polyimides.

A similar reaction was observed with polyurethane derivatives (4-26) which, at 150°C, undergo an intramolecular cyclization to form poly-imidazolidione derivatives (4-27). These polymers, reported by Imai,[35] were prepared by the condensation of diisocyanates and diglycinate esters.

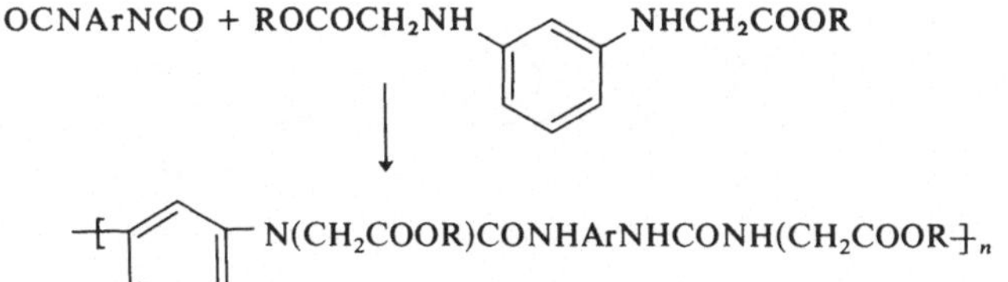

$$OCNArNCO + ROCOCH_2NH \qquad NHCH_2COOR$$

$$N(CH_2COOR)CONHArNHCONH(CH_2COOR)_n$$

4-26

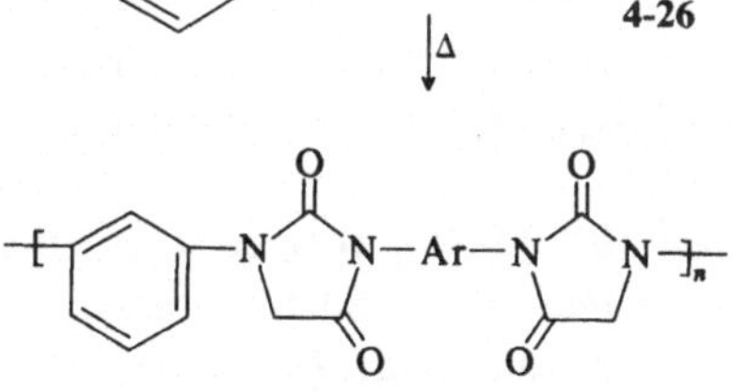

$\Delta$

4-27

Other polyurethanes (**4-28**), prepared from bis($\alpha$-hydroxycarboxylic acid) derivatives, also undergo an intramolecular cyclization with the release of polyoxazolidiones[36] (**4-29**) at 80–200°C

$$\text{+NHCOOC(R}^1\text{)(COOR}^2\text{)}-\text{R}-\text{C(R}^1\text{)(COOR}^2\text{)OCONHR+}_n$$

**4-28**

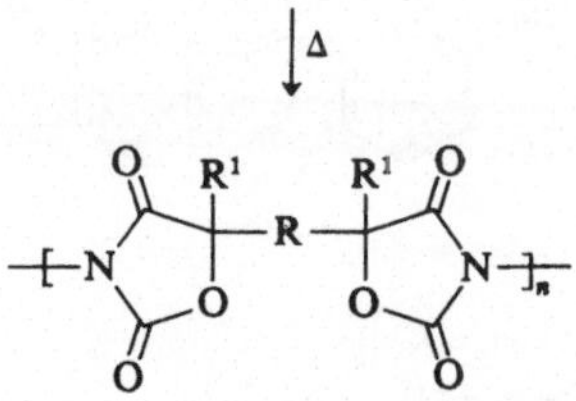

$\downarrow \Delta$

**4-29**

The release of polyimides (**4-30**) from polymers (**4-31**), containing a cyclic urethane structure, has also been reported.[37-40] These polymeric precursors (**4-31**) were prepared from the condensation reaction of diisocyanates with tetraacids or anhydrides.

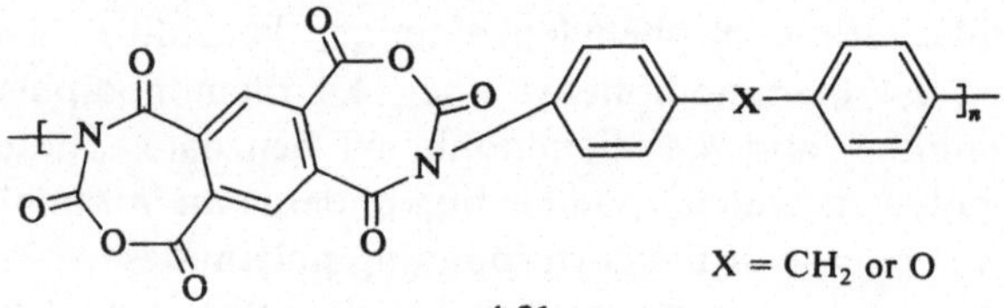

X = CH₂ or O

**4-31**

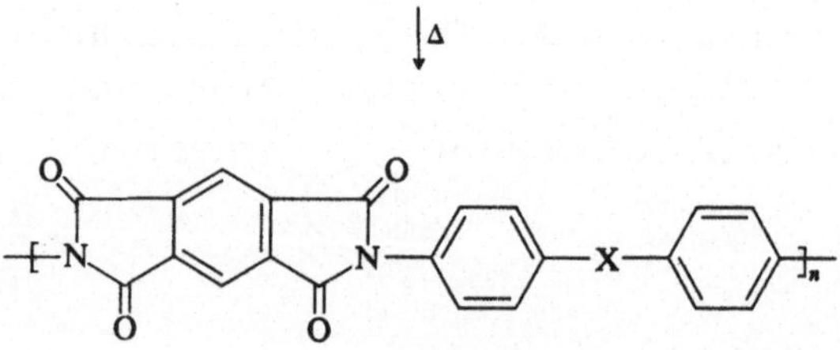

$\downarrow \Delta$

**4-30**

Polyimides (**4-32**), with a spiro ring structure, are formed from polyamide (**4-33**), derived from 1,2,3,4-butanetetracarboxylic dianhydride and aromatic diamines.[41]

Other poly(heterocyclic imides) (**4-34**), formed from the thermally sensitive polyamides (**4-35**), have been reported by Dokoshi *et al.*[42] They can be prepared by the reaction of dianhydrides of triacids and ortho-substituted aminobiphenyls. Similarly, Yoda[43] published the release reaction

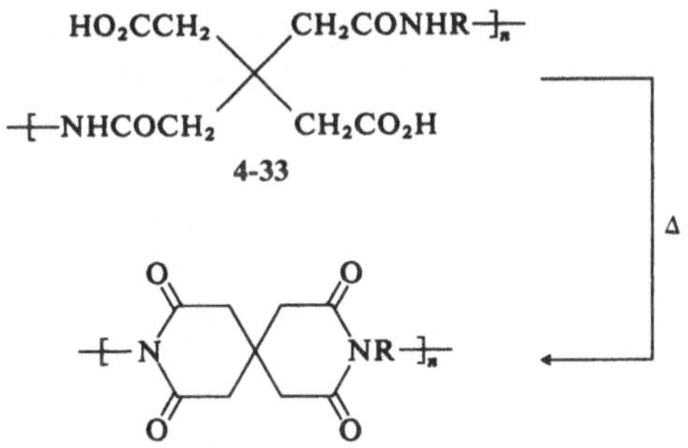

4-33

4-32

of polyimides with benzoxazinone rings (**4-36b**) from the corresponding polyamide derivatives (**4-36a**) at 18–300°C.

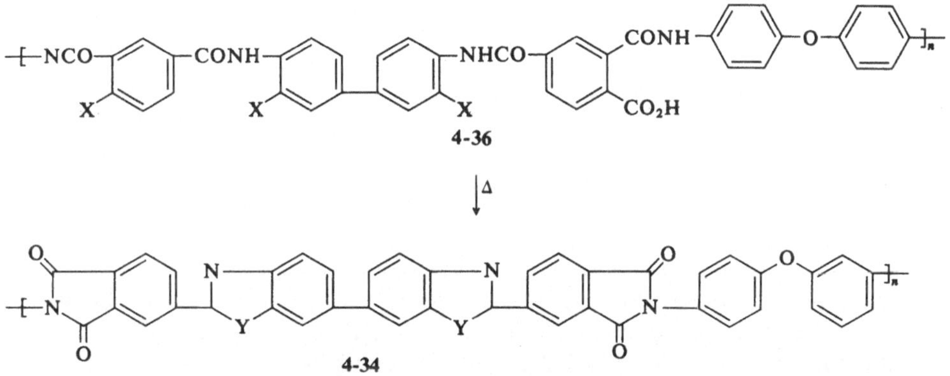

4-36

$\downarrow \Delta$

4-34

X = OH, Y = O (benzoxazole)
X = CO$_2$H, Y = COO (benzoxazinone)
X = NH$_2$, Y = NH (benzimidazole)
X = SH, Y = S (benzothiazole)

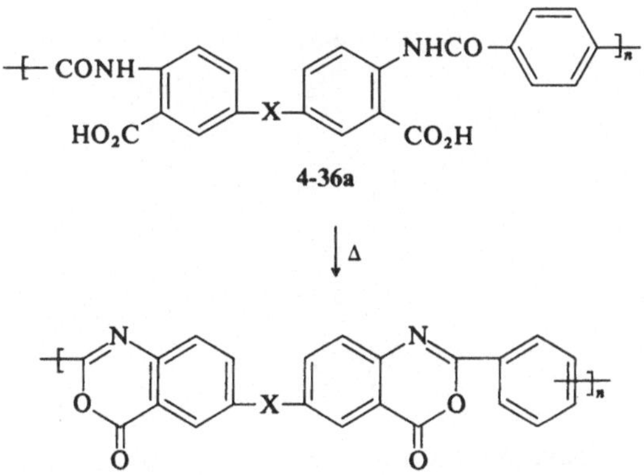

4-36a

$\downarrow \Delta$

4-36b

Thiazole rings have been incorporated[44] in the polyimide structure (4-37b) released at 100–160°C from the corresponding polyamides (4-37a).

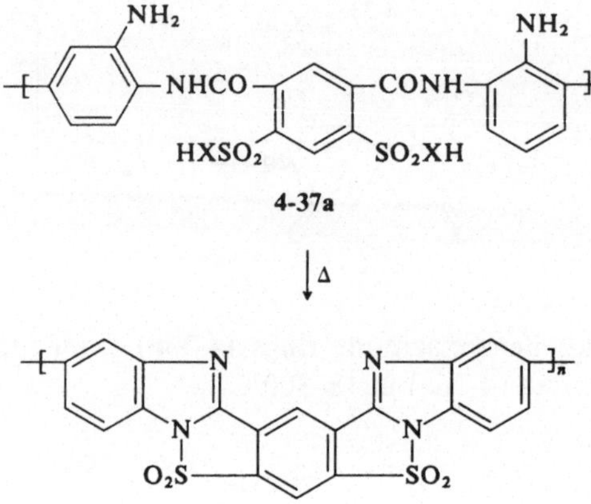

4-37a

↓ Δ

4-37b

## 4.2.1.2. Polyheterocycles from Polyoximes, Polyhydrazides, and Their Derivatives

Oxime derivatives (4-38) of poly(vinyl ketones) have been known since the 1930s. These derivatives thermally release[45] polypyridines (4-39) via an intramolecular cyclization reaction. The yields, 10–80% depending on substituents, are acceptable and the reaction found value as a route to the release of basic polymers.

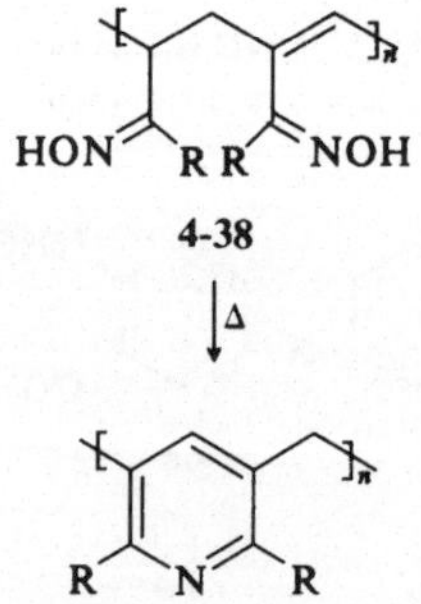

4-38

↓ Δ

4-39

$R = C_6H_5$, 10% yield
$R = CH_3$, 85% yield

Poly(ether-$O$-acylamidoximes) (4-40), obtained from solution polymerization of aromatic bisamidooximes (4-42) and aromatic dicarboxylic acid chlorides, were studied by Yoda et al.[46,47] Under thermal triggering conditions at 150–160°C, polyether-co-poly(1,2,4-oxadiazole) (**4-41**) was formed by cyclodehydration. The cyclodehydration reaction can also be triggered in the solid state at 150–180°C under 1 mm Hg pressure. The presence of poly(phosphoric acid) as a catalyst or a polar organic solvent helps the reaction to proceed at 180–220°C.

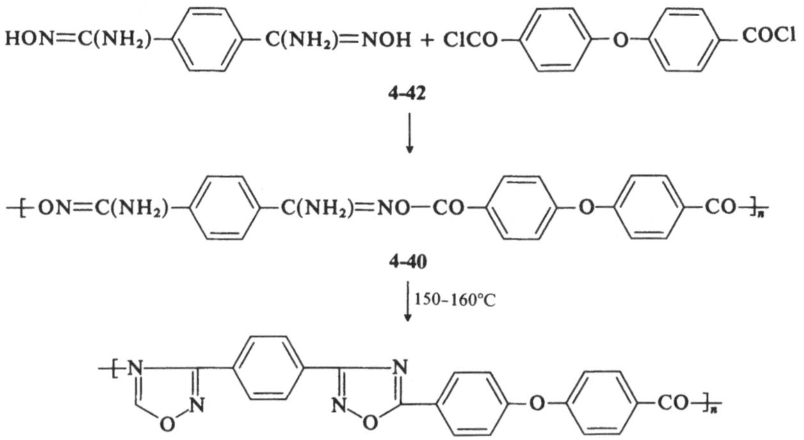
HON=C(NH₂)
C(NH₂)=NOH + ClCO
O
COCl
4-42
ON=C(NH₂)
C(NH₂)=NO—CO
CO
4-40
150–160°C
N
O
CO

**4-41**

Polyaminotriazole (**4-43**) is produced from a dicarboxylic hydrazide and hydrazine.[48] Heating (**4-43**) to its melting point triggers a cross-linking reaction with the release of an infusible resin (**4-44**). The cross-linking is believed to occur between the amino groups on the triazole ring and the terminal carboxylic groups. Other polyaminotriazole derivatives from dicarboxylic acids containing ether, thioether, or sulfone groups have also been prepared.[49]

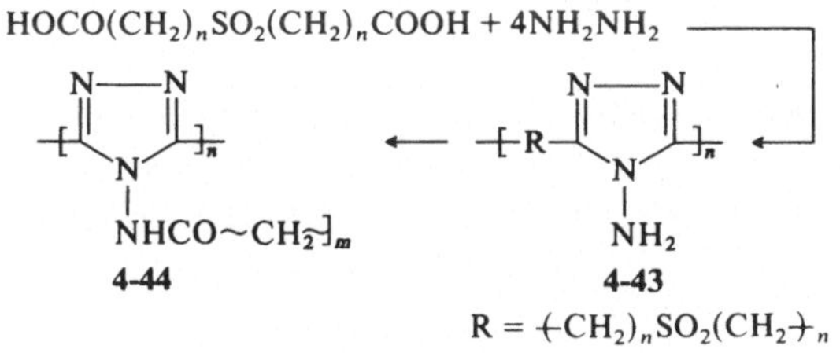
HOCO(CH₂)ₙSO₂(CH₂)ₙCOOH + 4NH₂NH₂
N—N
NHCO~CH₂
4-44
R
NH₂
4-43
R = (CH₂)ₙSO₂(CH₂)ₙ

| $n$ | m.p. (°C) (**4-43**) |
|---|---|
| 1 | dec. |
| 2 | dec. |
| 3 | 258 |
| 4 | 220–225 |
| 6 | 213–215 |
| 7 | 212 |

Similarly, film-forming polyhydrazides (**4-45**) undergo cyclodehydration at 170–280°C with the release of polyoxadiazoles (**4-46**).[50-55] The reaction is based on the preparation of 2,5-diaryl-1,3,4-oxadiazoles from diaryl hydrazide reported by Stolle[56] in 1903. The polyhydrazides were prepared from the corresponding diacid chloride and dihydrazide.

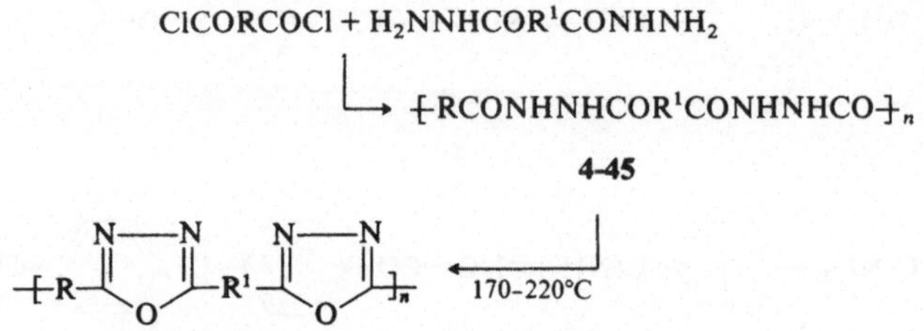

$$ClCORCOCl + H_2NNHCOR^1CONHNH_2$$

$$\longrightarrow \; \text{+RCONHNHCOR}^1\text{CONHNHCO+}_n$$

**4-45**

**4-46**

| R | R$^1$ |
|---|---|
| 1,4-Tetramethylene | 1,7-Heptamethylene |
| 1,8-Octamethylene | 1,8-Octamethylene |
| 1,3-Phenylene | 1,8-Octamethylene |
| 1,3-Phenylene | 1,4-Cyclohexyl |
| 1,3-Phenylene | 1,3-Phenylene |
| 1,3-Phenylene | 1,4-Phenylene |
| 1,3-Phenylene | 2,5-Dichloro-1,4-phenylene |
| 1,3-Phenylene | 2,6-Pyridine |
| 1,3-Phenylene | 4,4'-Phenylether |
| 1,3-Phenylene | 1,4-Tetramethylene |

Polyaminoamides (**4-47**), prepared from the condensation of diethylenetriamine with diacid derivatives, release a heterocyclic polymer (**4-48**) at 180–200°C via intramolecular cyclization.[57]

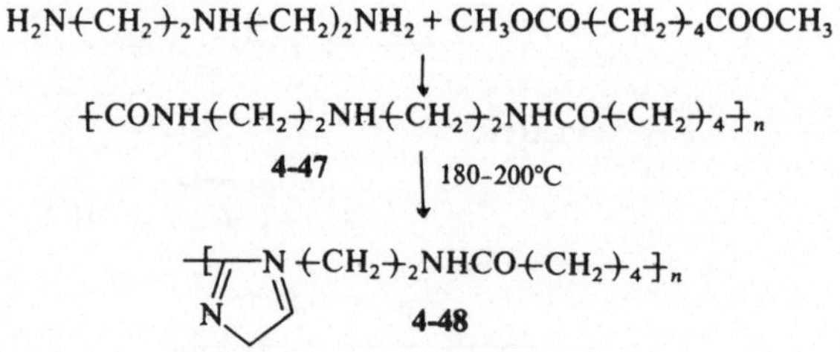

$$H_2N\text{+CH}_2\text{+}_2NH\text{+CH}_2)_2NH_2 + CH_3OCO\text{+CH}_2\text{+}_4COOCH_3$$

$$\text{+CONH+CH}_2\text{+}_2NH\text{+CH}_2\text{+}_2NHCO\text{+CH}_2\text{+}_4\text{+}_n$$

**4-47** | 180–200°C

**4-48**

### 4.2.1.3. Macro Radicals from Polyperoxides

Polymeric peroxides (**4-49**) are formed by the copolymerization of oxygen and monomers. These polymers are potentially explosive and are recommended to be handled as such. They, like other peroxides, thermally decompose to release aldehydes in quantitative yields by a radical mechanism. For example, polyperoxide of styrene releases benzaldehyde and formaldehyde in 80% yield.[58] Bovey and Kothoff[59] reported that at 100°C this polymer tends to explode. Miller and Mayo[60] published that a clean reaction with quantitative yields was observed when the polymer was slowly heated. This polymer is stable under cold and dark conditions, and the decomposition rate increases with increase in temperature.

$$\{OCH(R)-CH(R^1)O\}_n \xrightarrow{\Delta} n RCHO + n R^1CHO$$

**4-49**

$$R = R^1 = C_6H_5, 80\% \text{ yield}$$

Poly(1,4-peroxides) of 2,5-dimethyl-2,4-hexadiene release diepoxide, acetone, and $\beta$-methylcrotonaldehyde.[61] Several polyperoxides have been prepared from olefins, cyclic dienes, substituted phenylethylene acrylonitrile, and acrylic esters.

The polymeric nature of phthalyl peroxide was proposed in the early literature by Baeyer and Villiger[62] to explain its insolubility in a number of common solvents. Phthalyl peroxide is prepared from phthalyl chloride and sodium peroxide as described by Von Pechmann and Vanino.[63]

Common use of polyperoxides is in block polymer syntheses and the introduction of macro groups.

Thermally sensitive perester polymers (**4-50**) can be formed from precursor polymers or copolymers, by treatment with phosphorus pentachloride followed by a hydroperoxide.[64-66]

Polymeric hydroperoxides are closely related to the previously discussed polyperoxides and can be prepared from the reaction of an acyl peroxide and a polymer such as poly(4-isopropylstyrene) at about 80°C. The reaction results in a polyhydroperoxide with about 2–4 moles of active hydroperoxide groups per 100 moles of monomer units (Table 4.1), depending on the reaction conditions.[66] The peroxidation is easier[65] in the case of poly(4-isopropylstyrene) than polystyrene.

*Table 4.1.* Preparation of Polymeric Hydroperoxide by Peroxidation

| Polymer[a] | Solvent | Moles of hydroperoxide per 100 moles of monomer units |
|---|---|---|
| Polystyrene | Chlorobenzene | 0.1 |
| Polystyrene | Xylene | 0.15 |
| Polystyrene | Cumene | 0.13 |
| Polystyrene + 34% IPPS | Chlorobenzene | 0.2 |
| Polystyrene + 34% IPPS | Cumene | 3.0 |
| Polystyrene + 90% IPPS | Cumene | 3.5 |
| Polystyrene + 75% IPPS | Cumene | 0.6 |

[a] IPPS = poly(4-isopropylstyrene).

### 4.2.1.4. Macro Radicals from Azo Polymers

Azo polymers are characterized by the thermal lability due to the azo groups. Two types of polymers are known: those with azo groups in the main chain (**4-51**) and those with azo groups in the side chain (**4-52**).

$$\{R-N=N-R\}_n \quad \text{or} \quad \{R-N=N-R\}_n\{R^1\}_m$$

**4-51**

$$\{R\}_n \quad \text{or} \quad \{R\}_n\{R^2\}_m$$
$$\underset{N=NR^1}{|} \qquad\qquad \underset{N=NR^1}{|}$$

**4-52**

The patent literature contained references to azo polymers[67] as early as 1951. The synthesis of polymers with azo groups in the main chain commonly involves the polymerization of azo units with comonomers such as diamines,[68,69] diols,[71,72] or bisphenol A. Polymers with azo groups as side chains are prepared either by the modification of an arylamino group in a polymer into an azo group using diazonium salt coupling[73] or the copolymerization of an azo-containing monomer with other monomers,[73-78] for example, the azo-containing polyesters (**4-53**) reported by Waltz and Heitz.[79] These polymers thermally decompose to release macro radicals useful in block polymer syntheses.[80,81] These azo-containing polyesters can be prepared[81] from the reaction of azobisisobutyronitrile and poly(ethylene oxide) in an acid medium to give polymers with about ten azo groups per molecule.

$$\{(OR)_n OCOC(CH_3)_2-N=N-C(CH_3)_2CO\}_m$$

**4-53**

R = poly(ethylene oxide), poly(propylene oxide)

The macro radicals released from the thermal decomposition of azo polymers show a low degree of recombination to form homopolymers, making these polymers useful sources of stable macro radicals. The recombination phenomenon is more significant in low molecular azo compounds and is attributed to the cage effect.[82] These polymers (4-54) are prepared by the condensation of 4,4'-azobis-4-cyanopentanoyl chloride (4-55) with a bis-alcohol or bis-amine (4-56).

$$HXRXH + (ClCOCH_2CH_2C(CH_3)(CN)-N\!\!+)_2$$

4-56 4-55

$$\begin{array}{cc} CN & CN \\ | & | \end{array}$$
$$+COCH_2CH_2C(CH_3)-N=N-C(CH_3)CH_2CH_2COXRX+_n$$

4-54

| X—R—X (4-56) | Viscosity[a] | |
| --- | --- | --- |
| | Before heating | After heating |
| 1,4-Cyclohexanebismethylamine | 0.52 | 0.47 |
| Bisphenol | 0.30 | 0.21 |
| Ethylenediamine | 0.46 | 0.06 |
| Hexamethylendiamine | 0.68 | 0.11 |

[a] In dimethylacetamide solution at 2.5 g/liter concentration at 25°C.

Bis-(hydroxyalkyl)-2,2'-azodiisobutyrates (4-57), reported by Waltz et al.,[81] are prepared from the reaction of bishydroxy alkanes, such as ethylene glycol, 1,4-butanediol, 1,6-hexanediol, and 3,6-dioxaoctane-1,8-diol, poly(ethylene oxides), and poly(propylene oxides) with an excess of azobisisobutyronitrile. These polymers show similar lability to thermally triggered decomposition with the release of macro radicals and nitrogen gas.

$$HO+R+_nCOC(CH_3+_2N=N-C(CH_3+_2CO(R+_nOH$$

4-57

$$\xrightarrow{\Delta} [HO+R+_nCO+CH_3+_2C]^\cdot + N_2$$

$$+R+_n = +CH_2+_n \quad n = 2, 4, 6$$
$$= +(CH_2+_2-O+_3$$
$$= +CH_2+_2+OCH_2CH_2+_{4.4}+OCH_2CH_2+$$
$$= +CH_2+_2+OCH_2CH_2+_{20.3}+OCH_2CH_2+$$
$$= -CH_2CH(CH_3)+OCH_2CH(CH_3)+_5-OCH_2CH(CH_3)-$$
$$= -CH_2CH(CH_3)+OCH_2CH(CH_3)+_{32}-OCH_2CH(CH_3)-$$

### 4.2.1.5. Monomers from the Depolymerization of Polymers

The triggered depolymerization of poly(olefin sulfones), at 100–150°C with the release of sulfur dioxide and an olefin, was described by Bowmer and O'Donnell.[83] The rapid decomposition was initiated at the weak sulfur–carbon bond. These copolymers are prepared by the addition reaction of sulfur dioxide and an olefin. Daiton and Vin[84] described the rapid chain scission of poly(1-butene sulfone) (**4-58**) at 70–130°C to yield mainly sulfur dioxide and 1-butene. The yields were measured gravimetrically by Naylor and Anderson[85] to be within 80–90% at 179–250°C.

$$\{SO_2-CH_2CH(CH_2CH_3)\}_n$$

**4-58**

$$\longrightarrow nSO_2 + nCH_2=CHCH_2CH_3$$

The depolymerization of poly(methylene sulfones) (**4-59**), described by Gipstein et al.,[86,87] involves a $\beta$-hydrogen elimination leading to the chain scission. These polymers are prepared by free radical bulk polymerization of equimolar mixtures of olefins and sulfur dioxide using $t$-butylhydroperoxide as an initiator at 0–80°C. Examples are given in Table 4.2, together with the yields of the depolymerization reaction measured from the released sulfur dioxide and olefin.

$$\{SO_2\{CH_2\}_m\{R_1)CH(R^1)\}CH(R^2)\}_n \rightarrow nSO_2 + nR(CH_2)_mC(R^1)=C(R^2)$$

$$m = 0, 1, 2, 3, \ldots$$

**4-59**

*Table 4.2.* Thermally Triggered Depolymerization Reactions of Poly(olefin sulfones)

| | Percent yields at 150°C[a] | |
|---|---|---|
| Poly(olefin sulfone) (**4-59**) | Sulfur dioxide | Olefin |
| Poly(ethylene sulfone) | 60 | 25 |
| Poly(propene sulfone) | 48 | 51 |
| Poly(1-butene sulfone) | 25 | 70 |
| Poly(isobutene sulfone) | 47 | 52 |
| Poly(1-hexene sulfone) | 47 | 52 |
| Poly(4,4'-dimethyl-1-butene sulfone) | 56 | 43 |
| Poly(3-methyl-1-butene sulfone) | 90 | 7 |
| Poly(2-butene sulfone) | 47 | 45 |
| Poly(cyclohexene sulfone) | 57 | 41 |

[a] Percent yield of total volatile products.

Poly(D-$\beta$-hydroxybutyric acid) (**4-60**), described by Grassie *et al.*,[88] undergoes thermally triggered depolymerization at or above its melting point, within the 170–200°C range. The principal reaction involves chain scission, which results in the rapid decrease of the molecular weight, and the polymer is completely volatilized at 250–300°C.

$$\text{+COOCH(CH}_3\text{)CH}_2\text{COOCH(CH}_3\text{)CH}_2\text{COO+}_n$$

**4-60**

↓

$$\text{+COOCH(CH}_3\text{)CH}_2\text{COOH]}_n + \text{[CH}_3\text{CH=CHCOO+}_n$$

This depolymerization is delayed in the early stages of the reaction by a competing condensation reaction between the terminal hydroxyl and carboxyl groups present in the original polymer, but then the depolymerization rapidly increases with time. The poly(D-$\beta$-hydroxybutyric acid) is prepared via a biosynthetic route involving a culture of *Azotobacter beijewekii*.[88]

Poly(diallyldimethylpiperidinium hydroxide) (**4-61**), reported by Butler *et al.*,[89] undergoes a thermally triggered ring opening reaction with the release of a polymer containing free amino groups (**4-62**). The mechanism is analogous to the Hoffmann degradation reaction of quaternary salts.

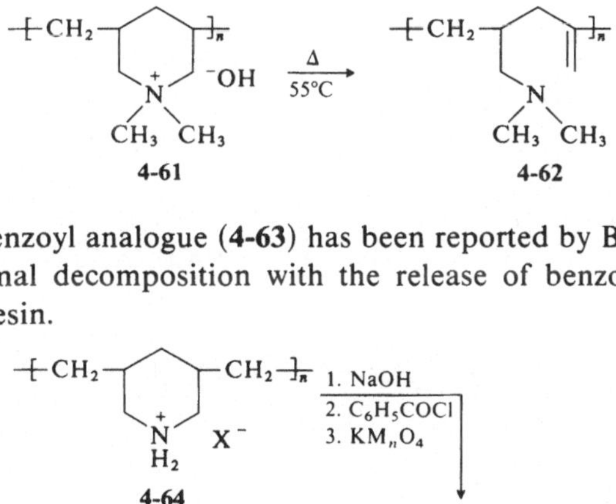

The *N*-benzoyl analogue (**4-63**) has been reported by Butler *et al.*[89] to undergo thermal decomposition with the release of benzoic acid and a cross-linked resin.

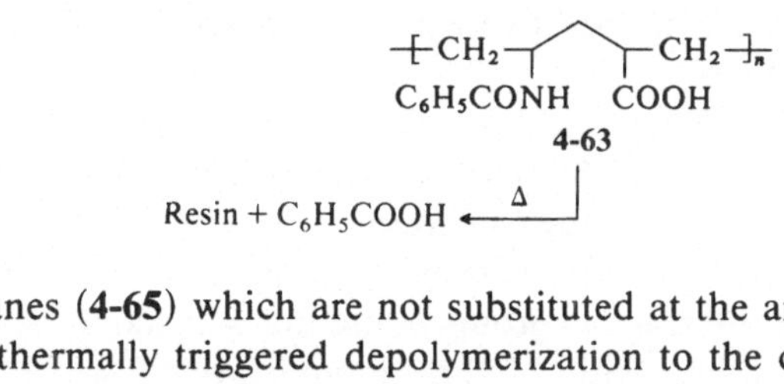

Polyurethanes (**4-65**) which are not substituted at the amide nitrogen atom undergo thermally triggered depolymerization to the corresponding monomers.[90] This reaction is accelerated by the presence of acidic groups. These polymers can be prepared by the reaction of methylenebis(4-phenyl isocyanate) with poly(ethylene glycol).

$$\text{-}\!\!\!\left[\text{R}\!\!-\!\!\!\left\langle\!\!\!\bigcirc\!\!\!\right\rangle\!\!\!-\text{NHCOOCH}_2\text{R}^1\right]_{\!\!n}$$

**4-65**

$$\downarrow \Delta/\text{H}^+$$

$$n\text{HOCH}_2\text{R}^1\text{R}\!\!-\!\!\!\left\langle\!\!\!\bigcirc\!\!\!\right\rangle\!\!\!-\text{NCO}$$

Heterocyclic polymers such as 3,5-poly-(1,2,4-oxadiazoles) (**4-66**) have been reported[91,92] to be sensitive to heating, which triggers the chain decomposition into nitriles and azo oxides.

$$\text{-}\!\!\!\left[\text{R}\!\!-\!\!\!\left\langle\!\!\!\begin{array}{c}\text{O}\!\!-\!\!\text{N}\\\text{N}\end{array}\!\!\!\right\rangle\!\!\!-\text{R}^1\right]_{\!\!n} \xrightarrow{\Delta} \text{R}^1\text{CN} + \overset{\overset{\displaystyle\text{O}^-}{|}}{\text{R}\text{C}}\!\!=\!\!\text{N}^+$$

**4-66**

Polyethylenemonothiocarbonate (**4-67**), a solid polymer with a molecular weight within 3000 units and a melting point between 85 and 115°C, has been reported to undergo thermally triggered depolymerization into ethylene sulfide and carbon dioxide[93] at 100–170°C. These polymers are prepared from the reaction of ethylene oxide and carbonyl sulfide.

$$\overset{\overset{\displaystyle\text{O}}{\diagup\!\!\diagdown}}{\text{H}_2\text{C}\!\!-\!\!\text{CH}_2} + \text{COS} \longrightarrow \text{-}\!\!\!\left[\text{CH}_2\text{CH}_2\text{OCOS}\right]_{\!\!n}$$

**4-67**

$$_n\text{CO}_2 + {}_n\text{CH}_2\!\!-\!\!\text{CH}_2 \xleftarrow[n = 4\text{-}30]{\Delta}$$

## 4.2.1.6. Polymers with Reactive Pendant Groups from Protected Derivatives

Polymers which can be released with reactive functional groups, such as isocyanate, are few in the literature.

The generation of the isocyanate group after polymer formation is necessary since the reactive group would not withstand the polymerization conditions. Polymers with N-ylide groups (**4-68**), which decompose at 130°C, are good precursors for polymers with free isocyanate groups.

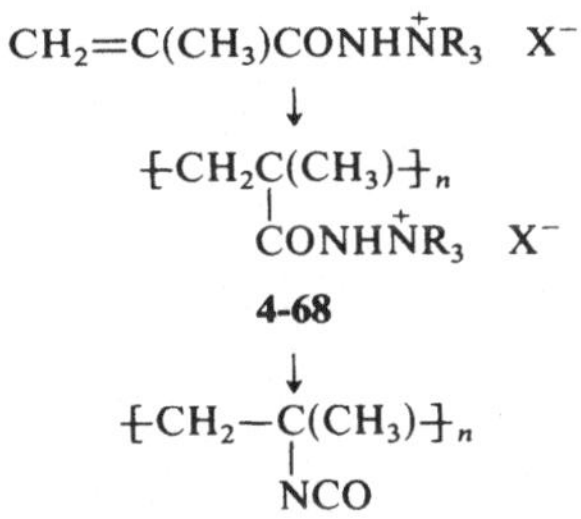

$$CH_2=C(CH_3)CONH\overset{+}{N}R_3 \quad X^-$$

$$\downarrow$$

$$+CH_2C(CH_3)+_n$$
$$\underset{\overset{|}{CONH\overset{+}{N}R_3 \quad X^-}}{}$$

**4-68**

$$\downarrow$$

$$+CH_2-C(CH_3)+_n$$
$$\underset{\overset{|}{NCO}}{}$$

Heterocyclic polymers, such as poly(2-vinyldioxazal-5-one)-co-poly-styrene (**4-69**), are reported[94] as precursors for the release of free-isocyanate-containing polymers (**4-70**). These released copolymers are stable to moisture even at high temperatures because the isocyanate groups are surrounded by the hydrophobic styrene structures.

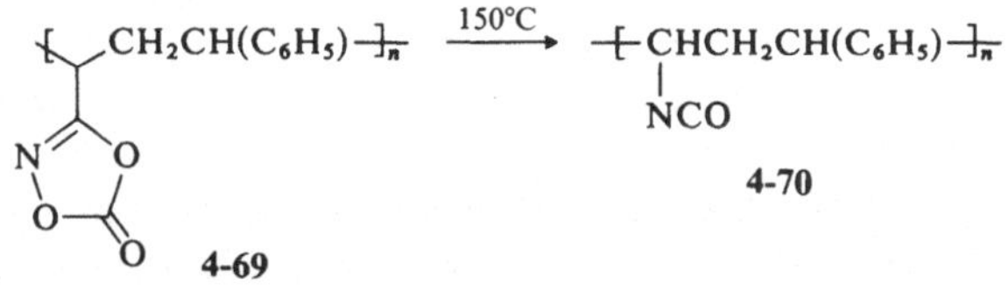

Thermally and photochemically sensitive polydiacetylenes undergo an intermolecular 1,4-addition which provides a route to the release of highly conjugated cross-linked polymers. The main discussion of this type of polymer is included in the photochemical section of this chapter, due to the common nature of the chemistry involved in the photochemical and thermal reactions.

### 4.2.2. Photochemically Triggered Release

#### 4.2.2.1. Cross-linked Polymers via Intermolecular Addition

A prime example of polymers which undergo intermolecular addition is polydiacetylenes (**4-71**). Diacetylenes have been known for more than a

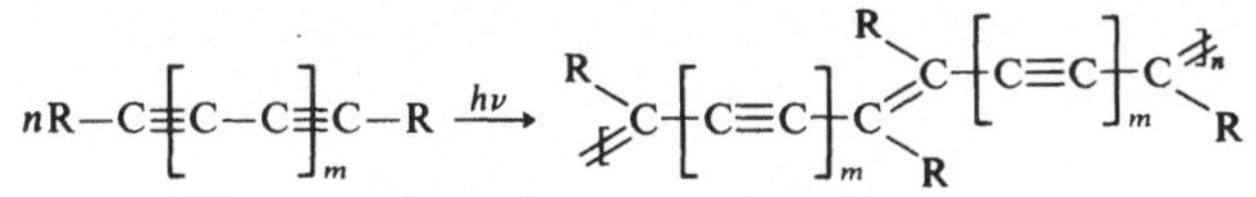

$$R = -CH_2OSO_2-(4-CH_3)C_6H_4$$
$$= +CH_2+_4OSO_2-(4-CH_3)C_6H_4$$
$$= +CH_2+_4OCONHC_6H_5$$
$$= +CH_2+OCONHC_6H_5$$
$$= +CH_2+_nOCONHCH_2COOC_4H_9$$

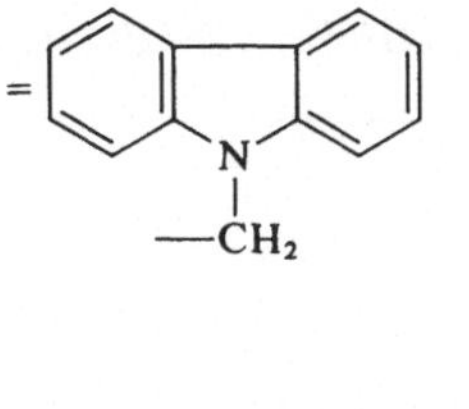

century to undergo dramatic color change upon prolonged storage. Wagner[95-99] concluded that the color change is the result of a triggered polymerization across carbons-1 and -4 of adjacent diacetylene moieties in a molecular stack.

The molecular stacking of diacetylenes in the solid state is crucial for the polymerization to proceed. This phenomenon is termed topo-polymerization.

Polydiacetylenes are unique among polymers due to their conjugated backbone. This property produces color changes that depend on the degree of cross-linking. Typically the non-cross-linked polymers are colorless solids, the partially cross-linked are either blue, purple, or red, and the fully cross-linked are metallic gold or copper in color, due to the extensive electron delocalization along the conjugated backbone.

Raman spectral monitoring of the 1,4-addition[100] of diacetylene monomers shows a gradual change of the C,C-triple bonds into C,C-double bonds accompanied by formation of the triple bonds of the cross-linked polymer (4-72).

A review of the syntheses of these polymers was published by Baughman.[101] Wagner[102,103] has reviewed the solid-state cross-linking mechanisms and the relation between molecular structure and reactivity. The mesomeric or inductive properties of substituents have no effect on the reaction rate. However, the size of the substituents can prevent the molecules from approaching each other and thus hinder the cross-linking.

Spectral sensitization[104] was examined using dyes carrying long aliphatic substituents which can be incorporated in the molecular layers without destroying the architecture.[106] The following are examples of polydiacetylenes which have been reported by Wagner,[105a] Hay,[105b,c] and Patil et al.[105d]

$$\left[\!C\!\equiv\!C\!-\!CH_2OArOCH_2\!-\!C\!\equiv\!C\right]_n \qquad \text{(Ref. 105c)}$$

4-73

(Ref. 105b)

$$\left[ OR{-}C{\equiv}C{-}C{\equiv}C{-}ROCONH{\left(CH_2\right)}_6NHCO \right]_n \qquad \text{(Ref. 105a)}$$

$$R = CH_2, \qquad CH_2{\left(OCH_2CH_2\right)}_2$$

$$\left[ OCH_2{-}C{\equiv}C{-}C{\equiv}C{-}CH_2OCOR^1CO \right]_n \qquad \text{(Ref. 105a, d)}$$

$$R^1 = {\left(CH_2\right)}_n; \qquad n = 2, 3, 4, 7, 8$$

Polymers with activated olefinic groups incorporated in their molecular structure exhibit photochemical sensitivity toward cycloaddition between adjacent polymeric chains. The reaction is less dependent on the molecular stacking and thus can be triggered in both the solid and solution phases. The cross-linking of polymeric molecules containing olefinic groups has been reviewed by Delzenne,[107] Kosar,[108] Reiser and Egerton,[109a] Williams,[109b] and Tazuke.[110]

Poly(vinyl cinnamate) (**4-74**) and its derivatives are the most studied structures but others have also been reported,[110–113] for example, polyesters of 4-carboxycinnamic acid (**4-75**), 4-phenylenediacrylic acid (**4-76**), poly(vinylcinnamylidene acetate) (**4-77**), and polydiphenylcyclopropene derivatives (**4-78**).

$$\left[ \underset{\displaystyle |\atop OCOCH{=}CHC_6H_5}{CH}{-}R \right]_n$$

**4-74**

**4-75**

**4-76**

$$\left[ \underset{\displaystyle |\atop OCOCH{=}CH{-}CH{=}CHC_6H_5}{CH}{-}R \right]_n$$

**4-77**

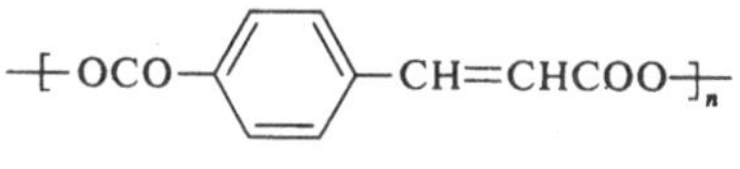

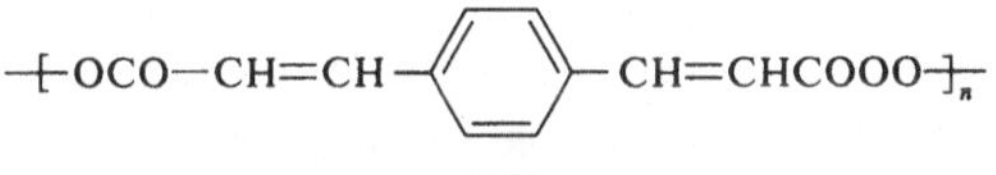

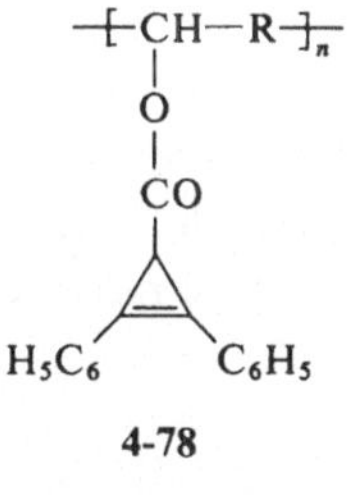

**4-78**

The quantum yields of these intermolecular cycloadditions were determined by their exposure, as films, to a monochromatic radiation of known intensity while monitoring the disappearance of the C,C-double bond band in the infrared spectrum. This provides a guide to the ease of the photocycloaddition.

Tanaka *et al.*[112] studied the photochemistry of the cinnamylidene acetyl group in a free monomer and within a polymeric structure. This inter-

*Table 4.3.* Quantum Yields of Photochemically Triggered Crosslinking

| Polymer/group | $\phi^a$ |
|---|---|
| Cinnamylidene acetic acid, crystals | 1.2 |
| Poly(vinylcinnamylidene acetate), film | 0.48 |
| Poly(vinylcinnamylidene acetate), dioxane solution | 0.091 |
| 1,4-Butanediol dicinnamylidene acetate, isopropanol solution | 0.073 |
| Poly(vinyl cinnamate), film | 0.33 |

$^a$ Quantum yields for the disappearance of the cinnamylidene acetate and the cinnamoyl groups. The irradiation was carried out at 20 cm distance using a 100-W super-high-pressure mercury lamp (Tosiba SHL 100-UV).

molecular cycloaddition exhibited a range of quantum yields depending on the physical form of the polymer studied (Table 4.3).

Williams *et al.*[111] described the synthesis of a range of poly(vinyl cinnamates) with pendant phenanthryl and cinnamyl groups (**4-79**) from poly(vinyl alcohol-co-acetate) and phenanthroyl and cinnamoyl chlorides.

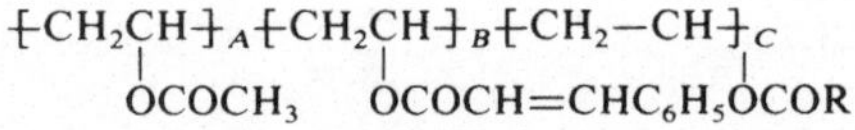

$$+CH_2CH\}_A+CH_2CH\}_B+CH_2-CH\}_C$$
$$\quad OCOCH_3 \quad OCOCH=CHC_6H_5OCOR$$

**4-79**

R = 2-, 3-, and 9-phenanthroate
A = 12, B = 78–88, C = 10

Polymers derived from poly(3-vinylpyridine) containing pendant olefinic groups (**4-80**) have also been reported with spectral absorbances between 300 and 400 nm. These were prepared to extend the spectral absorbance of the olefinic polymers into the near-visible region.

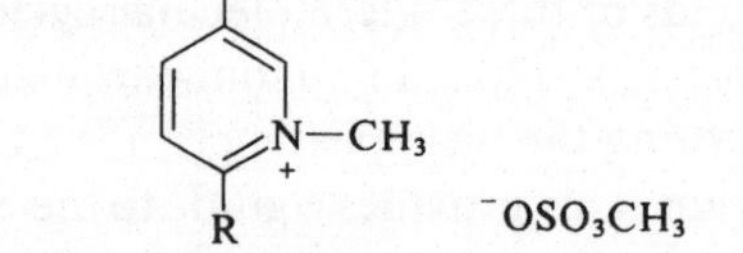

$$-(-CH_2-CH-CH_2-)_n$$

$$^-OSO_3CH_3$$

**4-80**

| R | $\lambda_{max}$ (nm)$^a$ |
|---|---|
| $C_6H_5CH{=}CH$ | 344 |
| 4-$CH_3OC_6H_4CH{=}CH$ | 380 |
| 3,4-$(CH_3O)_2C_6H_3CH{=}CH$ | 388 |
| 4-$[(CH_3)_2N]C_6H_4CH{=}CH$ | 467 |

| | |
|---|---|
| 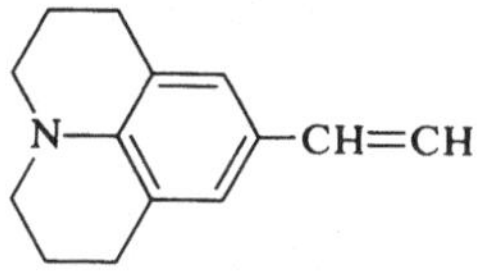 (structure)—$CH{=}CH$ | 486 |
| 4-$NO_2C_6H_4CH{=}CH$ | 453 |
| 2-$NO_2C_6H_4CH{=}CH$ | 345 |
| 4-$BrC_6H_4CH{=}CH$ | 351 |
| $C_6H_5{+}CH{=}CH{+}_2$ | 371 |

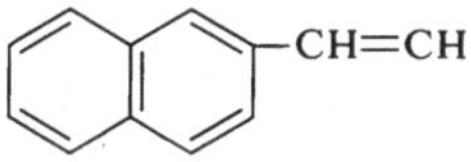

—$CH{=}CH$    365

—$CH{=}CH$    383

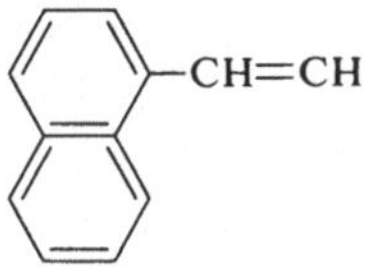

—$CH{=}CH$    309–408

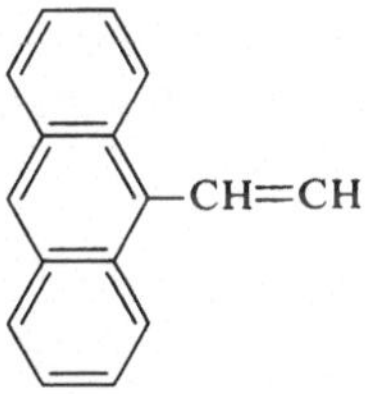

$^a$ In methanol.

Chalcone-containing polymers (**4-81**) also undergo a photochemically initiated intermolecular cycloaddition with the release of a cross-linking insolubilized resin.

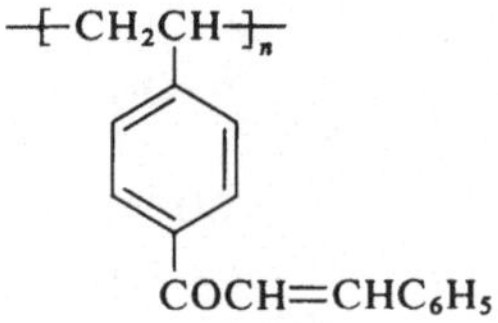

$+CH_2CH+_n$

$COCH{=}CHC_6H_5$

**4-81**

Unruh[114,115] prepared a series of polyvinylarylacetophenones by condensing polyvinylacetophenone with aldehydes.[116,118] Other copolymers from 4'-(β-hydroxyethoxy)chalcone with a styrene–maleic anhydride copolymer (4-82) were also reported.[115,119] Similarly, light-sensitive polymers such as polyvinylbenzophenone, poly(vinyl-4-chlorobenzophenone), poly(vinyl-4-methoxybenzophenone), and poly(vinyl-2-naphthophenone) were described by Merrill.[120]

$$\{CH(C_6H_5)CH(COOH)CH\}_n$$
$$\underset{\textbf{4-82}}{COOCH_2CH_2O}\!-\!\!\langle\!\langle\ \rangle\!\rangle\!-\!COCH{=}CH\!-\!\!\langle\!\langle\ \rangle\!\rangle\!-\!OCH_3$$

## 4.2.2.2. Polymers via Polymeric Rearrangements

Polycarbonates were reported by Bellus *et al.*[121] to undergo photochemically triggered rearrangement when exposed to radiation from a 900-W Hanovia xenon arc. The mechanism is believed to involve a two-step rearrangement of the carbonate structure.[122] For example, irradiation of poly[2,2-propanebis(4-phenyl carbonate)] (4-83), as a chloroform solution in a reactor with a 100-W medium-pressure mercury lamp in a quartz jacket at 150°C, triggered a first rearrangement with the release of phenylesters of salicylic acid groups (4-84) (315 nm absorbance). Further irradiation caused a second isomerization via a photo-Fries reaction and the formation of 2,2'-dihydroxybenzophenones (4-85) (330–360 nm absorbance).[124]

$$+C(CH_3)_2\!-\!\!\langle\!\langle\ \rangle\!\rangle\!-\!OCOO\!-\!\!\langle\!\langle\ \rangle\!\rangle\!\rightarrow_n$$
$$\textbf{4-83}$$

$$\downarrow h\nu$$

$$+C(CH_3)_2\!-\!\!\langle\!\langle\ \rangle\!\rangle\!-\!OH$$
$$COO\!-\!\!\langle\!\langle\ \rangle\!\rangle\!\rightarrow_n$$
$$\textbf{4-84}$$

$$\downarrow h\nu$$

$$+C(CH_3)_2\!-\!\!\langle\!\langle\ \rangle\!\rangle\!-\!OH$$
$$CO$$
$$HO\!-\!\!\langle\!\langle\ \rangle\!\rangle\!\rightarrow_n$$
$$\textbf{4-85}$$

Maerov[123] observed an analogous photochemical rearrangement of aromatic polyesters, such as poly(4,4'-diphenylolpropane isophthalate) (4-86) prepared as described by Eareckson,[124] with the release of a polymer (4-87) which contained 2-hydroxybenzophenone groups.

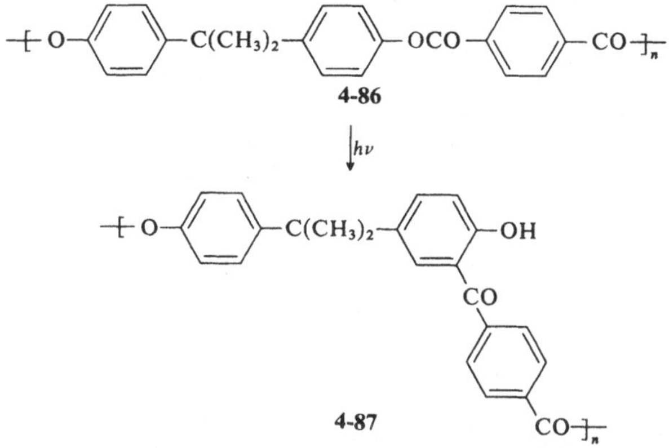

**4-86**

**4-87**

Carlsson *et al.*[125] reported a similar rearrangement with polyamides (4-88) under exposure from a 500-W super-pressure mercury lamp. Based on the absorption spectra and model compounds, the researchers concluded the occurrence of a photo-Fries rearrangement with the release of 2-aminobenzophenone groups in the polymer backbone (4-89).

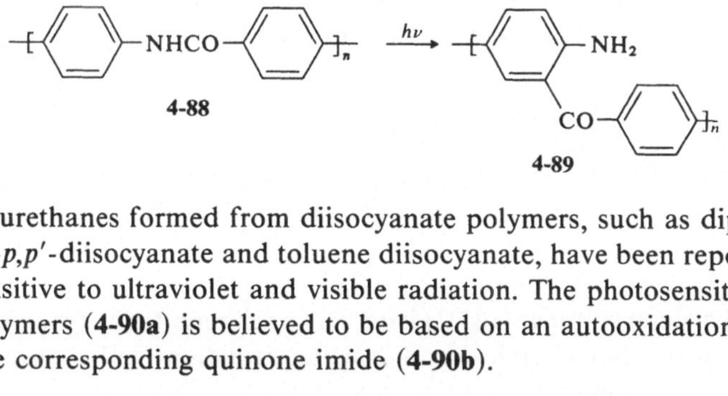

**4-88**

**4-89**

Polyurethanes formed from diisocyanate polymers, such as diphenyl-methane-*p,p'*-diisocyanate and toluene diisocyanate, have been reported[126] to be sensitive to ultraviolet and visible radiation. The photosensitivity of these polymers (4-90a) is believed to be based on an autooxidation which forms the corresponding quinone imide (4-90b).

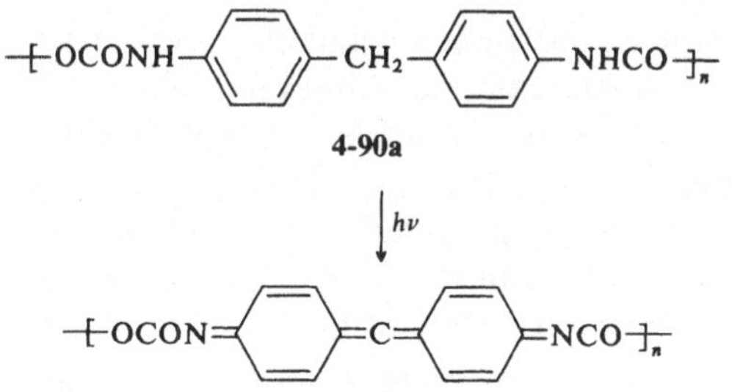

**4-90a**

**4-90b**

### 4.2.2.3. Polymeric Fragments via Molecular Fragmentation

The survey of molecular fragmentation in this section will encompass two general types: (1) the fragmentation of groups pendant from the polymer backbone and the subsequent changes in the polymer structure and properties and (2) the fragmentation of the polymer backbone and the formation of smaller molecular weight units.

Fragmentation of the backbone structure of poly[oxy(2,6-dimethyl)-1,4-phenylene] (4-91) has been examined. This polymer absorbs ultraviolet radiation at about 290–350 nm and upon exposure to sunlight undergoes photochemically triggered oxidation involving scission of the polymer chain.[127]

**4-91**

Analogously, poly(arylether sulfone) (4-92) undergoes photochemically triggered depolymerization upon exposure to sunlight.[128,129] The polymer is believed to undergo scission at the oxygen–carbon bond followed by the release of phenyl radicals and sulfur dioxide. These polymers are also characterized by their high thermal stability.

**4-92**

Poly(olefin sulfones) (4-93) have also been reported to fragment, under the action of photochemical and ionizing radiation, with cleavage of the carbon–sulfur bond.[130-133]

Photosensitization of such degradation reactions was studied by Gipstein et al.[130] For example, poly(5-hexen-2-one sulfone) was reported to be sensitized with benzophenone and poly(1-butene sulfone) with pyridine N-oxide.[131]

These polymers can be prepared by the addition of sulfur dioxide to a variety of olefins in a radical copolymerization to give a regular 1 : 1 alternating composition regardless of the monomer ratio.[132] A few exceptions do exist, such as the addition of styrene and sulfur dioxide which gives the corresponding polymer with varying styrene content, since styrene can compete effectively with sulfur dioxide in the addition to its own radical.[133] $\beta$-Methylstyrene[134] and 4-vinylpyridine[135] also fail to form the corresponding alkene sulfone copolymers via simple addition in liquid sulfur dioxide. However, these addition polymers can be prepared by alternative routes described by Wilson et al.[136]

Bowmer et al.[137-141] have studied the degradation of a variety of poly(olefin sulfones) with different olefin structures. The extent of depolymerization was correlated with the ceiling temperature of the copolymer.[138] This temperature is defined as that above which the equilibrium lies increasingly to the monomer side and depropagation dominates propagation. These poly(olefin sulfones) also undergo facile thermal degradation, which is attributed to the weak carbon–sulfur bond in the main chain and is also related to the ceiling temperature.

A number of poly(olefin sulfones) (**4-93**) have been examined,[138] for example, poly(1-butene sulfone), poly(2-methyl-1-pentene sulfone), poly(5-hexen-2-one sulfone), and poly(cyclopentene sulfone).

$$\{CHR^1CR^2R^3SO_2\}_n \xrightarrow{h\nu} R^1CH{=}CR^2R^3 + SO_2$$

**4-93**

| $R^1$ | $R^2$ | $R^3$ |
|---|---|---|
| H | H | H |
| H | H | $CH_3$ |
| H | H | $CH_2CH_3$ |
| H | H | $(CH_2)_3CH_3$ |
| H | H | $CH(CH_3)_2$ |
| $CH_3$ | H | $CH_3$ |
| $-CH_2-(CH_2)_2CH_2-$ | | H |
| H | H | $CH_2C(CH_3)_3$ |
| H | $CH_3$ | $CH_3$ |

Terpolymers from 1-butene and cis-2-butene with sulfur dioxide were also reported by Gipstein et al.[142] These provide a route to the plasticization of the polymer structure.

Terblocked polymers containing methyl methacrylate and methacrylic acid improved the solubility characteristics.

Photochemically triggered fragmentation of groups pendant from the polymer backbone is the other route for the release of molecular fragments. Azide polymers have been reported to undergo triggered cross-linking upon exposure to light with the formation of macro molecules.[143] Examples of these azide polymers, together with the wavelengths at which they absorb, are given in Table 4.4. Other azide polymers include poly(vinyl alcohol p-azidobenzoate) (**4-94**) and poly(vinyl azide) (**4-95**), which can be prepared

Table 4.4. Spectral Properties of Azide Polymers

| Azide polymer[a] | Spectral absorption (nm) |
|---|---|
| Poly(vinyl acetate-3-azidophthalate) | 310 |
| Poly(vinyl acetate-4-azidophthalate) | 310 |
| Poly(vinyl acetate-3,4-azidophthalate) | 320 |
| Poly(vinyl acetate-4-azidobenzoate) | 310 |
| Cellulose acetate-3-azidophthalate | 320 |
| Gelatin-3-azidophthalate | Broad |

[a] Ref. 143.

by the esterification of poly(vinyl alcohol) with $p$-azidobenzoyl chloride, as described by Merrill and Unruh[144] and Sano and Hasegawa.[143]

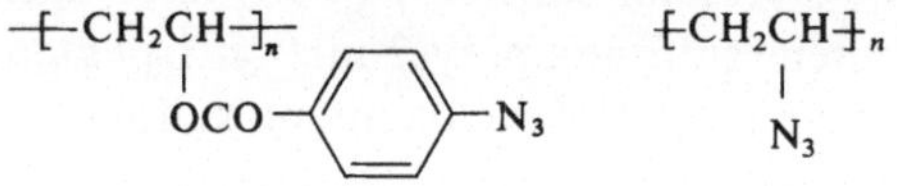

Poly[1-($\beta$-oxyethyl-4'-sulfonylazidocarbanilate)-carbonyl-2-hydroxy-carbonyl-3-methoxycarbonyl-4-tetramethylene] (4-96), a light-sensitive polymer reported by Stuber et al.,[145] is prepared from the reaction of a maleic anhydride–vinyl ether copolymer and hydroxy-terminated arenesulfonyl azide. The polymer spectral sensitivity can be extended into the visible region by using appropriate dye sensitizers.

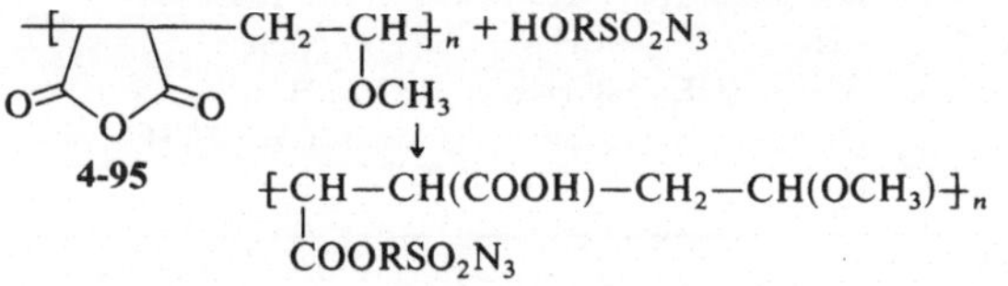

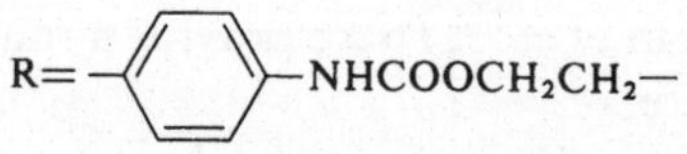

Polymers with diazonium groups are sensitive to photochemical triggering which causes the loss of their ionic character and the formation of cross-linked water-insoluble resins. An example is polymer (4-97), which can be prepared by the condensation of formaldehyde and a diazonium salt.[146]

This reaction was first described by Schmidt and Zahan[147] and is commonly carried out in a manner similar to the condensation of formaldehyde and phenol in the presence of an acid catalyst.[148,149]

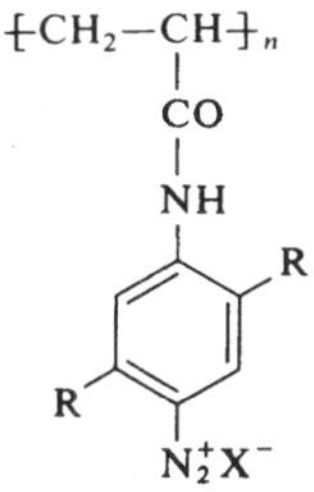

$$\begin{array}{c} \{CH_2-CH\}_n \\ | \\ CO \\ | \\ NH \end{array}$$

**4-98**

Diazopolyacrylamides (**4-98**) have also been found to exhibit photochemical properties which allow the release of cross-linked macromolecules upon exposure to light.[146]

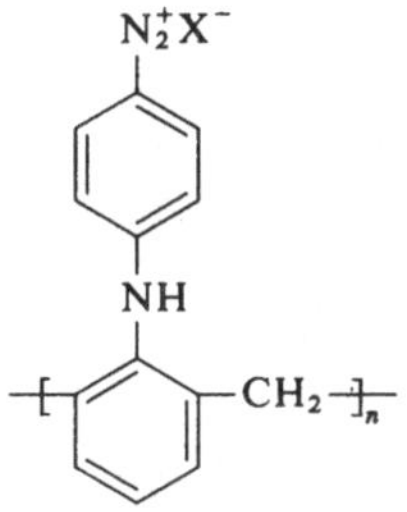

**4-97**

Polymeric sulfonium salts are also photochemically sensitive. Triaryl sulfonium salts, discussed in Chapter 2, are known to release radicals and acids under photochemical triggering conditions. Their polymeric analogues (**4-99**), examined by Crosby et al.,[150] can be prepared from poly(4-bromostyrene) and alkyl or aryl sulfide followed by quaterization.

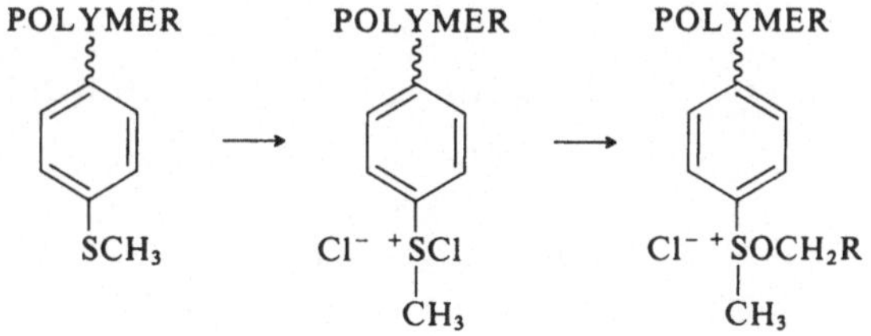

**4-99**

Dorman and Love[151] described a range of dialkyl- and cycloalkyl(arylmethylene)sulfonium resins (**4-100**) which were prepared from the reaction of the corresponding benzyl chloride polymer with dialkyl sulfide.

Other examples include polymers (**4-101**), described by Crosby *et al.*[150]

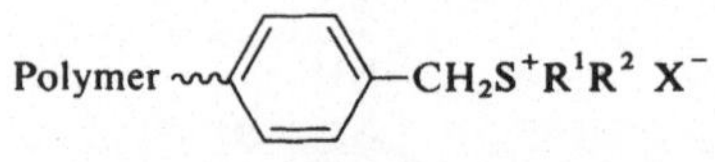

$$\text{Polymer} \text{---}\!\!\!\langle\!\!\!\bigcirc\!\!\!\rangle\!\!\!\text{---} CH_2S^+R^1R^2 \ X^-$$

**4-100**

| R¹ | R² | X⁻ |
|---|---|---|
| $CH_3$ | $CH_3$ | $HCO_3^-$ |
| $-CH_2CH_2CH_2-$ | | $HCO_3^-$ |
| $CH_3$ | $CH_3$ | $(CO_3)^{--}$ |

$$\text{Polymer}\text{---}(CH_2\text{---})_n S^+CH_3R \ \ X^-$$

**4-101**

| R | X⁻ | $n$ |
|---|---|---|
| $CH_3$ | $CH_3SO_4^-$ | 0 |
| Cl | $Cl^-$ | 0 |
| $CH_3$ | $FSO_3^-$ | 0 |
| $CH_3$ | I | 1 |
| Alkyl | $HCO_3^-$ | 1 |

Polymers containing the acylimino ester chromophore (**4-102**) have a strong absorption at about 225 nm and undergo rapid photochemical decomposition at the nitrogen–oxygen bond with the release of macro radicals.

$$\begin{array}{cc} \text{---}\!\!\!+CH_2C(CH_3)CH_2C(CH_3)\!\!+_n \\ \ \ | \qquad\qquad\quad | \\ CH_3OCO \qquad\quad COON{=}C(CH_3)OCH_3 \end{array}$$

**4-102**

$$\downarrow h\nu$$

$$\begin{array}{cc} +CH_2C(CH_3){-}CH_2C(CH_3)\!\!+_n + n[CH_3COC(CH_3){=}N^{\cdot}] \\ \ \ | \qquad\qquad\quad | \\ COOCH_3 \qquad COO^{\cdot} \end{array}$$

$$\downarrow h\nu$$

Polymer degradation

The degradation of $\alpha$-ketooximinomethacrylate esters (**4-103**) in solution upon irradiation with 365-nm light was reported by Delzenne *et al.*[152] and in the solid phase by Reichmanis and Wilkins.[153] Copolymers such as poly(methyl methacrylate-co-3-oximino-2-butanone methacrylate-co-methacrylonitrile) were prepared using free radical solution polymerization initiated by benzoyl peroxide.

$$CH_3C=CH_2 + CH_3C(R)=CH_2$$
$$| \quad\quad\quad\quad\quad\quad\;$$
$$COON=C(CH_3)COCH_3$$

$$\downarrow \Delta$$

$$+CH_2-CR(CH_3)-CH_2C(CH_3)+_n$$
$$\quad\quad\quad\quad\quad\quad\quad\quad | $$
$$\quad\quad\quad\quad\quad\quad COON=C(CH_3)COCH_3$$

**4-103**

Copolymers of methyl methacrylate and acyloximino methacrylate esters have been reported[154] to absorb in the deep ultraviolet region at 200–250 nm. The acyl group of the 3-oximino-2-butanone substituent was found to be more effective than the *t*-butyl, phenyl, or benzoyl analogues in promoting the photochemically triggered depolymerization.

The polymers have also been shown sensitive toward electron beam radiation. Delzenne *et al.*[152] have reported a number of *O*-acylated oximinoketones (**4-104**) synthesized from the corresponding $\alpha$-diketones.

$$R^1COC(R^2)=NOCOR^3$$

**4-104**

| $R^1$ | $R^2$ | $R^3$ |
|-------|-------|-------|
| $C_6H_5$ | H | $C_6H_5$ |
| $C_6H_5$ | $CH_3$ | $C_6H_5$ |
| $C_6H_5$ | $CH_3$ | $4\text{-}N_3C_6H_4$ |
| $CH_3$ | H | $C_6H_5$ |
| $CH_3$ | $CH_3$ | $C_6H_5$ |
| $CH_3$ | $CH_3$ | $2\text{-}ClC_6H_4$ |
| $CH_3$ | $CH_3$ | $2\text{-}CH_3OC_6H_4$ |
| $CH_3$ | $CH_3$ | $4\text{-}NO_2C_6H_4$ |
| $CH_3$ | $CH_3$ | $C_6H_5CH=CH$ |
| $CH_3$ | $CH_3$ | $(CH_3)_2CH=CH$ |

Copolymers of 1-phenyl-1,2-propanedione-2-*O*-methacroyloxime with methyl methacrylate and those from 1-(4'-hydroxyphenyl)-1,2-propane-dione-2-oxime, *p*-hydroxyphenylglyoxal aldoxime, or 2,2-(4,4'-dihydroxy-diphenyl)propane and isophthaloyl, terphthaloyl, and sebacoyl chlorides have also been reported.[154] The latter route forms polymers with the oximino group built into the polymer backbone (**4-105**).

$$+O-\!\!\!\left\langle\!\!\!\bigcirc\!\!\!\right\rangle\!\!\!-COC(R^1)=NOCOR^2CO+_x+O-\!\!\!\left\langle\!\!\!\bigcirc\!\!\!\right\rangle\!\!\!-C(CH_3)_2-\!\!\!\left\langle\!\!\!\bigcirc\!\!\!\right\rangle\!\!\!-OCOR_2CO+_y$$

**4-105**

$R^1 = H \text{ or } CH_3$
$R^2 = +CH_2+_8$, *M*- and *O*-phenylenes

These polymers can be triggered in the near ultraviolet at 300–380 nm, although their main absorption is in the region of 250 nm. Werner and Piguet[155] reported the decomposition of the $O$-benzoylbenzyl monooxime by sunlight. Recently, Vermes and Bengelmans[156] described the slower decomposition of $O$-oxime acetates.

An increase in the polymer sensitivity in the 230–260-nm region can be accomplished[154] by the incorporation of 3-oximino-2-butanone methacrylate[168] into poly(methyl methacrylate). The incorporation of acrylonitrile further enhances the polymer photosensitivity. The light energy used was from a 1000-W mercury lamp focused through quartz condenser optics.

Poly($N$-alkyl-2-nitroamides) (4-106) have been reported to be sensitive to photochemical and thermal cleavage of the nitro amide group. The photosensitivity of these polymers is a function of the photolabile aromatic amide linkages. Upon light exposure, these polymers undergo a triggered rearrangement with the release of the corresponding carboxylic polymer. This chemistry is based on the 1973 research of Amit and Patchornik[157] dealing with light-sensitive $N$-substituted-2-nitroanilides as protective groups for carboxylic acids.

4-106

These polymers were exposed as films to light from a 1-kW mercury-xenon lamp through quartz optics. The released acid functionality was monitored by solubility differentials between exposed and unexposed areas in an ammonium hydroxide solution.

Two types of polymers were reported, those with the 2-nitroanilide group as a pendant group (4-106) and those where the 2-nitro-$N$-methyl-anilide group is part of the backbone chain of a homo- or copolymer (4-107). Upon light exposure, the latter polymer underwent depolymerization with the release of benzoxadiazoles (4-108).

4-107                                   4-108

Synthesis of the homopolymer (**4-107**) by thermal polymerization of the monomer 3-nitro-4-(N-methylamino)benzoyl chloride proceeds in the solid phase at the melting point of the monomer or in a dimethylacetamide solution.[158]

The analogous copolymer (**4-111**) is formed from the reaction of 3,3'-dinitro-4,4'-di-N-methylaminodiphenyl ether[170] (**4-110**) with aromatic diacylchlorides (**4-109**) at room temperature.

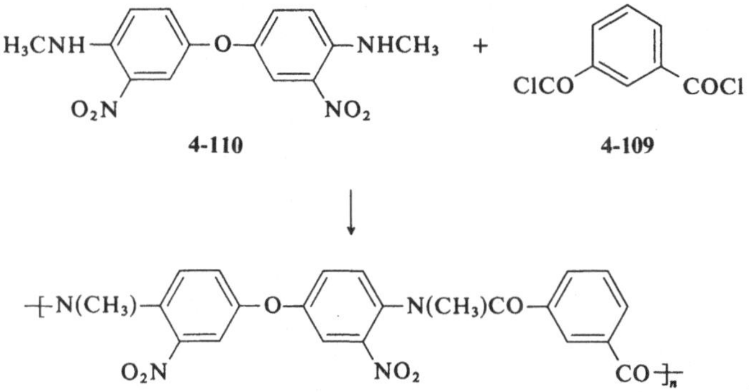

**4-110**                                                    **4-109**

**4-111**

Zehavi and Patchornik[160] reported the synthesis of a light-sensitive polymer (**4-112**), which absorbs at 320 nm, from 6-nitrovanillin and chloromethylstyrene–divinylbenzene copolymer. This synthetic route follows earlier work by Frechet and Shuerch.[161]

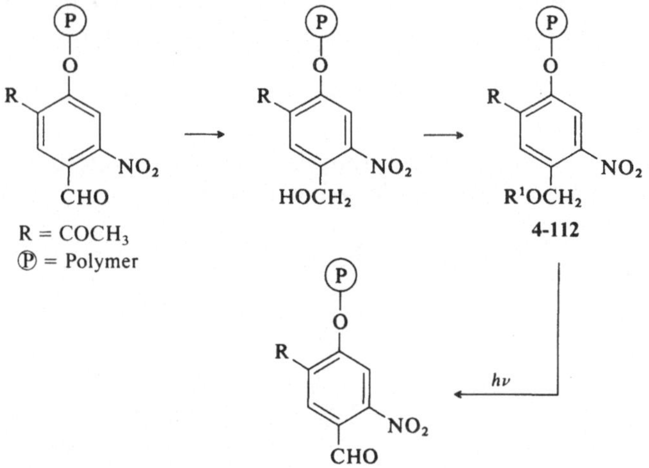

R = COCH₃
Ⓟ = Polymer

**4-112**

Tjoeng et al.[162] reported a 3-nitro-4-(bromomethyl)benzoylpoly(ethylene glycol) (**4-113**) which is used in protective group synthesis of peptides

*Table 4.5.* Photochemically Triggered Release of Peptides from Polymers with Protective Groups

| Released peptides[a,b] | Percent yield |
|---|---|
| Boc—Lys(2)—Leu—Glu(OBzl)—Ala | 95 |
| Boc—Lys(2)—Leu—Glu(OBzl)—Ala—Leu—Glu(OBzl)—Ala | 87 |
| Boc—Lys(2)—Ala—Glu(OBzl)—Ala—Leu—Glu(OBzl)—Ala | 96 |
| Boc—Lys(2)—Leu—Glu(OBzl)—Ala—Ala—Glu(OBzl)—Ala | 90 |
| Boc—Lys(2)—Ala—Glu(OBzl)—Ala—Ala—Glu(OBzl)—Ala | 92 |

[a] Ref. 173.
[b] Amino acids abbreviations follow IUPAC-IUB rules; see *J. Biol. Chem. 247*, 997 (1972).

possessing free terminal carboxylic groups. The peptide is released in high yields from the support polymer by irradiation at 350 nm (Table 4.5).

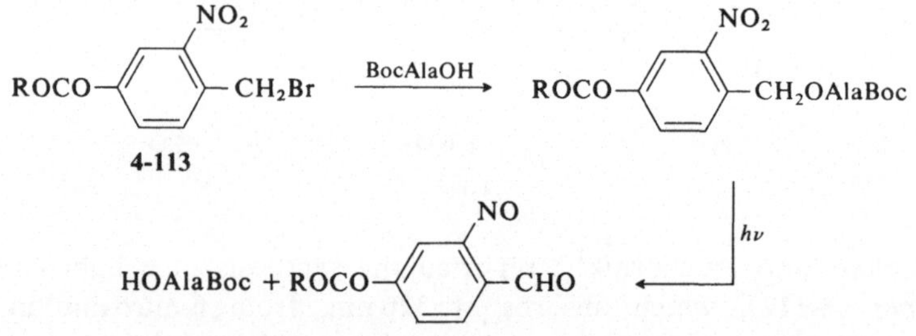

$$R = HOCH_2CH_2(OCH_2CH_2)_nOCH_2CH_2-$$

The polymer in turn was prepared from poly(ethylene glycol) and 3-nitro-4-(bromomethyl)benzoic acid via a polymeric condensation reaction.

Similarly, Rich and Gurwar[163] reported 3-nitro-4-bromomethylbenzoyl amide polystyrene resin (**4-114**) to be useful in the protective group synthesis of peptides possessing terminal carboxylic groups. The protected acids were released by photochemical irradiation at 350 nm.

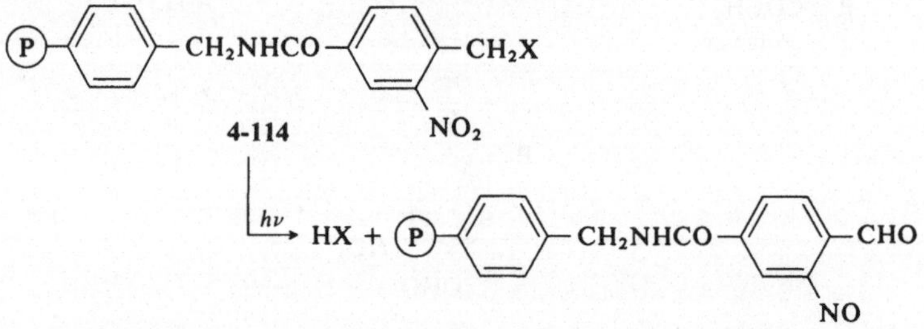

Ⓟ = Polymer
X = Cl, Boc—Gly—O—Boc—Ser(Bzl)—Tyr(Bzl)—Glu—O—

## 4.2.2.4. Photochromic Polymers

Photochromism was first reported by Marckwald[164] in 1899 and the common monomeric candidates are surveyed in Chapter 5.

The incorporation of photochromic monomers in a polymeric matrix has shown a reduction in the thermal self-bleaching reaction, which is an advantage over the monomeric analogues and valuable in certain applications such as image recording.[165]

Two types of photochromic polymers have been described in the literature: those which are photochromic as a result of their backbone structure and those which contain pendant photochromic units. For example, the reaction of bis-aldehydes and 4-aminoaniline forms azo polymers which are photochromes[166] of the first type. An example of the second type of photochromic polymers may be prepared by the copolymerization of acrylamido or vinyl-substituted azobenzenes with vinyl monomers.[165] A review of the solid-state photochromism of polymers was published by Smets[167] in 1983.

Photochromic polymers based on the chemistry of spirobenzopyran and its photochemical isomerization are known; for example, copolymer (4-115) from ethyl acrylate and 8,8'-bisacryloxymethylenebisphotochrome D was described by Smets.[168]

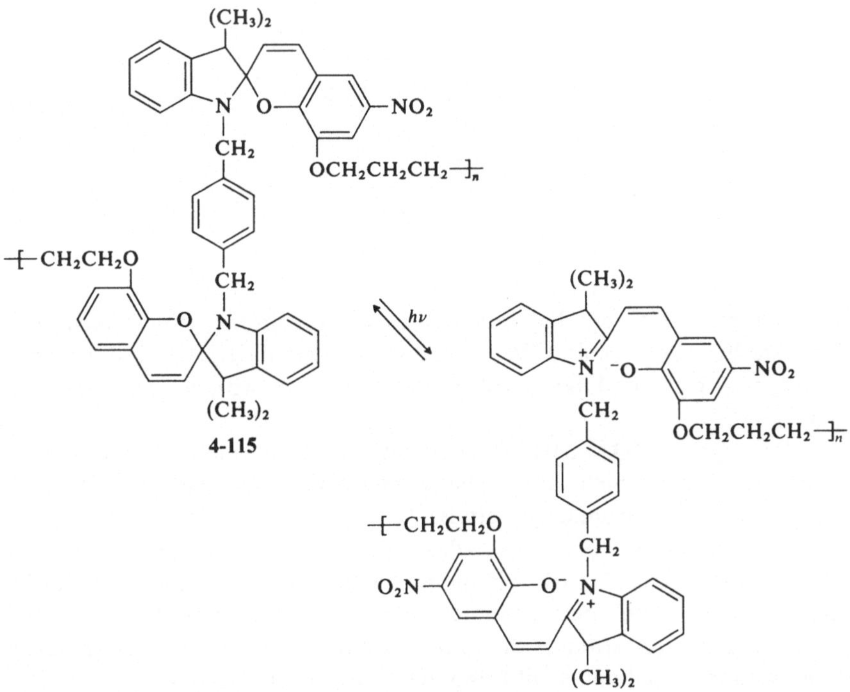

Similarly, Eisenbach[169] and Smets *et al.*[170] reported polymers (**4-116**) and (**4-117**), respectively, both with spirobenzopyran as the photochromic nucleus.

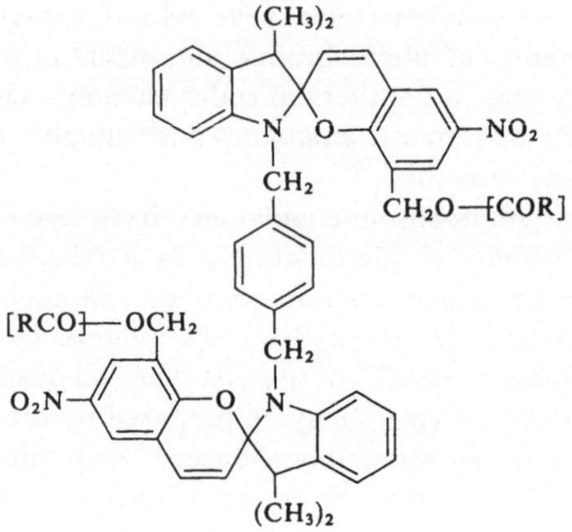

**4-116**   RCO = —{CO(CH₂)ₙ—COO—⟨benzene⟩—C(CH₃)₂—⟨benzene⟩—O—}ₙH

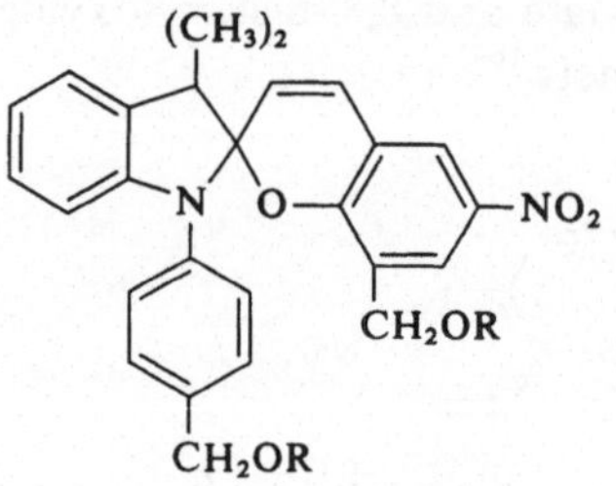

**4-117**   R = —{CO(CH₂)₅COO—⟨benzene⟩—C(CH₃)₂—⟨benzene⟩—O—}ₓH

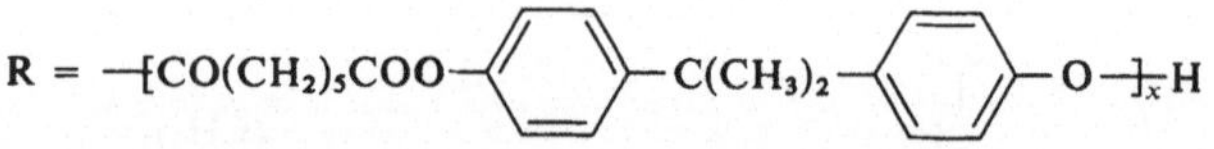

Polymers with pendant photochromic azobenzene groups have also been studied. Lovrien and Waddington[171] in 1964 described photochromic copolymers of 4-acrylamidoazobenzene derivatives with acrylic and methacrylic acids.

Polystyrene (**4-18**) partially substituted at the aromatic ring para position with bis(4-dimehylaminophenyl)methanol was also described as possessing photochromic properties.[172]

Kamogawa *et al.*[173] described polyvinylazobenzenes, polyvinylhydroxyazobenzenes, polyarylamidomethylaminoazobenzenes, and polyacrylamidomethylthione as photochromic polymers. These were prepared by either polymerization of the photochromic vinyl monomer or chemical modification of polymers with photochromic chromophores.

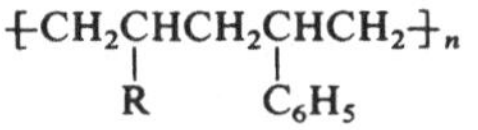

$$\{CH_2CHCH_2CHCH_2\}_n$$
$$\quad\ \ |\qquad\quad |$$
$$\quad\ \ R\qquad\quad C_6H_5$$

**4-118**

| R | $\lambda_{max}$ (nm) | Irradiation time (min) | 50% Recovery time (min) |
|---|---|---|---|
| —C6H4—N=N—C6H4—N(CH3)2 | 415 (solution)<br>415 (film) | 10<br>10 | 30<br>150 |
| 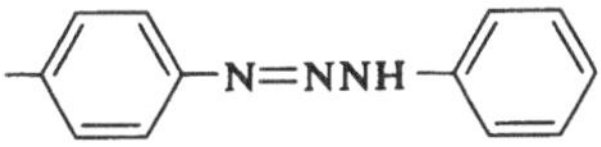 —C6H4—N=NNH—C6H5 | 365 (solution) | 45 | 30 |
| (CH3)2N—C6H4—N=N—C6H4(CH3) | 420 (solution)<br>420 (film) | 30<br>10 | 15<br>7 |
| (CH3)2N—C6H3—N=N—C6H4—CH3 | 424 (solution)<br>424 (film) | 10<br>10 | 12<br>10 |
| (CH3)2N—C6H3—N=N—C6H4—C6H4—NH2 | 398 (solution)<br>400 (film) | 10<br>— | 2<br>— |
| 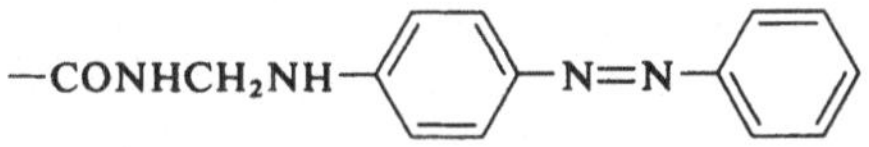 —CONHCH2NH—C6H4—N=N—C6H5 | 412 (solution) | 10 | 3 |
| HO—C6H3—N=N—C6H5 | 358 (solution)<br>358 (film) | 60<br>60 | —<br>— |
| HO—C6H3—N=N—C6H4—Cl | 362 (solution)<br>364 (film) | 60<br>60 | 60<br>— |
| HO—C6H3—N=N—C6H4—CH3 | 357 (solution)<br>359 (film) | 60<br>60 | —<br>— |

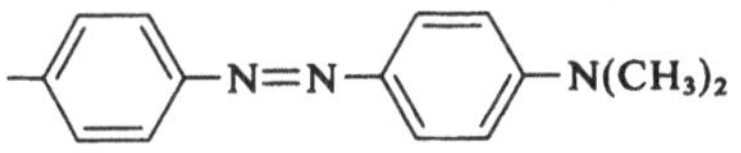

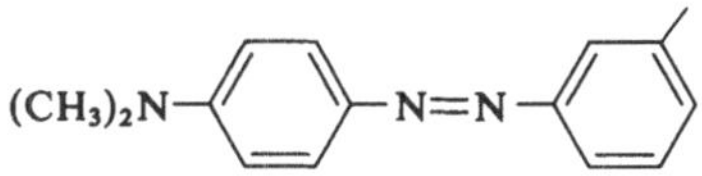

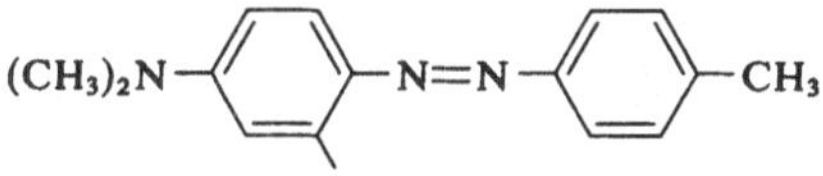

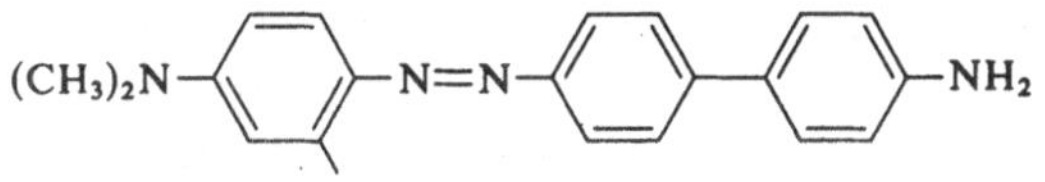

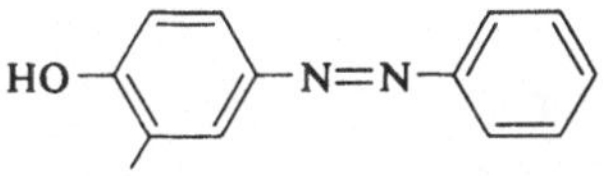

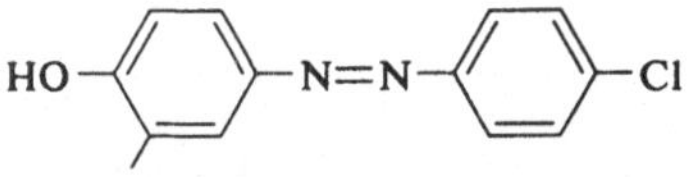

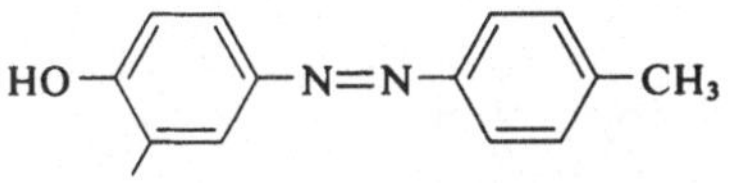

The photochromic properties of such polymers were examined as films or in solution under irradiation from a 100-W projection lamp.

Photo-isomerization, the basis of the photochromic properties of these azo polymers, involves the *cis* and *trans* configurations of the azo group. The former configuration is the most sterically hindered, and thus small differences in the steric structure of the pendant groups have a large influence on the polymer photochromic properties.[173]

Polymers with the photochromic chromophore in the backbone chain are best exemplified by azo and Schiff's base polymers, such as polyanil (**4-119**), formed from the condensation of a bis-aldehyde with 4-aminoaniline, and polyazo, prepared from diazonium salts and poly(2-hydroxystyrene).[173]

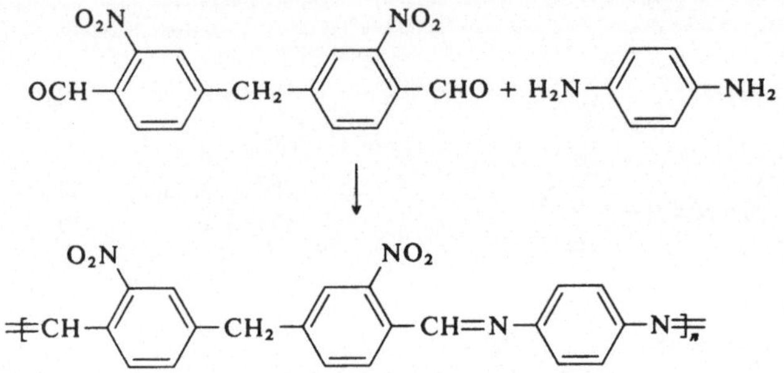

**4-119**

Blair *et al.*[174] described photochromic properties of polyamides (**4-120**) and (**4-121**) containing azo groups, such as 3,3'- and 4,4'-azodibenzoyl-*trans*-2,5-dimethylpiperazine. The photomechanical conversion energy of polyamides (**4-122**) containing a stilbene chromophore in the backbone was examined by Osada and Katsumura.[175]

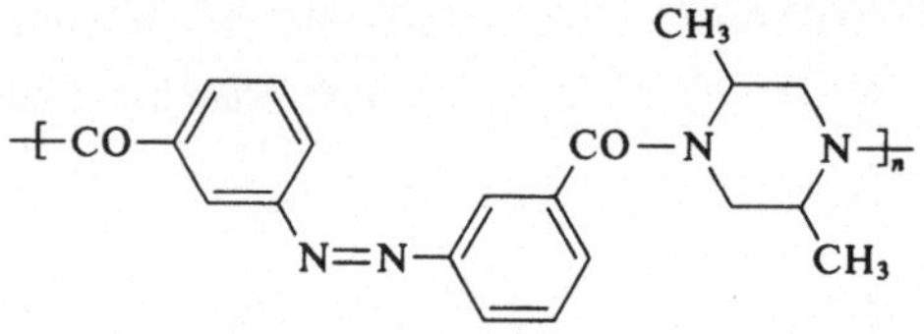

**4-120**

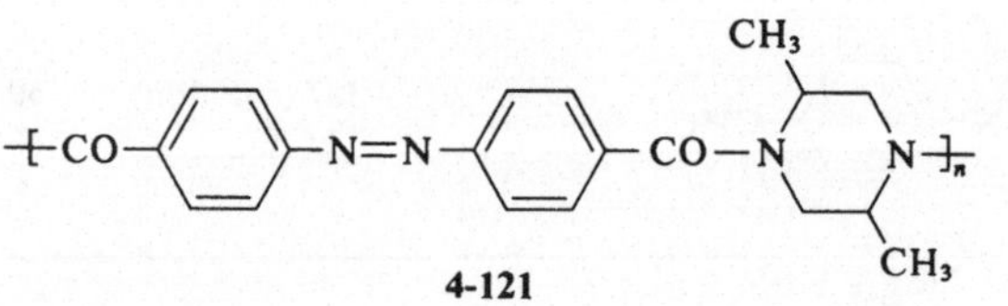

**4-121**

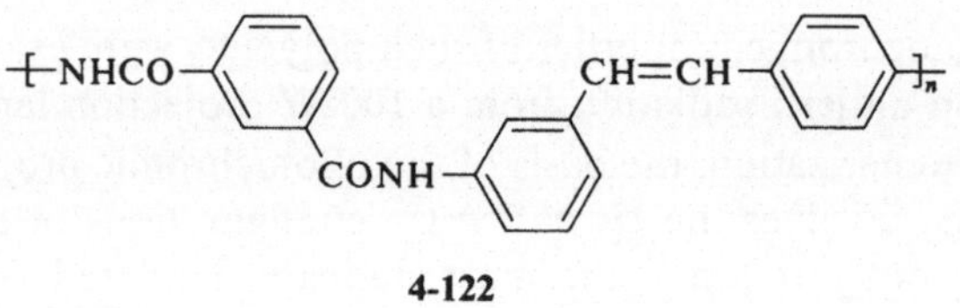

**4-122**

# REFERENCES

1. L. J. Miller, J. D. Margerum, and J. B. Rust, *J. SMPTE 71*, 1177 (1968).
2. D. W. Woodward, V. C. Cambers, and A. B. Cohen, *Photogr. Sci. Eng. 7*, 360 (1963).
3. A. B. Cohen, SPSE Symp. on Unconventional Photographic Systems, Washington, p. 122 (1967).
4. S. Levinos, SPSE Symp. on Unconventional Photographic Systems, Washington, p. 145 (1964).
5. F. A. Stuber, H. Ulrich, B. V. Rao, and A. A. R. Sayigh, *Photogr. Sci. Eng. 17*, 446 (1973).
6. K-I. Shimazu, *Photogr. Sci. Eng. 7*, 33 (1973).
7. C. L. Jewett and J. M. Case, U.S. Pat. 2,714,066 (1955).
8. J. Kosar, *Light Sensitive Systems*, p. 137, J. Wiley, New York (1965).
9. D. G. Borden and J. L. R. Williams, *Makromol. Chem. 178*, 3035 (1977).
10. H. Hiraoka and L. W. Welsh, Jr., in: *Polymers in Electronics* (T. Davidson, ed.), p. 55, ACS Symposium Series, No. 242, Washington, D.C. (1984).
11. L. Stillwagon, in: *Polymer Materials for Electronic Applications* (E. D. Fiet and C. Wilkins, Jr., eds.), p. 19, ACS Symposium Series, No. 184, Washington, D.C. (1982).
12. H. Bamford and A. D. Jenkins, *Nature 176*, 78 (1955).
13. D. M. French, *Rubber Chem. Technol. 42*, 71 (1969).
14. S. F. Reed, *J. Polym. Sci., Part A-1, 9*, 2029 (1971).
15. R. Waltz, B. Bomer, and W. Heitz, *Makromol. Chem. 178*, 2527 (1977).
16. R. Waltz and W. Heitz, *J. Polym. Sci., Polym. Chem. Ed. 16*, 1807 (1978).
17. W. M. Edwards and I. M. Robinson, U.S. Pat. 2,710,853 (1955).
18. W. M. Edwards and I. M. Robinson, U.S. Pat. 2,867,609 (1959).
19. E. I. Valko, G. C. Tesoro, and E. D. Szubin, U.S. Pat. 2,914,427 (1959).
20. H. R. Ing and R. H. F. Manske, *J. Chem. Soc.*, 2348 (1926).
21. R. Rubner, *Siemans Forsch.-U. Entwickl.-Ber. 5*, 92 (1976).
22. J. Duran and N. S. Viswanathan, in: *Polymers in Electronics* (T. Davidson, ed.), p. 2395, ACS Symposium Series, No. 242, Washington, D.C. (1984).
23. M. L. Wallach, *J. Polym. Sci., Part A-2, 5*, 653 (1967).
24. V. L. Bell and G. F. Pezdirtz, *J. Polym. Sci., Part B, 3*, 977 (1965).
25. V. L. Bell and A. R. Jewell, *J. Polym Sci., Part A-1, 5*, 3043 (1967).
26. V. L. Bell, *J. Polym. Sci., Part B, 5*, 941 (1967).
27. F. Dawans and C. S. Marvel, *J. Polym. Sci., Part A, 3*, 3549 (1965).
28. J. G. Colson, R. H. Michel, and R. M. Paufler, *J. Polym. Sci., Part A-1, 4*, 59 (1966).
29. E. N. Teleshov and A. N. Pravednikov, *Dokl. Akad. Nauk SSSR 192*, 1347 (1964).
30. R. L. Van Deusen, *J. Polym. Sci., Part B, 4*, 211 (1966).
31. (a) V. L. Bell, *Polym. Lett. 5*, 941 (1967); (b) V. L. Bell and R. A. Jewell, *J. Polym. Sci., Part A-1, 5l*, 3013 (1967).
32. F. Dawans and C. S. Marvel, *J. Polym. Sci., Part A, 3*, 3549 (1965).
33. R. L. Van Deusen, *Polym. Lett. 4*, 211 (1966).
34. G. M. Bower and L. W. Frost, *J. Polym. Sci., Part A, 1*, 3135 (1963).
35. Y. Imai, *J. Polym. Sci., Part A-1, 5*, 2289 (1967).
36. Farbenfrabriken Bayer, Netherlands Pat. Appl. 67/05, 983 (1967).
37. L. M. Azberino, W. J. Farrisey, and J. S. Rose, U.S. Pat. 3,708,458 (1973).
38. W. M. Alvino and L. E. Edleman, *J. Appl. Polym. Sci. 19*, 2961 (1975).
39. P. S. Carleton, W. J. Farrisey, & J. S. Rose, *J. Appl. Polym. Sci. 16*, 2983 (1972).
40. R. A. Mayers, *J. Polym. Sci., Part A-1, 7*, 2757 (1969).
41. D. G. Locrini and J. M. Witzel, *J. Polym. Sci., Part A-1, 7*, 2185 (1969).
42. N. Dokoshi, S. Tohyama, S. Fujita, M. Kurihara, and N. Yoda, *J. Polym. Sci., Part A-1, 8*, 2197 (1970).

43. N. Yoda, *Encycl. Polym. Sci. Technol. 10*, 682 (1962).
44. G. F. D'Alelio, D. M. Feigl, T. Ostdick, M. Saha, and A. Chang, *J. Macromol. Sci. Chem. A6*, 1 (1972).
45. C. S. Marvel and D. S. Casey, *J. Org. Chem. 24*, 957 (1959).
46. N. Yoda, R. Nakanishi, M. Kurihara, Y. Bamba, S. Tohyama, and K. Ikeda, *J. Polym. Sci., Part B, 4*, 11 (1966).
47. N. Yoda, M. Kurihara, K. Ikeda, S. Tohyama, and R. Nakanishi, *J. Polym. Sci., Part B, 4*, 551 (1966).
48. Y. Iwakura, T. Kurosaki, and H. Watamoto, *Yuki Gosei Kagaku Kyokai Shi 19*, 327 (1961).
49. J. W. Fisher, *Fibers (Nat. Syn.) 14*, 365 (1954).
50. A. H. Frazer, S. Sweeny, and F. T. Wallenberger, *J. Polym. Sci., Part A, 2*, 1157 (1964).
51. A. H. Frazer and F. T. Wallenberger, *J. Polym. Sci., Part A, 2*, 1137 (1964).
52. A. H. Frazer and F. T. Wallenberger, *J. Polym. Sci., Part A, 2*, 1139 (1964).
53. A. H. Frazer and F. T. Wallenberger, *J. Polym. Sci., Part A, 2*, 1147 (1964).
54. A. H. Frazer and F. T. Wallenberger, *J. Polym. Sci., Part A, 2*, 1181 (1964).
55. A. H. Frazer and F. T. Wallenberger, *J. Polym. Sci., Part A, 2*, 1825 (1964).
56. R. Stolle, *J. Prakt. Chem. 68*, 30 (1903).
57. E. I. Valko, G. C. Tesoro, and E. D. Szubin, U.S. Pat. 2,914,427 (1959).
58. F. R. Mayo and A. A. Miller, *J. Am. Chem. Soc. 78*, 1023 (1956).
59. E. A. Bovey and I. M. Kothoff, *J. Am. Chem. Soc. 69*, 2143 (1947).
60. A. A. Miller and F. R. Mayo, *J. Am. Chem. Soc. 78*, 1017 (1956).
61. G. Bernhardt and F. Korte, *Angew. Chem. 77*, 133 (1965).
62. A. Baeyer and V. Villiger, *Ber. 34*, 762 (1901).
63. H. Von Pechmann and L. Vanino, *Ber. 27*, 1511 (1894).
64. T. Saegusa, T. Nozaki, and R. Oda, *J. Chem. Soc. Japan (Ind. Chem. Sect.) 57*, 233 (1954).
65. W. Hahn and H. Lechtenbohmer, *Markromol. Chem. 16*, 50 (1955).
66. D. J. Metz and R. B. Mesrobian, *J. Polym. Sci. 16*, 345 (1955).
67. J. Hill, U.S. Pat. 2,556,876 (1951).
68. W. Heitz, H. G. Stahl, and R. Dick, German Pat. 3,005,889 (1981).
69. S. Nagai, Y. Hidaka, and A. Ueda, Japanese Pat. 74 17,897 (1974).
70. D. A. Smith, *Makromol. Chem. 103*, 301 (1967).
71. K. B. Chandalia and P. J. Preston, U.S. Pat. 4,094,868 (1978).
72. J. J. Laverty and Z. G. Gardlund, *Polym. Prepr., Am. Chem. Soc., Div. Polym. Chem. 15*, 306 (1974).
73. W. Hahn and A. Fischer, *Makromol. Chem. 21*, 77 (1956).
74. R. Kerber, O. Nuyken, and R. Steinhausen, *Makromol. Chem. 177*, 1357 (1976).
75. R. Kerber, O. Nuyken, and R. Steinhausen, *Makromol. Chem. 178*, 1833 (1977).
76. R. Kerber, O. Nuyken, and M. Dorn, *Makromol. Chem. 179*, 2083 (1978).
77. O. Nuyken, M. Dorn, and R. Kerber, *Makromol. Chem. 180*, 1651 (1979).
78. O. Nuyken, R. Ren Gel, and R. Kerber, *Makromol. Chem. 181*, 1565 (1980).
79. R. Waltz and W. Heitz, *J. Polym. Sci., Polym. Chem. Ed. 16*, 1807 (1977).
80. G. Smets and A. E. Woodward, *J. Polym. Sci. 14*, 126 (1954).
81. R. Waltz, B. Bomer, and W. Heitz, *Makromol. Chem. 178*, 2527 (1977).
82. D. A. Smith, *Makromol. Chem. 103*, 301 (1967).
83. T. N. Bowmer and J. H. O'Donnell, *Polym. Deg. Stab. 3*, 87 (1981).
84. F. S. Daiton and K. J. I. Vin, *Proc. Roy. Soc. (London) A212*, 207 (1952).
85. M. A. Naylor and A. W. Anderson, *J. Am. Chem. Soc. 76*, 3962 (1954).
86. E. Gipstein, E. Wellisch, and O. J. Sweeting, *J. Org. Chem. 29*, 207 (1964).
87. E. Wellisch, E. Gipstein, and O. J. Sweeting, *J. Appl. Polym. Sci. 8*, 1623 (1964).
88. N. Grassie, E. J. Murray, and A. A. Homles, *Polym. Deg. Stab. 6*, 47 (1984).
89. G. B. Butler, A. Crawshaw, and W. L. Miller, *J. Am. Chem. Soc. 80*, 3615 (1957).

90. N. Grassie and G. A. Perdomo Mendoza, *Polym. Deg. Stab. 11*, 145 (1985).
91. D. A. Bochvar, I. V. Stankevich, E. S. Krongruz, A. L. Rusanov, and V. V. Korshak, *Vysokomol. Soedin, Ser. A9*, 1429 (1967); *Chem. Abstr. 68*, 2943, 30114m (1968).
92. N. Dokoshi, Y. Bamba, M. Kurihara, and N. Yodt, *Makromol. Chem. 108*, 170 (1967).
93. S. W. Osborn and E. Broderick, U.S. Pat. 3,213,108 (1965).
94. O. Makoto, in: *An Introduction to Specialty Polymers* (N. Ise and I. Tabushi, eds.), p. 27, Cambridge University Press, Cambridge (1983).
95. G. Wagner, *Z. Naturforsch. 24B*, 824 (1969).
96. G. Wagner, *Makromol. Chem. 145*, 85 (1971).
97. J. Kaiser, G. Wagner, and E. W. Fischer, *Israel J. Chem. 10*, 157 (1972).
98. G. Wagner, in: *Molecular Metals* (W. E. Hatfield, ed.), p. 209, Plenum Press, New York (1979).
99. G. Wagner, *Makromol. Chem. 154*, 35 (1972).
100. A. J. Melveber and R. H. Baughman, *J. Polym. Sci., Polym. Phys. Ed. 11*, 603 (1973).
101. R. H. Baughman, *J. Polym. Sci., Polym. Phys. Ed. 12*, 1551 (1974).
102. G. Wagner, *Pure Appl. Chem. 49*, 443 (1977).
103. (a) G. Wagner, *J. Polym. Sci., Chem. Lett. B9*, 133 (1971); (b) G. Wagner, *Makromol. Chem. 145*, 85 (1971).
104. (a) J. P. Fouassier, B. Tieke, and G. Wagner, *Israel J. Chem. 18*, 227 (1979); (b) C. Bubeck, B. Tieke, and G. Wagner, *Ber. Bunsenges Phys. Chem. 86*, 499 (1982).
105. (a) G. Wagner, *Makromol. Chem. 134*, 219 (1970); (b) A. S. Hay, *J. Polym. Sci., Part A-1, 8*, 1022 (1970); (c) A. S. Hay, D. H. Bolon, K. B. Leimer, and R. F. Clark, *J. Polym. Sci., Part B, 8*, 97 (1970); (d) A. O. Patil, D. D. Deshpande, S. S. Talwar, and A. B. Biswas, *J. Polym. Sci. 19*, 1155 (1981).
106. H. Kuhn and D. Mobius, *Angew. Chem. 83*, 672 (1971).
107. (a) G. A. Delzenne, in: *Encyclopedia of Polymer Science and Technology* (H. F. Mark and N. M. Bikales, eds.), Suppl. Vol. 1, p. 401, Interscience, New York (1976); (b) G. A. Delzenne, *Adv. Photochem. 11*, 1 (1979).
108. J. Kosar, *Light-Sensitive Systems: Chemistry and Applications of Nonsilver Halide Photographic Processes*, p. 140, J. Wiley and Sons, New York (1965).
109. (a) A. Reiser and P. L. Egerton, *Photogr. Sci. Eng. 23*, 144 (1979); (b) J. L. R. Williams, *Fortschr. Chem. Forsch. 13*, 227 (1969).
110. S. Tazuke, in: *Developments in Polymer Photochemistry—3* (N. S. Allen, ed.), p. 53–93, Applied Science Publishers, London (1982).
111. J. L. R. Williams, S. Y. Farid, J. C. Doty, R. C. Daly, D. P. Specht, R. Searle, D. G. Borden, H. J. Chang, and P. A. Martic, *Pure Appl. Chem. 49*, 523 (1977).
112. H. Tanaka, M. Tsuda, and H. Nakanishi, *J. Polym. Sci., Part A-1, 10*, 1729 (1972).
113. C. D. Beboer, *J. Polym. Sci., Lett. Ed. 11*, 25 (1973).
114. C. C. Unruh, *J. Appl. Polym. Sci. 2*, 358 (1959).
115. C. C. Unruh, *J. Appl. Polym. Sci. 3*, 310 (1960).
116. C. C. Unruh and C. F. N. Allen, U.S. Pat. 2,716,102 (1955).
117. C. C. Unruh and C. F. N. Allen, U.S. Pat. 2,716,097 (1955).
118. A. C. Smith, J. L. R. Williams, and C. C. Unruh, U.S. Pat. 2,816,091 (1957).
119. C. C. Unruh and C. F. N. Allen, U.S. Pat. 2,706,725 (1955).
120. S. H. Merrill, U.S. Pat. 2,831,768 (1958).
121. D. Bellus, P. Hrdlovic, and Z. Manasek, *J. Polym. Sci., Part B, 4*, 1 (1966).
122. P. A. Mullen and N. Z. Searle, *J. Appl. Polym. Sci. 14*, 765 (1970).
123. S. B. Maerov, *J. Polym. Sci., Part A-3*, 487 (1965).
124. W. M. Eareckson III, *J. Polym. Sci. 40*, 399 (1955).
125. (a) D. J. Carlsson, L. H. Gan, and D. M. Wiles, *J. Polym. Sci., Polym. Chem. Ed. 16*, 2353 (1978); (b) J. D. Carlsson, L. H. Gan, and D. M. Wiles, *J. Polym. Sci., Polym. Chem. Ed. 16*, 2365 (1978).

126. C. S. Schollenberger and F. D. Stewart, in: *Advances in Urethane Science & Technology* (K. C. Frisch and S. L. Reegen, eds.), Vol. III, Technomic Pub., Westport, Connecticut (1973).
127. N. S. Allen and J. F. McKellar, *Makromol. Chem. 108*, 2875 (1979).
128. B. D. Gesner and P. G. Kelleher, *J. Appl. Polym. Sci. 12*, 1199 (1968).
129. B. D. Gesner and P. G. Kelleher, *J. Appl. Polym. Sci. 13*, 2183 (1969).
130. E. Gipstein, W. Moreau, G. Chiu, and O. Need, *J. Appl. Polym. Sci. 21*, 677 (1977).
131. R. J. Himics, M. Kaplan, N. V. Desai, and E. S. Poliniak, *Tech. Paper Soc. Plastics Eng.*, Mid-Hudson Section, Oct. 13–15 (1976).
132. K. J. Ivin and J. B. Roste, *Adv. Macromol. Chem. 1*, 335 (1968).
133. R. E. Cais, J. H. O'Donnell, and F. A. Bovey, *Macromolecules 10*, 254 (1977).
134. M. Matsuda, M. Lino, and N. Tokura, *Makromol. Chem. 65*, 232 (1963).
135. C. Schneider, J. Denaxas, and D. Hummel, *J. Polym. Sci., Part C, 16*, 2203 (1967).
136. C. G. Wilson, J. M. Frechet, and M. J. Farrall, in: *Polymer Science and Technology—Modifications of Polymers* (C. E. Carraher, Jr. and J. A. Moore, eds.), Vol. 21, p. 25, Plenum Press, New York (1983).
137. T. N. Bowmer and J. H. O'Donnell, *J. Polym. Sci., Polym. Chem. Ed. 19*, 45 (1981).
138. T. N. Bowmer and J. H. O'Donnell, *J. Macromol Sci. Chem. A17*, 243 (1982).
139. T. N. Bowmer, J. H. O'Donnell, and P. R. Wells, *Makromol. Chem., Rapid Commun. 1*, 1 (1980).
140. T. N. Bowmer, J. H. O'Donnell, and P. R. Wells, *Polymer Bulletin 2*, 103 (1980).
141. T. N. Bowmer and J. H. O'Donnell, *Polymer 22*, 71 (1980).
142. E. Gipstein, W. Moreau, G. Chiu, and O. U. Need III, *J. Appl. Polym. Sci. 21*, 677 (1977).
143. R. Sano and K. Hasegawa, *Photogr. Sci. Eng. 15*, 309 (1971).
144. S. H. Merrill and C. C. Unruh, *J. Appl. Polym. Sci. 7*, 273 (1963).
145. F. A. Stuber, H. Ulrich, D. V. Rao, and A. A. R. Sayigh, *Photogr. Sci. Eng. 17*, 446 (1973).
146. K. Shimazu, *Photogr. Sci. Eng. 17*, 33 (1973).
147. M. P. Schmidt and R. Zahan, U.S. Pat. 2,003,631 (1936).
148. C. L. Jewett and J. M. Case, U.S. Pat. 2,714,066 (1955).
149. I. Mellan, U.S. Pat. 3,050,502 (1962).
150. G. A. Crosby, N. M. Weinshenker, and H-S. Uh, *J. Am. Chem. Soc. 97*, 2232 (1975).
151. L. C. Dorman and J. Love, *J. Org. Chem. 34*, 158 (1969).
152. G. A. Delzenne, U. Laridon, and H. Peeters, *Eur. Polym. J. 6*, 933 (1970).
153. E. Reichmanis and C. W. Wilkins, Jr., in: *Polymers Materials for Electronic Applications* (E. D. Feit and C. W. Wilkins, Jr., eds.), p. 29, ACS Symposium Series, No. 184, Washington, D.C. (1982).
154. C. W. Wilkins, Jr., E. Reichmanis, and E. A. Chandross, *J. Electochem. Soc. 127*, 2510 (1980).
155. A. Werner and A. Piguet, *Chem. Ber. 34*, 4295 (1904).
156. J. P. Vermes and R. Bengelmans, *Tetrahedron Lett.*, 2091 (1969).
157. B. Amit and A. Patchornik, *Tetrahedron Lett.*, 2205 (1973).
158. A. P. Foken, T. N. Gerasenova, K. E. Matoshena, V. E. Soklenko, and L. N. Ogneva, *Siberian Chem. J.*, Ser. 4, 89 (1972).
159. R. T. Foster and C. S. Marvel, *J. Polym. Sci., Part A, 3*, 417 (1965).
160. U. Zehavi and A. Patchornik, *J. Am. Chem. Soc. 95*, 5673 (1973).
161. J. M. Frechet and C. Shuerch, *J. Am. Chem. Soc. 93*, 4492 (1971).
162. F. S. Tjoeng, E. K. Tong, and R. S. Hodges, *J. Org. Chem. 43*, 4190 (1978).
163. D. H. Rich and S. K. Gurwar, *J. Am. Chem. Soc. 97*, 1571 (1975).
164. W. Marckwald, *Z. Phys. Chem. (Leipzig) 30*, 140 (1899).
165. G. A. Delzenne, in: *Proceedings of IIIrd International Congress on Reprography*, 1971, p. 169, IPC Science and Technology Press, Guildford (1971).

166. H. Kamogawa, M. Kato, and H. Sagiyama, *J. Polym. Sci., Part A, 6,* 2967 (1968).

167. G. Smets, *Advances in Polymer Sciences*, p. 17, Springer-Verlag, Berlin (1983).

168. (a) G. Smets, *Pure Appl. Chem. 42,* 516 (1975); (b) G. Smets, *J. Polym. Sci., Polym. Chem. Ed. 13,* 2223 (1975).

169. C. Eisenbach, *Ber. Bunsenges. Phys. Chem. 84,* 680 (1980).

170. (a) G. Smets, J. Thoen, and A. Aerts, *J. Polym. Sci., Polym. Symp. 51,* 119 (1975); (b) G. Smets and E. Evans, *Pure Appl. Chem., Suppl. Macromol. Chem. 8,* 357 (1973).

171. R. Lovrien and J. C. B. Waddington, *J. Am. Chem. Soc. 86,* 2315 (1964).

172. P. Friut, French Pat. 1,365,308 (1964).

173. H. Kamogawa, M. Kato, and H. Sugiyama, *J. Polym. Sci., Part A-1, 6,* 2967 (1968).

174. H. S. Blair, H. I. Pogue, and E. Riordan, *Polymer 21,* 1195 (1980).

175. Y. Osada and E. Katsumura, *Makromol. Chem., Rapid Commun. 2,* 241, 411 (1981).

# 5

# Triggered Isomerization and Color Change

## 5.1. PROCESSES UTILIZING TRIGGERED ISOMERIZATION AND COLOR CHANGE

The triggered displacement of the equilibrium between isomers of a compound has been frequently used in industrial applications. This section will deal with highlights of such processes.

The most frequent use of a triggered isomerization is found in the field of imaging science, where isomerization is accompanied by a visible change in color. Such a change when reversible is termed chromism. Polymer applications, on the other hand, make use of the higher reactivity of *cis* over *trans* isomers toward double-bond additions during the cross-linking of polymeric chains.

### 5.1.1. Imaging Science

Photochromic indolinospiropyrans have been used by the NCR Corporation as a basis for forming colored images.[1] They also reported[2] the use of spiro compounds encapsulated in microcapsules, which on imaging rupture to form an image.

Fulgides have also been reported[3] useful for making photographic proofs. The process is sensitive to blue light and forms positive images from the required negatives with contrasts comparable to a continuous-tone image. Their use has also extended into holographic data storage media.[4]

3-Pyridylsydnone and 2-(4'-nitrobenzylpyridine) were both found useful in litho plate applications.[5] Their color change was used to record the images on an exposed printing plate.

Triphenylmethane leuco dyes were also used in recording the images on an exposed printing plate. Furthermore, the leuco forms of triphenylmethane compounds[7] such as 4,4'-dimethylaminotriphenylmethyl cyanide have been utilized in heat-sensitive copying paper. The high sensitivity of triphenylmethane leuco cynanides to ultraviolet light aroused interest in their use as actinometers,[8-10] where UV dosage can be monitored by a color formation reaction.

Photochromic compounds have also found applications in the field of photochromic ophthalmic lenses.[11a] Spiropyrans, triphenylmethane leuco compounds, and polynuclear aromatic hydrocarbons were candidates for incorporation into plastic lenses. In general, little commercial success was achieved due to the high fatigue rate of such organic compounds under continuous cycling.

Hammond et al.[11b] have described an interesting use of photochromic compounds and their triggered color change as chemical gears. The authors proposed the use of the light-triggered color change of 2,3-diphenylindenone oxide (5-1) from colorless to red as a design variable in a working system.

In principle, light was used both to do work and as a control signal. Both the color formation and bleaching depended upon radiant energy to do the work in a phototransformation, while the analyzing light was the signal and control device.

Colorless                                    Colored
5-1

Nitrones (5-2) have also found a place in imaging.[12,13] A UV-triggered intramolecular cyclization of a nitrone molecule was the basis for a dye-forming imaging system. The cyclized nitrone, now an oxaziridine molecule formed by the imagewise exposure, decomposes on heating (the image fixing step) to give benzaldehyde and a yellow azo dye. Reviews dealing with the decomposition reaction have been published by Hammer and Macaluso[14] and Delpierre and Lamchen.[15]

5-2

Irreversible decolorization processes other than those utilizing the previously described nitrone chemistry have found interest in imaging science, where the dye does not constitute the formed image but is an aid toward image formation. In such applications the dye color is generally not required following the imaging step. For example, Eastman Kodak Co.[16] describes the use of Stenhouse salts as sensitizing dyes for the photoconductor in an electrophotographic process. Following the imaging step, the dye is decolorized by heating. In a different context, 3M Company[17a] describes the use of the Stenhouse salts as antihalation dyes in a silver-based photothermographic process. In this case the dye reduces the amount of scattered light within the light-sensitive layer during imaging and is decolorized during the heat developing step. Dyes with an N-alkoxypyridinium group have been published by Eastman Kodak[17b] as antihalation dyes for photothermographic processes. Such dyes undergo a color change when exposed to intense heat or light due to the triggered cleavage of the alkoxy group off the pyridinium ring.

$$R^2HN\overset{+}{=}\underset{X^-}{\underset{}{\underset{}{}}}\overset{R}{\underset{}{(\quad)_n}}NHR^1$$

Stenhouse salts

## 5.1.2. Polymer Science

Triggered cross-linking of polymeric molecules has made use of the *cis–trans* equilibrium as a route to the cross-linking reaction. Compounds capable of *cis–trans* photoisomerization, for example, maleic and fumaric acids, azo compounds, and oximes, yield isomeric structures with different characteristics. Polymerization of such unsaturated compounds through photocycloaddition reactions was reviewed by Dilling in 1983.[18]

Kodak[19] used triggered isomerization of stilbene to switch on a cross-linking only through the *cis*-isomer. This is formed from the more stable *trans* form upon UV exposure.

The same concept of the triggered *cis–trans* isomerization of stilbene (5-3) is used in the sensitization of polymeric cross-linking for printing plates.[20,21]

$$C_6H_5C{:}CC_6H_5 \xrightarrow{h\nu} C_6H_5C{:}CC_6H_5 \xrightarrow{O_2}$$

*trans* Isomer                    *cis* Isomer

**5-3**

## 5.2. CHEMICAL REACTIONS FOR COLOR CHANGE

### 5.2.1. Thermally Triggered Release

The reversible thermal triggering of a color change is termed thermochromism. This is defined as the noticeable change in color of compounds on heating and the reversion to the origional color on cooling. Such a reversible color change is an essential aspect of thermochromism, as compared to other color changes which are irreversible.

The mechanism of the thermochromic process varies with molecular structure and is covered in a number of reviews.[22,23] Theilacker *et al.*[24] published articles dealing with thermochromism of ethylenic compounds. Mustafa[25] reviewed spiropyrans and their thermochromic properties, while Pullman and Pullman[26] examined the relation between structure and thermochromism.

In this section we will limit our discussion to organic compounds which display thermally triggered color changes. Inorganic compounds are also known to undergo thermochromism and a good review by Day[23] covers this subject.

Irreversible color changes have been also reported for certain classes of compounds. These changes usually accompany thermal reactions where the conjugation of a colored species is altered or destroyed. Such a process is not reversed when the triggering energy is withdrawn. An example of such a class of colored compounds is the Stenhouse salts, reviewed by Lewis and Mulquiney.[27]

#### 5.2.1.1. Thermochromic Spiropyrans

Early observations of the thermally triggered color change of spiropyrans were made in 1926. Since then over seventy compounds have been reported to show similar properties. A review of the chemistry of spiropyrans published in 1948 by Mustafa[25] outlines these early results.

The color change of a spiro structure from colorless to blue or violet in solution or in the solid phase involves a thermally triggered equilibrium between the colorless spiro (**5-4**) and colored open-chain structures (**5-5**).

|  Colorless  |  Colored  |
| 5-4 | 5-5 |

X = O, N, S; Y = O

General requirements for such thermochromism can be summarized as follows:

1. Two possible spiro structures may involve the ortho or para-linkages.

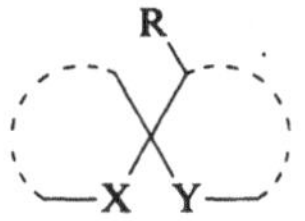

**5-6**
*ortho*-Link

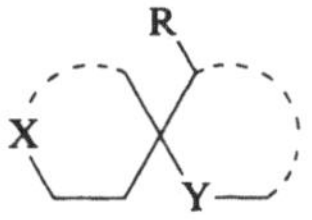

**5-7**
*para*-Link

2. The ring containing the heteroatom, Y, must be part of a naphthopyran ring.
3. The R-group should always be a hydrogen; substituents hinder the opening of the ring.
4. The extent of the chromism is related to the electron-donating properties of the heterocyclic ring.
5. Substituents will enhance the thermochromism if they stabilize the positive charge on the heterocyclic ring or the negative charge on the pyran ring of the open structure.

The open-chain colored form of the spiro compound is easily identified as a member of the merocyanine class of dyes. The question of whether it is a quinoidal form or a polar one has been extensively studied, as summarized by Knott[28] and Dilthey and Wubken.[29] In general, the nature of the hetero-ring would dictate the structure of the open-chain colored form. The following are selected structures of classes of spiro compounds reported in the literature.

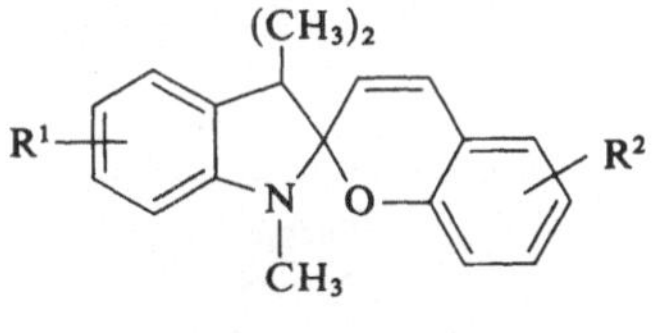

**5-8** Colorless to purple-green[30]

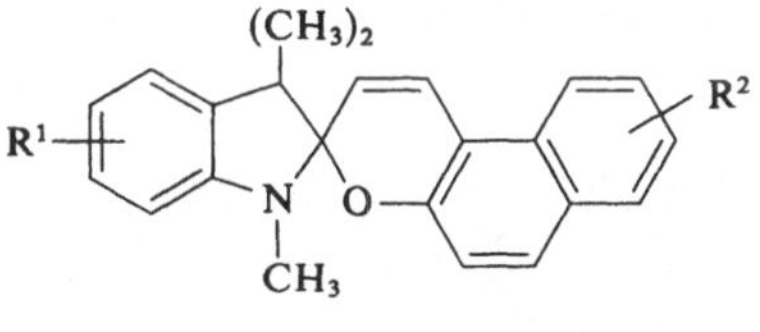

**5-9** Colorless to blue-red[31,32]

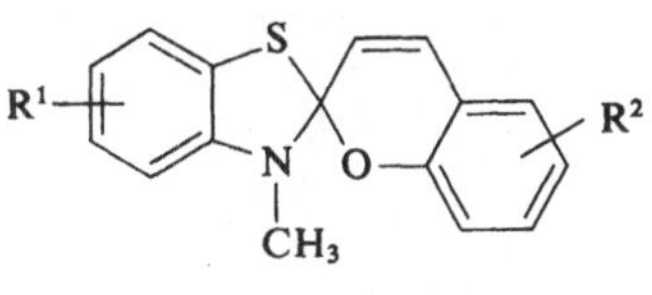

**5-10** Colorless to purple-red[32]

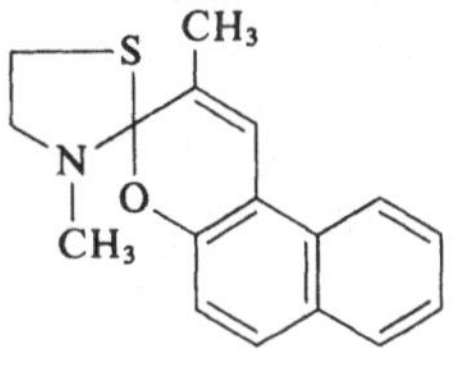

**5-11** Colorless to yellow[33]

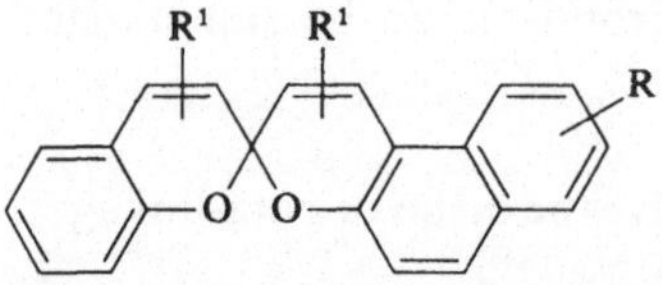

**5-12**  Colorless to red-purple[32]

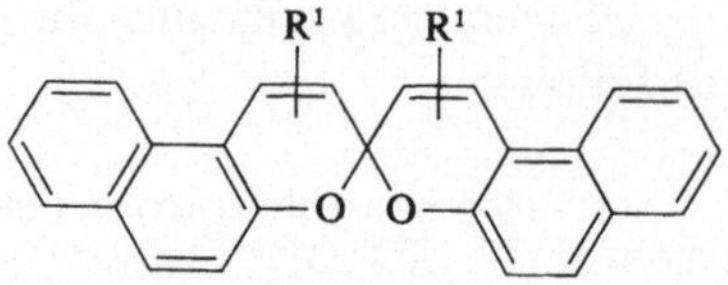

**5-13**  Ref. 32

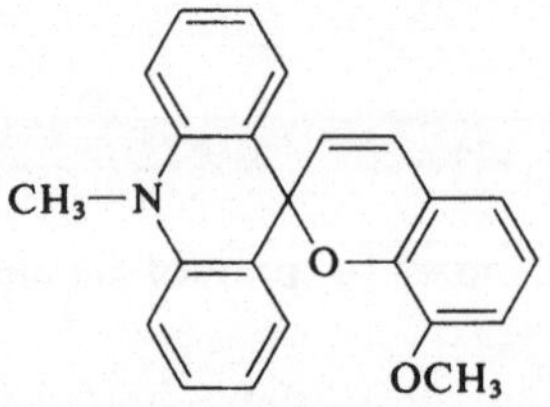

**5-14**  Colorless to violet

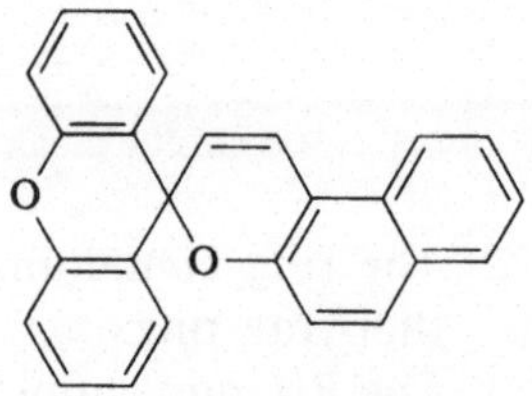

**5-15**  Colorless to deep red[32]

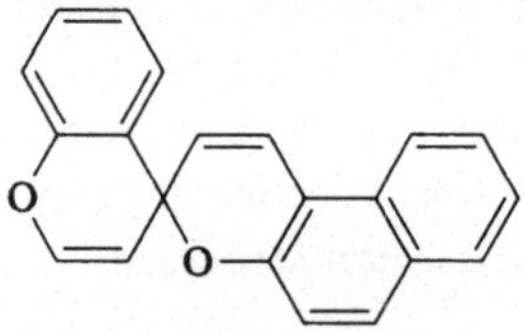

**5-16**  Ref. 35

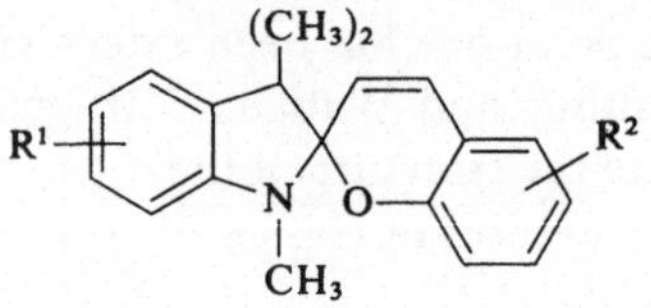

**5-17**

| $R^2$ | Triggering temperature[a] (m.p., °C) | Triggered color |
|---|---|---|
| 6-$NO_2$ | 176–177 | Purple |
| 7-$NO_2$ | 107–108 | Orange |
| 7-Cl | 116–117 | Purple |
| 6,8-$Br_2$ | 115–116 | Blue |
| 6-$NO_2$-8-CH:$CH_2$ | 108–109 | Purple |
| 6-$NO_2$-8-Br | 252–256 | Purple |
| 6-$NO_2$-8-F | 302–303 | Purple |
| 6-Cl-8-$NO_2$ | 134–136 | Purple |
| 6-Br-8-$NO_2$ | 122–123 | Purple |
| 6-$NO_2$-8-$CH_3$O | 152–153 | Dark blue |
| 6-$CH_3$-8-$NO_2$ | 116–117 | Green |
| 5-$NO_2$-8-$CH_3$ | 166–167 | Dark green |

[a] Ref. 30.

5.2.1.2. Thermochromic Ethylenic Compounds

These compounds are characterized by a double bond linking two sets of rings. Such compounds include bianthrone (**5-18**), dixanthylene (**5-19**), xanthylideneanthrone (**5-20**), and diphenylmethyleneanthrone (**5-21**).

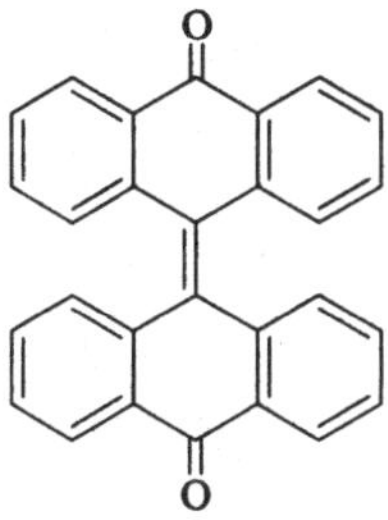

**5-18**

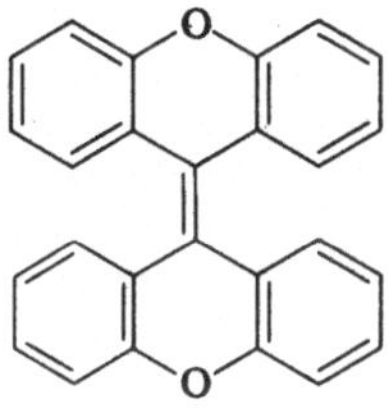

**5-19**

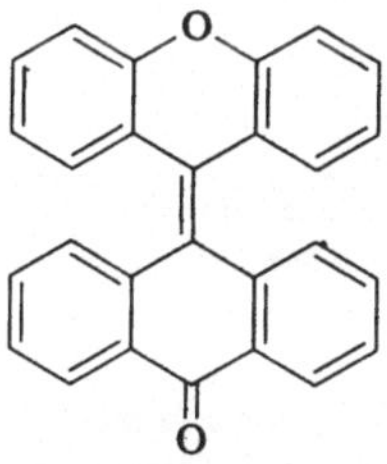

**5-20**

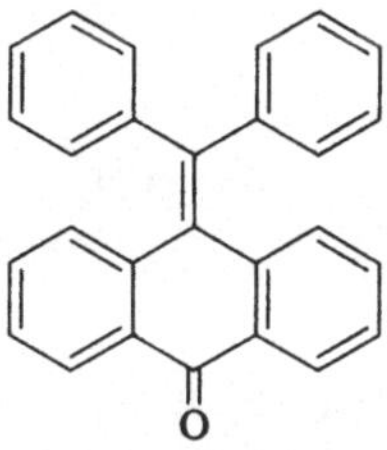

**5-21**

Various discussions and postulates as to the mechanism of the thermally triggered color change of these multi-cyclic ethylenes cited in the literature include theories of biradical, ionic, and steric basis. Day, in his review,[22] has summarized chronologically these different postulates. The main line of thought behind most of these theories[36] is the relief of the steric strain.

Examples of bianthrones (**5-18**), dixanthylenes (**5-19**), and diphenyl-methyleneanthrones (**5-21**) reported in the literature will be covered here with a brief reference to their thermochromic properties under various conditions.

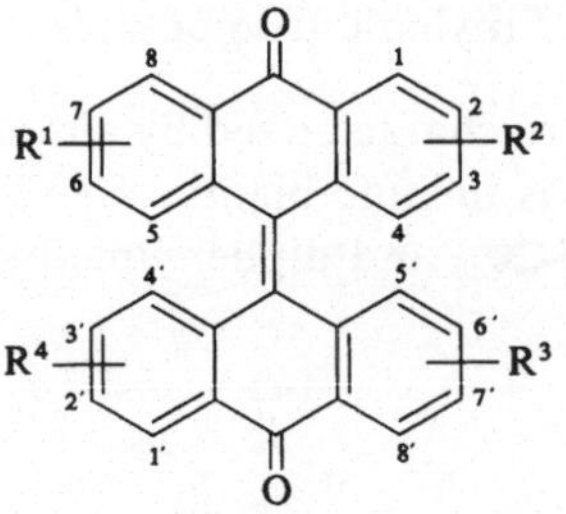

**5-18**

| $R^1, R^2, R^3, R^4$ | Triggered color and medium[a] |
|---|---|
| H, H, H, H | Colorless at $-180°C$<br>Blue-green at melting point<br>Deep blue-green in hot solution |
| $2,2'-Br_2$ | Colorless in cold solution<br>Blue-green in hot solution |
| $4,4'-Br_2$ | Colorless in cold solution<br>Blue-green in hot solution |
| $2,3,2',3'-(CH_3)_4$ | Colorless in cold solution<br>Blue-green in hot solution |
| $2,4,2',4'-(CH_3)_4$ | Colorless in cold solution<br>Blue-green in hot solution |

[a] Ref. 37.

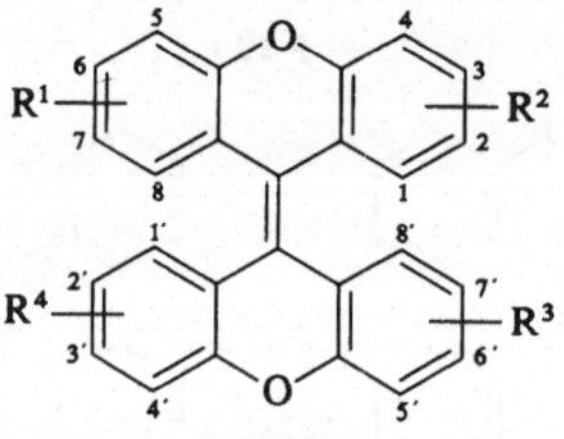

**5-19**

| $R^1, R^2, R^3, R^4$ | Triggered color and medium[a] |
|---|---|
| H, H, H, H | Pale yellow at $-180°C$<br>Lemon yellow at $25°C$<br>Green at melting point<br>Yellowish green in solid solution<br>Deep green in hot solution |
| $1,3,1',3'-(CH_3)_4$ | Yellow in cold solution<br>Olive green in hot solution |
| $2,3,2',3'-(C_6H_5)_2$ | Yellow in cold solution<br>Olive green in hot solution |
| $1,2,1',2'-(C_6H_5)_2$ | Yellow in cold solution<br>Deep green in hot solution |

[a] Ref. 22.

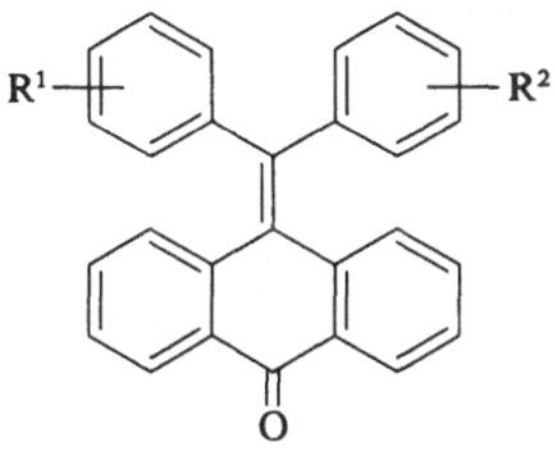

**5-21**

| $R^1, R^2$ | Triggered color and medium[a] |
|---|---|
| H, H | Pale yellow at −180°C |
| | Yellow at 25°C |
| | Ruby red at melting point |
| | Yellow in cold solution |
| 4-CH₃O | Yellowish green at −60°C |
| | Yellowish orange at 25°C |
| 4,4′-(CH₃O)₂ | Yellow at 25°C |
| | Red at melting point |
| | Reddish orange in hot solution |

[a] Refs. 38 and 39.

Other compounds of related sterically crowded structures have been reported to show thermochromic properties. The following are a few examples of their parent ring structures.

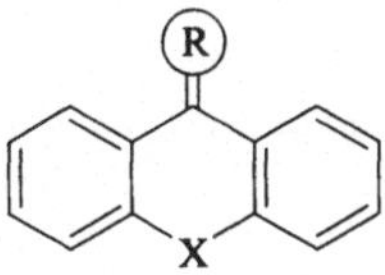

Diphenyl methylenexanthene (**5-22**)
R = (C₆H₅)₂C:, X = 0
Colorless at 25°C, reddish orange at melting point[40,41]

N-Methylacridan (**5-23**)
R = (C₆H₅)₂C:, X = NCH₃
Deep red at melting point[42]

10-(9-Flurylidene)thioxanthene (**5-24**)

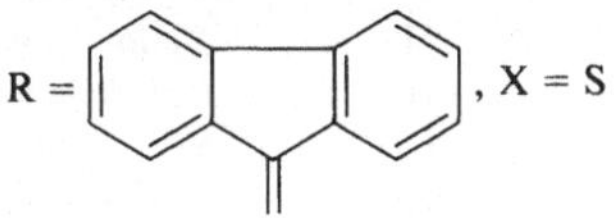

Yellow solid, green at melting point[43]

Dimethyldiacridine[44] (**5-25**)

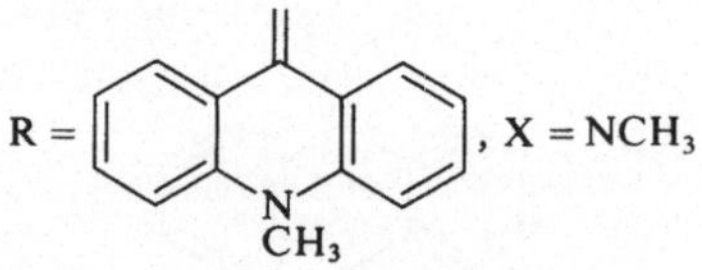

R = , X = NCH$_3$

2-(9-Xanthylidene)-indan-1,3-dione (**5-26**)

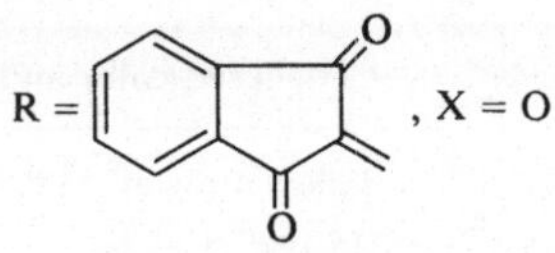

R = , X = O

Violet in cold solution, reddish brown in hot solution[45]

## 5.2.1.3. Thermochromic Disulfide Derivatives

A number of disulfides (**5-27**) have been reported[22] to be thermochromic. The following are a few examples.

$$R-S-S-R$$
**5-27**

Diphenyldisulfide[46]
R = C$_6$H$_5$; colorless solid, yellow at melting point

$\beta$-Dinaphthyldisulfide[47]
R = $\beta$-C$_{10}$H$_7$; colorless solid, yellow at melting point

9,9'-Dianthryldisulfide[47]
R = C$_{14}$H$_{10}$; orange solid, red at melting point

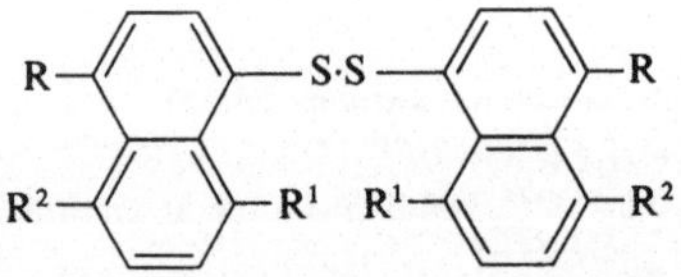

**5-28**

| R | R$^1$ | R$^2$ | Triggered color[a] |
|---|---|---|---|
| H | H | H | Colorless solid, yellow at m.p. |
| C$_2$H$_5$COO | H | H | Pale yellow solid, deep yellow at m.p. |
| H | OH | H | Yellow solid, orange at m.p. |
| H | C$_2$H$_5$COO | H | Colorless solid, yellow at m.p. |
| H | OH | CH$_3$ | Yellow solid, orange at m.p. |
| H | C$_6$H$_5$SO$_2$NH | H | Colorless solid, yellow at m.p. |

[a] Ref. 47.

## 5.2.1.4. Miscellaneous Thermochromic Compounds

A number of molecular structures have been reported to undergo thermally induced color change. These include carbocyclics, polycarbocyclics, and heterocyclics such as the following examples:

9,10-Di-*p*-anisylanthracene[48]
Colorless solid, red in toluene solution

Acecyclone[54] and tetracyclone[45]
Blue and violet in cold solutions, violet and reddish violet in hot solutions, respectively

4-Triphenyl-1,2-quinone[45]
Orange in cold solution, deep green and brown in hot solutions

1,1′-Methylenedi-2-hydroxyphenazine[49]
Yellow solid, orange at about 200°C

1,1′-Methylenedi-2-hydroxyacridine[49] (**5-29**)
Yellow in cold solution, red in hot solution

2-Hydroxy-1-naphthylidene compounds[50]

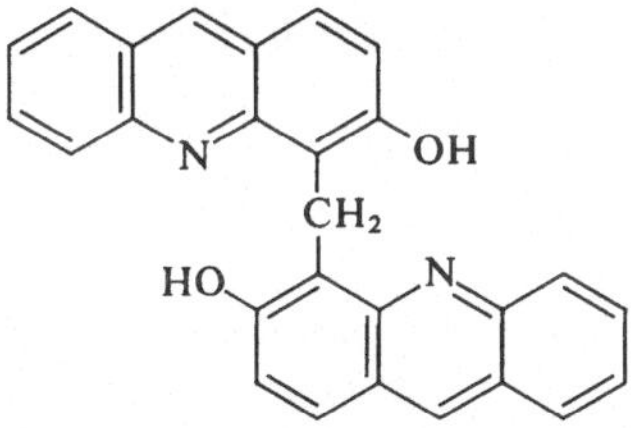

**5-29**

## 5.2.1.5. Isomerization about Aliphatic and Aromatic C,C-Double Bonds

Most studies of the *cis–trans* isomerization of aliphatic olefins involve activated double bonds such as those of carotenoids. These are colored pigments with a conjugated chain of eight C,C-double bonds. The *cis–trans* isomerization is accompanied by a change in the absorption spectrum of the compound. Three reviews[51-53] are cited in the literature covering the isomerization of carotenoids. Axeropthene[54] (**5-30**) for example was found to undergo thermally and photochemically triggered *cis* to *trans* isomerization.

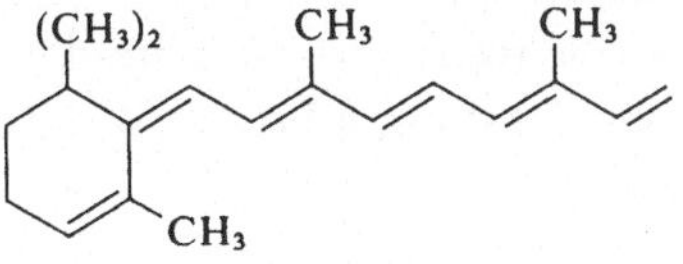

**5-30**

Other conjugated olefins which have also been reported to undergo similar *cis–trans* isomerization include the dimethyl ester of corticrocin (**5-31**), cinnamelazine (**5-32**), and phenyl pentadienealazine (**5-33**). All these *trans* isomers were observed to isomerize into *cis* forms on exposure to sunlight or an incandescent lamp. The *cis* isomers can be converted into the more stable *trans* isomers photochemically, thermally, or catalytically.

$$HOOCH:CHCH:CHCH:CHCOOH$$

**5-31**

$$C_6H_5CH:CHCH:N–N:CHCH:CHC_6H_5$$

**5-32**

$$C_6H_5CH:CHCH:CHCH:N–N:CHCH:CHCH:CHC_6H_5$$

**5-33**

Stilbene and its 4-methoxy, 4-nitro, 4-methoxy-4'-nitro, and 2-cyano derivatives also undergo *cis–trans* isomerization and this was examined by a number of kinetic methods.[55-57] Similarly, heating the *cis* isomer of benzylidene-3,3-dipenyl-1-hydroindone (**5-34**) to 160°C causes it to isomerize to the more stable *trans* isomer.[58]

**5-34**

Isomerization mechanisms for substituted or conjugated double bonds were reported to be unimolecular in nature and with an activation energy of about 24 kcal, a value very close to that for stilbene.[56]

### 5.2.1.6. Isomerization about C,N- and N,N-Double Bonds

A number of compounds with C,N- and N,N-double bonds are classified as dyes. Merocyanine dyes, which contain the C,N-double bond, show a thermally triggered *cis–trans* isomerization about their double bond, as observed by spectroscopic studies.[59] Azo compounds, which include structures with N,N-double bonds, display a facile *cis–trans* isomerization, with the *trans* isomer the more stable of the two. The activation energy for the isomerization of azobenzene was measured to be 23 kcal[60] as were energies of its 4-methoxy, 4-chloro, 4-bromo, 4-nitro, and 4-methyl derivatives.

A number of other compounds have been examined, for example, benzene diazocyanide (**5-35**) and its halo and nitro derivatives, which

$$C_6H_5N:NCN$$

**5-35**

showed activation energies between 21 and 26 kcal,[61] as well as azoxyben-zenes and $p,p'$-azoxytoluenes.[62]

### 5.2.1.7. Thermally Irreversible Color Change

The highly crystalline colored salts (**5-36**), formed from the reaction of two molecules of aromatic primary amines with a molecule of furfural-dehyde, were first isolated by Stenhouse.[63]

**5-36**

Extensions of this reaction were published in subsequent years by Schiff,[64] Zincke and Muhlhausen,[65] and Borsche *et al.*[66]

Schiff[64] reported no formation of colored salts when the Stenhouse reaction is conducted using the tertiary amines. Fischer[67a] and Renshaw and Naylor[67b] showed that dimethylaniline reacted with furfuraldehyde to form diphenylfurfurylmethane leuco which, upon oxidation, gave a colored compound analogous in structure to malachite green. Zincke and Muhlhausen[65] extended the Stenhouse reaction by reacting anilines with 1-(2,4-dinitrophenyl)pyridinium chloride, forming colored salts which he proposed to have an open-chain structure (**5-37**).

$$R = 2,4\text{-}(NO_2)_2C_6H_3$$                    **5-37**

His structural assignments were later favored by a number of workers[68] and accepted as common to Stenhouse type salts.

Since the first Stenhouse reaction, a number of reviews[69] have been published covering the formation and chemistry of the Stenhouse salts. These salts are sensitive to heat, which causes them to undergo a decoloriza-tion reaction.[70] Products of this reaction have been extensively examined[69b] and accepted to be the result of cyclization, where the color-forming conju-gation is interrupted. For example, salts formed from furfuraldehyde gave rise to 3-hydroxypyridinium salts (**5-40**)[70,71] and those of the type prepared by Zincke formed pyridinium salts (**5-39**).[72]

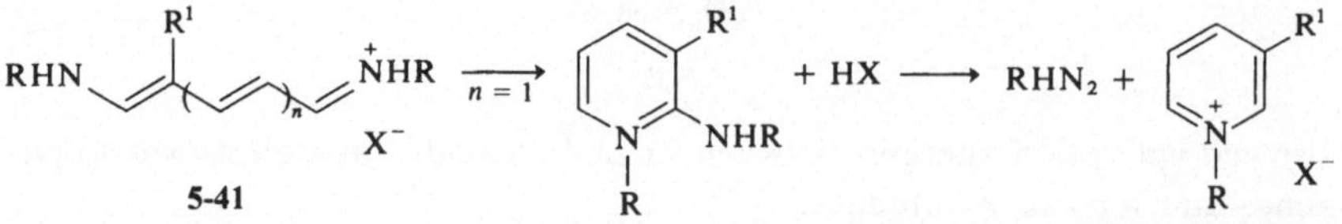

R¹ = H (5-39);
R¹ = OH (5-40)

The variation of the counterions of the Stenhouse salts and their effect on the heat-triggered cyclization reaction was examined by Chen *et al.*[73] Variation of substituents on the anilino ring has also been reported,[74,75] for example, in those anilines[75] with two nitrogen substituents and those with nonsymmetrical anilino portions of the molecule.[76]

Stenhouse type salts (5-41) with various lengths of the polymethine chain are also known for $n = 0$[74] and $n = 2$,[74,77] which extend their visible absorption characteristics from the blue into the green and red regions of the spectrum. Structures with short polymethine chains ($n = 0$) are less sensitive while those with longer chains ($n = 2$) are more sensitive to the thermally triggered decoloration. This can be rationalized as due to the ring strain which accompanies the cyclization of the shorter chain length salts. The following are some reported Stenhouse salts and their absorption wavelengths.

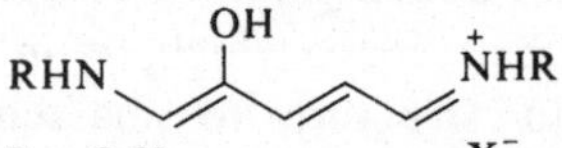

R = C₆H₅

| X⁻ | Visible absorption wavelength (nm) (extinction coefficient) |
|---|---|
| Trifluoroacetate | 518 (18,700) |
| Trichloroacetate | — |
| Oxalate | 518 (11,400) |
| Acetylenedicarboxylate | — |
| Maleate | 520 (11,400) |
| Chloroacetate | 520 (6,330) |
| Bromoacetate | — |
| Phthalate | 514 (4,300) |
| α-Bromopropionate | — |
| Salicylate | — |
| Tartarate | — |
| Fumarate | 520 (2,000) |
| 2-Furoate | 520 (2,300) |
| Formate | — |
| Glycolate | — |
| Benzenesulfonate | 520 (10,600) |
| Methanesulfonate | 520 (18,700) |
| Succinate | — |
| Benzoate | — |

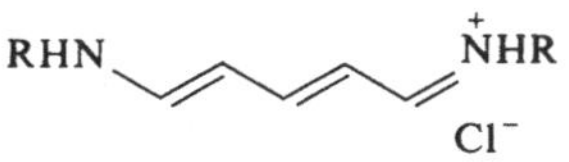

$$RHN \diagdown \diagup \diagdown \diagup \diagdown \overset{+}{N}HR$$

Cl⁻

**5-37**

| R | Visible absorption wavelength (nm)[a] (extinction coefficient) |
|---|---|
| 4-(CH₃)₂NC₆H₄ | 550 (9,300) |
| 4-CH₃OC₆H₄ | 489 (9,400) |
| 4-FC₆H₄ | 489 (10,000) |
| 4-ClC₆H₄ | 492 (11,200) |
| 4-NO₂C₆H₄ | 450 (5,400) |
| 3,4-(CH₃)₂C₆H₃ | 490 (8,400) |
| 3,5-(CH₃)₂C₆H₃ | 488 (6,400) |
| 3-ClC₆H₄ | 480 (2,900) |
| 4-BrC₆H₄ | 410 (3,500) |
| 3-BrC₆H₄ | 400 (1,900) |
| 4-HOOCC₆H₄ | 500 (1,200) |
| 4-CH₃SC₆H₄ | 510 (5,500) |
| 3,4-(CH₃O)₂C₆H₃ | 515 (5,500) |
| 3,5-(CH₃O)₂C₆H₃ | 498 (3,200) |
| 4-(CH₃)₃CC₆H₄ | 490 (7,800) |
| 4-CH₂(CH₂)₃C₆H₄ | 490 (10,000) |
| 4-C₆H₅OC₆H₄ | 495 (7,100) |
| 4-C₆H₅N:NC₆H₄ | 550 (9,200) |
| 4-C₆H₅NHC₆H₄ | 547 (9,200) |
| 4-(4'-CH₃)C₆H₄NHC₆H₄ | 557 (6,400) |

[a] Refs. 72 and 74.

## 5.2.2. Photochemically Triggered Release

Photochromism is a phenomenon involving triggering a change in the color of a chemical compound by the action of light. The process is reversible in the absence of light. The change can be between two colored states or a colored and a colorless state. The rate of change is a function of the chemical nature of the compound, the temperature, and the medium.

A variety of compounds undergo photochemically triggered changes in color. Extensive reviews on this subject have been published by Stobbe[78] in 1920, Chalkley[79] in 1929, Brown and Shaw[80] in 1961, Bhatnager et al.[81] in 1939, Van Overbeek[82] in 1939, Exelby and Grinter[83] in 1964, Dorion and Wiebe[84] in 1970, and Brown[85] in 1971.

### 5.2.2.1. Photochromic Anils or Schiff's Bases

Anils are anilino derivatives (**5-42**) of aldehydes. The majority of the photochromic examples (**5-43**) are derived from salicylaldehyde. Photo-

$R^1—C=N—R^2$

**5-42**

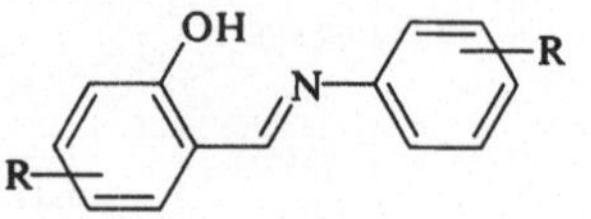

**5-43**

chromic behavior is observed only in the crystalline solid phase where colorless or slightly yellow crystals darken under the action of UV radiation to deeper yellow or brown shades. These revert to their original color thermally in the dark or by irradiation at their new wavelength of absorption.

It is concluded that the phenomenon is a function of the molecular arrangement in the crystal lattice of the solid phase, since different crystal forms of a given anil vary in their photosensitivity. Furthermore, ring substituents display no correlation with photosensitivity; an exception to this is the hydroxy group ortho to the aldehyde functional group.[86]

Photochromic Schiff's bases of this class vary greatly in their speed of decay after irradiation. Senier et al.[87] measured the decay time of a number of these compounds (**5-44**) during their study of the effect of temperature on the photochromic properties.

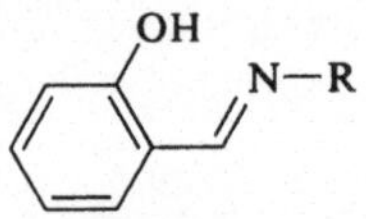

**5-44**

| R | Decay time (min)[a] |
|---|---|
| 4-HOOCC$_6$H$_4$ | 4.0 |
| 2-HOOCCH:CHC$_6$H$_4$ | — |
| 4-C$_2$H$_5$OOCC$_6$H$_4$ | — |
| C$_6$H$_5$ | 1.0 |
| 2-CH$_3$OC$_6$H$_4$ | 0.05 |
| 4-CH$_3$OC$_6$H$_4$ | — |
| 3-BrC$_6$H$_4$ | — |
| 2-BrC$_6$H$_4$ | 1.5 |
| 4-BrC$_6$H$_4$ | 1.3 |
| 2-ClC$_6$H$_4$ | 1.5 |
| 3-CH$_3$C$_6$H$_4$ | 80.0 |
| 3,4-(CH$_3$)$_2$C$_6$H$_3$ | 3.0 |
| 3-NH$_2$C$_6$H$_4$ | 0.017 |

[a] Refs. 84 and 86.

A number of references[89-91] cover these compounds and their properties. Senier and Forster[89] have also reported anils (**5-45**) with the hydroxy group on the aldehyde portion of the molecule in a para rather than an ortho position with respect to the aldehyde group. These still exhibit color changes in various crystalline and polymorphous forms.

**5-45**

| R | Color of crystalline state | Color of polymorphous state[a] |
|---|---|---|
| 4-CH$_3$C$_6$H$_4$ | Colorless | Greenish yellow |
| 3,4-(CH$_3$)$_2$C$_6$H$_3$ | Pale yellow | Deep yellow |
| 2,4-(CH$_3$)$_2$C$_6$H$_3$ | Straw yellow | Deep yellow |
| 3-BrC$_6$H$_4$ | Pale yellow | Canary yellow |
| 4-BrC$_6$H$_4$ | Pale yellow | Deep yellow |
| 2-HOOCC$_6$H$_4$ | Deep yellow | Brownish yellow |
| 4-HOOCC$_6$H$_4$ | Deep yellow | Brownish yellow |
| 2-CH$_3$OC$_6$H$_4$ | Off-white | Pale yellow |
| 3-CH$_3$OC$_6$H$_4$ | Off-white | Canary yellow |
| 4-CH$_3$OC$_6$H$_4$ | Bright yellow | Deep yellow |
| 1-C$_{10}$H$_7$ | Yellow | Deep yellow |
| 2-C$_{10}$H$_7$ | Deep yellow | Deep yellow |

[a] Ref. 89.

## 5.2.2.2. Photochromic Arylhydrazones

These compounds of the general formula (**5-46**) are formed through the condensation of arylhydrazines with aldehydes, ketones, or $\alpha$-diketones. The photochromic properties[92,93] are observed mainly in the solid phase and for the majority of the compounds, only if R$^2$ is a phenyl or naphthyl group.

$$R^1-CH:NNH-R^2$$

**5-46**

The mechanism of this photochromism is still unclear. A number of theories have been proposed but none have been fully proven. For example, a proposed mechanism involving isomerization with hydrogen transfer to the aldehyde carbon or the ring carbons was not corroborated by the nmr studies which followed. Gheorghiu and Arrventiu,[94] in 1931, postulated that the photochromic properties resulted from the photochemically excited $\pi$-electrons assuming a variety of positions in the molecule. Examples of hydrazones showing photochromic properties with a variety of substituents are compounds (**5-47**).

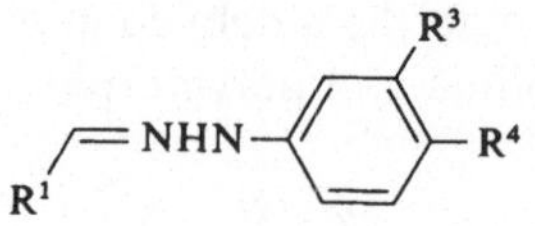

**5-47**

| $R^1$ | $R^3$ | $R^4$ | Reference |
|---|---|---|---|
| $2\text{-}NH_2C_6H_4$ | $-C_4H_4-$ | | 95 |
| $C_6H_5$ | $-C_4H_4-$ | | 96 |
| $C_6H_5$ | H | H | 97 |
| $C_6H_5CH{:}CH$ | H | H | 99 |
| $C_6H_5CH{:}CH$ | $-C_4H_4-$ | | 99 |
| $2\text{-}CH_3C_6H_4$ | $-C_4H_4-$ | | 98 |
| $3\text{-}CH_3C_6H_4$ | $-C_4H_4-$ | | 98 |
| $4\text{-}CH_3C_6H_4$ | $-C_4H_4-$ | | 98 |

### 5.2.2.3. Photochromic Osazones and Semicarbazones

Osazones of the general structure (**5-48**) are formed through the condensation of hydrazines with $\alpha$-diketones, where $R^1(R^2)$ and R can be alkyl groups and aryl groups, respectively. In general, the R-groups are phenyl or $\alpha$- or $\beta$-naphthyl derivatives. Few examples of this class of compounds are known in the literature.[99] However, all show chromism in the solid phase.

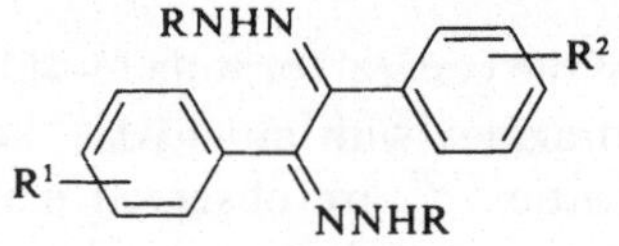

**5-48**

| $R^1$ | $R^2$ | R | Reference |
|---|---|---|---|
| H | H | $2\text{-}C_{10}H_7$ | 100 |
| $4\text{-}CH_3O$ | $4\text{-}CH_3O$ | $1\text{-}C_{10}H_7$ | 101 |
| $4\text{-}CH_3O$ | $4\text{-}CH_3$ | $2\text{-}C_{10}H_7$ | 104 |
| $4\text{-}(CH_3)_2CH$ | $4\text{-}(CH_3)_2CH$ | $C_6H_5$ | 99 |

Semicarbazones, characterized by the general formula (**5-49**), are formed through the condensation reactions of semicarbazide, thiosemicarbazide, and their aryl derivatives with ketones or aldehydes. $R^1$ and $R^2$ can be alkyl or aryl groups, preferably those substituted with electron-rich groups, which would extend conjugation and favor photochromism. For example, semicarbazones with $R^1$ as $CH_3$ and $R^2$ as $C(CH_3){:}CHC_6H_4[$ $p$-$(OCH_3)]$ show more photochromism than when $R^2$ is $C(CH_3){:}CHC_6H_5$.[102]

The following are examples of this family of compounds which generally displays photochromism in the solid phase. The color change is usually from colorless or pale yellow to a deep yellow.

$$R^1R^2C:NNHCXNHR^3$$

**5-49**

| R¹ | R² | R³ | X | Reference |
|---|---|---|---|---|
| $C_6H_5CH:CH$ | H | H | O | 103 |
| $3,4-(CH_3O)_2C_6H_3CH:CH$ | $CH_3$ | H | O | 104 |
| $C_6H_5CH:CH$ | H | H | S | 105 |
| $4-HOC_6H_4C(CH_3):CH$ | $CH_3$ | H | S | 106 |
| $2-HOC_6H_4CH:CH$ | $C_2H_5$ | H | O | 109 |
| $4-(CH_3)_2CHC_6H_4CH:CH$ | $CH_3(CH_2)_2$ | H | S | 109 |
| $4-(CH_3)_2CHC_6H_4CH:CH$ | $CH_3(CH_2)_2$ | $C_6H_5$ | O | 109 |
| $4-(CH_3)_2CHC_6H_4CH:CH$ | $C_2H_5$ | H | S | 109 |
| $4-(CH_3)_2CHC_6H_4CH:CH$ | $C_2H_5$ | $C_6H_5$ | O | 109 |
| $3-CH_3OC_6H_4CH:CH$ | H | H | O | 108 |
| $4-CH_3OC_6H_4CH:CH$ | H | H | O | 108 |
| $4-CH_3OC_6H_4CH:CH$ | H | $C_6H_5$ | O | 108 |
| $2-CH_3OC_6H_4CH:CH$ | H | H | O | 108 |
| $2-CH_3OC_6H_4CH:CH$ | H | H | S | 108 |
| $2-CH_3OC_6H_4CH:CH$ | H | $C_6H_5$ | O | 108 |

### 5.2.2.4. Photochromic Stilbene and Thiosulfonate Derivatives

Stilbene derivatives of the general structure (**5-50**) were discovered by Stobbe and Mallison[107] to be photochromic in the solid phase. Generally, the colorless or pale yellow crystals are photochemically converted into pink-colored crystals. Amino and sulfonic acid groups are necessary in all photochromic structures. The mechanism suggested by Stobbe involves an interaction with oxygen whereby the stilbene derivative takes up oxygen, forming a colored organic oxide. The following are examples reported to be photochromic.

**5-54**

$Y = H^+, Na^+, K^+, Ba^{2+}, Ca^{2+}, Mg^{2+}, Pb^{3+}$
$R = HCO, CH_3CO, 4-HCONHC_6H_4CO, 4-CH_3CONHC_6H_4CO,$
$\quad 4-(4'-CH_3CONHC_6H_4)CONHC_6H_4CO$

Photochromic thiosulfonates have the general structure (**5-51**). These compounds were found by Smiles *et al.*[108,109] to show photochromic properties in the solid phase and only when the crystals are solid solutions of the sulfonate with about 0.1% of the disulfide. The pure thiosulfonate is not photochromic nor are its solid physical mixtures with the disulfide.

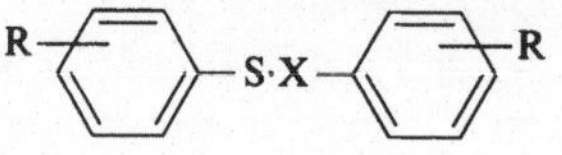

**5-51**   R = CH$_3$CONH, X = SO$_2$
           R = CH$_3$CONH, X = S

### 5.2.2.5. Fulgides and Azo Dye Derivatives

Fulgides are derivatives of succinic anhydride, acid, salts, or esters of the general formula (**5-52**). Most photochromic structures are anhydrides with R$^1$, R$^2$, R$^3$, and R$^4$ substituted phenyl rings. These undergo a color change from yellow or pale red to deeper shades under light exposure. Stobbe[110-112] and Hänel[113] have studied a variety of structures and have attributed the color change to an interconversion between stereoisomers.

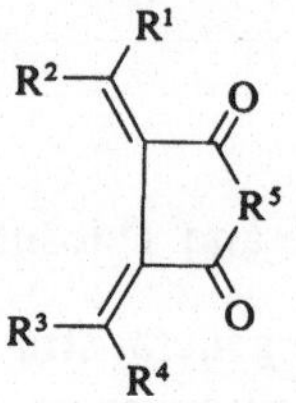

R$^5$ = O, (OH)$_2$, (OR)$_2$

**5-52**

| R$^5$ = O | R$^1$ | R$^2$ | R$^3$ | R$^4$ |
|---|---|---|---|---|
| 4-ClC$_6$H$_4$ | H | C$_6$H$_5$ | C$_6$H$_5$ |
| C$_6$H$_5$ | H | C$_6$H$_5$ | H |
| 2-CH$_3$OC$_6$H$_4$ | H | C$_6$H$_5$ | C$_6$H$_5$ |
| 2-NO$_2$C$_6$H$_4$ | H | C$_6$H$_5$ | C$_6$H$_5$ |
| 4-NO$_2$C$_6$H$_4$ | H | C$_6$H$_5$ | C$_6$H$_5$ |
| C$_6$H$_5$ | C$_6$H$_5$ | C$_6$H$_5$ | C$_6$H$_5$ |

Azo dyes of the general structure (**5-53**) have been studied[114-117] for their photochromism in solution. These photochromic properties are dependent on the structures of the dyes. In general, when the R-groups are electron-rich groups, such as hydroxyl or amino, the sensitivity to visible light increases with an increase in electron richness. For example, *p-*

aminoazobenzene shows stronger photochromism than $p$-hydroxyazoben-zene. The latter are stronger still than the naphthyl derivatives, which in turn are more sensitive than $o$-hydroxyazobenzene and its derivatives.

The mechanism of such a phenomenon was attributed to a $cis$-$trans$ equilibrium which is shifted to the $cis$ form by light exposure. Accordingly, the steric hindrance experienced by the ortho derivatives of azobenzene influences their photochromic sensitivity. Since the mechanism is a $cis$-$trans$ isomerization, the photochromic recovery rates are rapid.

Brode et al.[114] reported a number of azo dyes (5-53) and their absorption characteristics as a function of the ring substituents. The photochromism was found to be dependent on the type of solvent and medium tem-perature.[115] Their results showed an increase in the photochromic charac-teristics of a number of azo dyes with decrease in solvent polarity.

$$R^1 - \langle\text{ring}\rangle - N{:}N - \langle\text{ring}\rangle - R^2$$

**5-53**

| | | Visible absorption wavelength (nm) (extinction coefficient)[a] | |
|---|---|---|---|
| $R^1$ | $R^2$ | Before triggering | After triggering |
| 4-NH$_2$ | H | 377 (25,800) | 332 (9,200) |
| 4-(CH$_3$)$_2$N | H | 410 (28,300) | 362 (12,000) |
| 4-(CH$_3$)$_2$N | 4-(4'-NHCOC$_6$H$_4$)CH$_2$CH$_2$ | 410 (28,400) | 360 (10,700) |
| 4-(CH$_3$)$_2$N | 4-(4'-NH$_2$C$_6$H$_4$)CH$_2$ | 410 (30,000) | 360 (10,900) |
| 4-(CH$_3$)$_2$N | 4-C$_6$H$_5$ | 413 (35,000) | 362 (12,200) |
| 4-(CH$_3$)$_2$N | 4-(4'-NH$_2$)C$_6$H$_4$ | 426 (35,900) | 362 (12,200) |
| 4-(CH$_3$)$_2$N | 4-(4'-CH$_3$CONH)C$_6$H$_4$ | 425 (35,600) | 360 (11,900) |
| 4-OH | 4-CH$_3$ | 347 (25,600) | 304 (7,700) |
| 2-CH$_3$-4-OH | 2,4,6-(CH$_3$)$_3$ | 348 (20,000) | 305 (8,900) |
| 2,6-(CH$_3$)$_2$-4-OH | 2,4,6-(CH$_3$)$_3$ | 340 (18,100) | 301 (7,600) |
| 4-OH | 2,4-(Cl)$_2$ | 334 (16,800) | 311 (8,200) |

[a] In benzene solution.

## 5.2.2.6. Photochromic Indigo Dyes

Indigo dyes of the general formula (5-54) are found to undergo a reversible $cis$-$trans$ isomerization in solution[118,119] triggered by light. This isomerization involves a change in the light absorption characteristics and thus the color of the dye. The reversible isomerization is most pronounced, with a color change from blue to red, when X is a sulfur atom.

Brode and Wyman[118] have examined and reported absorption curves of ten purified thioindigo dyes with a variety of substituents. Their study centered on the direction of the $cis$-$trans$ equilibrium of (5-55) upon

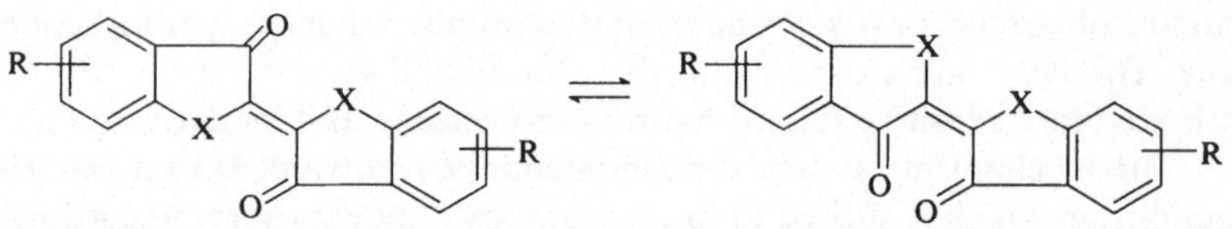

**5-54**

**5-55**   X = S;     **5-56,**   X = NH;     **5-57,**   X = O;     **5-58,**   X = Se

irradiation at different wavelengths. Some of their results are presented in Tables 5.1.

Wyman and Zenhausern[120] have examined spectroscopically the *cis–trans* isomerization of *N*-substituted indigo dyes (**5-56**). The substituents

*Table 5.1.* Visible Absorption Wavelength for the Photochemically Triggered *cis–trans* Isomerization of Thioindigo Dyes

| | Visible absorption wavelength (nm) | | Isomer distribution (%) | |
|---|---|---|---|---|
| R | Before triggering | After triggering | *trans* | *cis* |
| 5,5′,7,7′-(CH$_3$)$_4$ | 565 | — | 84 | 16 |
| | — | 575 | 29 | 71 |
| 4,4′,5,5′,7,7′-(Cl)$_6$ | 565 | — | 78 | 22 |
| | — | 575 | 45 | 55 |
| Vat scarlet G | 515 | — | 58 | 42 |
| | — | 520 | 28 | 72 |

*Table 5.2.* Quantum Yields for the Photochemically Triggered *cis–trans* Isomerization of Indigo Dyes

| X | Visible absorption wavelength (nm) (extinction coefficient) before triggering | Quantum yield (*trans* to *cis*) |
|---|---|---|
| Se (*trans*)[a] | 562 (14,000) | 0.025 |
| Se (*cis*) | 485 (9,000) | |
| O (*trans*)[b] | 413 (13,000) | 0.63 |
| O (*cis*) | 396 (13,100) | |
| S (*trans*)[c] | 542 (17,800) | 0.027 |
| S (*cis*) | 485 (23,000) | |
| NCH$_3$ (*trans*)[d] | 644 (23,000) | 0.004 |
| NCH$_3$ (*cis*) | 588 (8,300) | |

[a] Ref. 122.
[b] Ref. 121.
[c] Ref. 123.
[d] Ref. 124.

were designed to influence the isomerization in order to monitor the equilibrium spectroscopically.

Furthermore, Güsten[121] reported the *cis–trans* isomerization of the oxoindigo dye 2,2'-3-oxo-2,3-dihydrobenzo[*b*]furylidine (5-57). The UV irradiation (medium-pressure mercury immersion arc lamp, Osram 581) of the bright yellow *trans*-oxindigo (m.p. 276–278°C) for 3 hours gave the *cis*-oxindigo (m.p. 252–254°C), with absorptions at 413 nm ($\varepsilon = 1.38 \times 10^4$) and 396 nm ($\varepsilon = 1.31 \times 10^4$), respectively.

Selenoindigo dyes (5-58) have also been studied.[122] The quantum yields for the *cis–trans* isomerization were compared with those of other indigo dyes and constituted a basis for the measurement of the photochromism. Table 5.2 lists the *cis–trans* isomerization quantum yields of photochromic indigo dyes.

### 5.2.2.7. Photochromic Triphenylmethane and Camphor Derivatives

Triphenylmethane dye derivatives (5-59) undergo ionization triggered by light. Such a process manifests itself with a color change in solution. The photochromic properties are well apparent when X is a nitrile, hydroxy, or sulfite group.

$$(C_6H_5)_3C-X \underset{\text{Dark}}{\overset{\text{Light}}{\rightleftharpoons}} (C_6H_5)_3C^+ + X^-$$

**5-59**

Examples of the known nitrile derivatives include curamine, brilliant green, crystal violet, methyl violet,[125] malachite green, and pararosaniline in ethanol or ether[126] and benzaurine cyanide and phenolphthalein cyanide in water.[127]

Carbinol derivatives such as malachite green and crystal violet in alkaline ethanol,[126,128] benzaurine carbinol, and phenolphthalein carbinol[127] have been reported.

Sulfite derivatives, exemplified by crystal violet, malachite green, methyl violet, and pararosaniline[128,129] sulfites, are also known.

The mechanism of such solution photochromism involves dissociation of the molecule into a cation/anion pair. The increased conductivity of the solution containing these dyes with light exposure collaborates such a mechanism.

Harris *et al.*[130] examined the photolysis of malachite green leucocyanide and came to the same conclusion. Chalkley,[131] in a U.S. patent, described the synthesis of triarylmethane dyes by the photolysis of the nitrile derivatives.

Camphor derivatives of the general structure (**5-60**), with the camphor ring as part of the whole molecule, were shown to be photochromic in solution.[132]

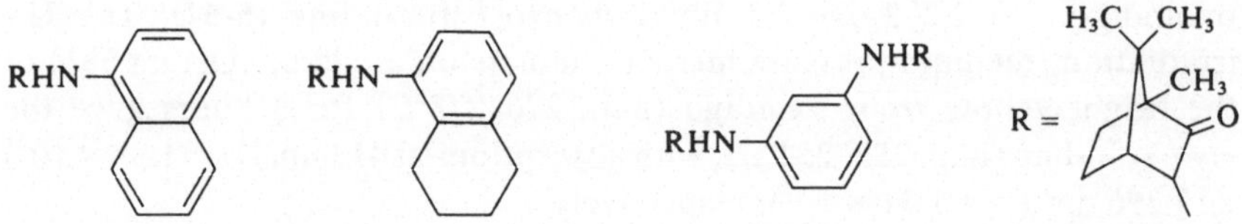

**5-60**

The color change involves shades of green from light to dark, in halogenated solvents (e.g., bromoform, chloral, or chloroform). The proposed mechanism involves the photochemical addition of the solvent to the camphor molecule, forming a triarylmethane type of dye.

Singh and Patt[133] have examined α-naphthylaminocamphor, which showed photochromism in a number of halogenated solvents. Their results added more support to a mechanism involving the interaction of the α-naphthylaminocamphor molecule with the halogenated solvent. They observed an increase in acidity of the irradiated solution containing the photochromic compound while acetylation of the amino group or the presence of a base (e.g., sodium ethoxide) quenched the color reaction.

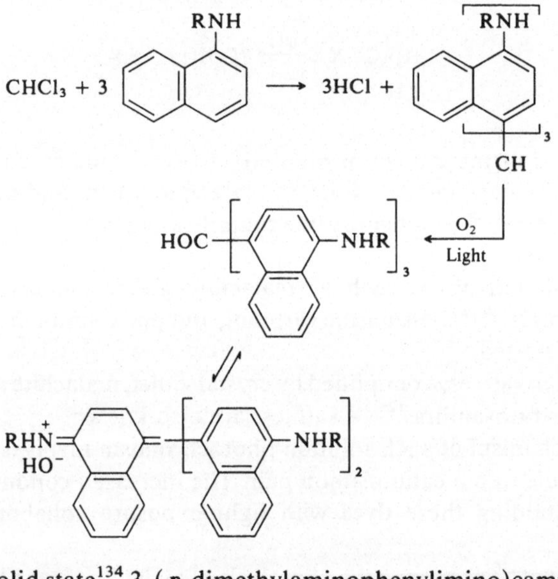

In the solid state[134] 3-(p-dimethylaminophenylimino)camphor and the diethyl analogue displayed photochromic properties. The color change was from yellow to a deep shade of yellow and scarlet. Such a solid-state photochromism involves the generation of radicals.

## 5.2.2.8. Photochromic *o*-Nitrobenzyl Derivatives

*o*-Nitrobenzyl derivatives (**5-61**) have been studied extensively for the past decade for their photochromic properties in the solid state[135-138] and in solution at low temperatures[139-141]

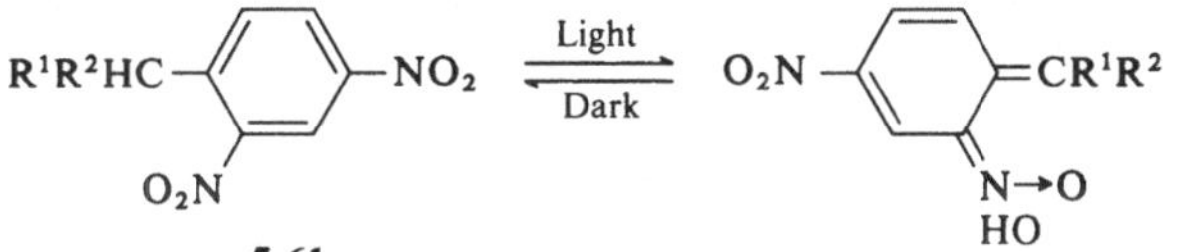

$R^1, R^2 = H, CH_3, C_6H_5$

The mechanism involves hydrogen transfer from the benzyl group to the ortho-nitro group. Such a process requires the presence of a nitro group in the position ortho to the benzylic hydrogens. The rate of photochromism is a function of temperature, structure, and solvent. The colors are mainly shades of purple, blue, and green. Several compounds have been examined and a few examples, together with the color of their irradiated forms, are given here.

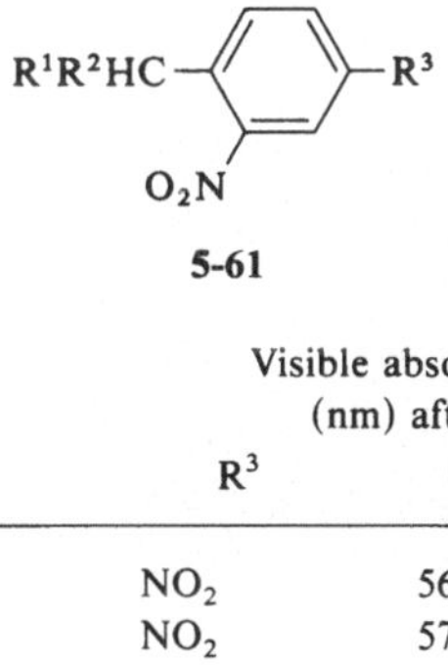

**5-61**

| $R^1$ | $R^2$ | $R^3$ | Visible absorption wavelength (nm) after triggering in ethanol | Color |
|---|---|---|---|---|
| | $-2\text{-}C_5H_4N-$ | $NO_2$ | 567 | Purple |
| | $-4\text{-}C_5H_4N-$ | $NO_2$ | 575 | Blue |
| | $-2\text{-}C_5H_4NO-$ | $NO_2$ | 501 | Pink |
| | $-4\text{-}C_5H_4NO-$ | $NO_2$ | 527 | Pink |
| H | $2\text{-}NO_2C_6H_4$ | $NO_2$ | — | Rose |
| H | $4\text{-}NO_2C_6H_4$ | $NO_2$ | 580 | Blue |
| H | $2,4\text{-}(NO_2)_2C_6H_3$ | $NO_2$ | 712 | Bluish green |
| $2,4\text{-}(NO_2)_2C_6H_3$ | $2,4\text{-}(NO_2)_2C_6H_3$ | $NO_2$ | 712 | Bluish green |
| H | $COONH_2$ | $NO_2$ | — | Green |
| H | $COOH$ | $NO_2$ | 425, 560 | Green |
| | $-2\text{-}C_5H_4N-$ | $COOCH_3$ | — | Bluish green |
| | $-2\text{-}C_5H_4N-$ | $COOC_2H_5$ | — | Bluish green |
| | $-2\text{-}C_5H_4N-$ | $CONH_2$ | 485, 560 | Green |
| | $-2\text{-}C_5H_4N-$ | $COOH$ | 405, 450, 470, 580 | Gray |

[a] Refs. 142 and 143.

## 5.2.2.9. Photochromic Spiro Compounds

In 1953 Chaudé and Rumpf[144] found that spirans, in solution, can exist in an equilibrium between two forms (**5-62**). The direction of the equilibrium can be altered by irradiation with light.

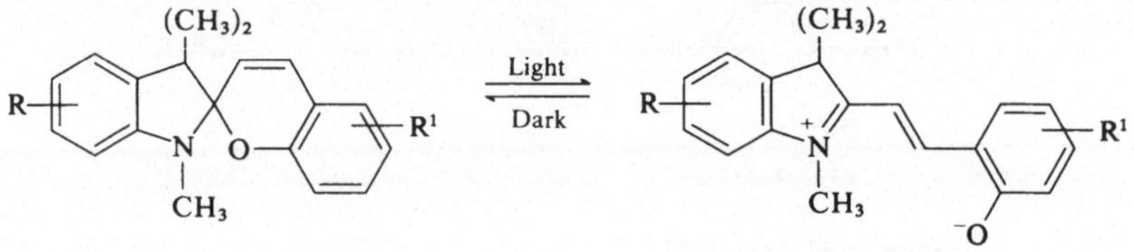

**5-62a**  Colorless isomer                          **5-62b**  Colored isomer

Hirshberg[145] has studied the photo- and thermochromic properties of spiro compounds. His observations have revealed that UV light triggers the formation of colored species. These species thermally revert to the thermo-dynamically more stable colorless form in the dark. At low temperatures the colored form is more stable but can be converted to the colorless form by irradiation with visible light of a specific wavelength. Furthermore, polar solvents are found to stabilize the colored form, which is additional proof of its polar structure.

The pH of solvents also affects the stability of the colored form. Hirshberg[145] has reported the effect of pH on the color of the open form. Spiropyrans (**5-63**), (**5-64**), and (**5-65**) showed a color change from yellow to red with a change in pH.

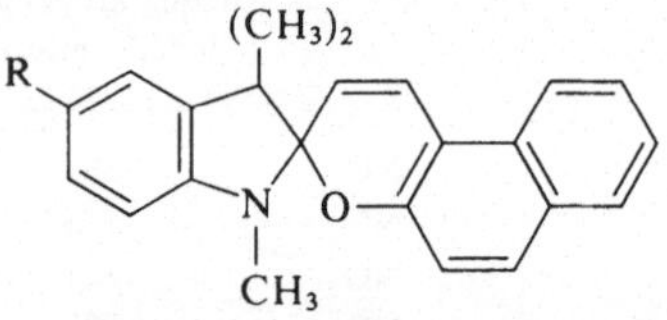

**5-63** R = H; **5-64** R = CH$_3$O

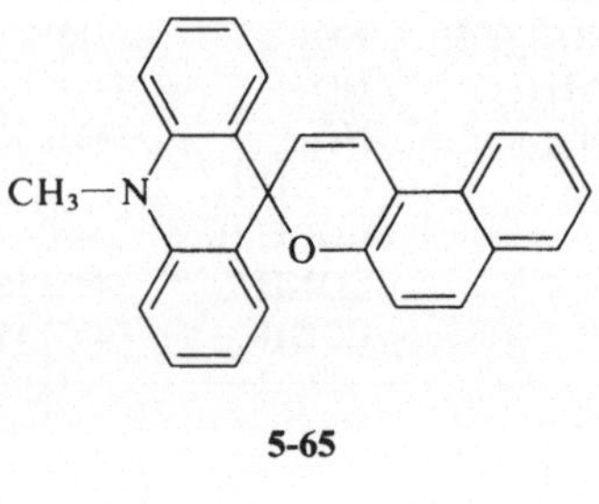

**5-65**

| Compound | Solvent | Visible absorption wavelength (nm) after triggering | Color |
|---|---|---|---|
| **5-63** | Ethanol | 570 | Reddish violet |
|  | +0.5 equiv. HCl | 570 | Rose |
|  | +1.0 equiv. HCl | 480 | Yellow |
|  | Ethanol + toluene | 570 | Reddish violet |
|  | +0.5 equiv. HCl | 490 | Orange |
|  | +1.0 equiv. HCl | 480 | Yellow |
|  | +10.0 pyridine | 570 | Reddish violet |
| **5-67** | Ethanol | 580 | Reddish violet |
|  | +0.1 equiv. HCl | 490, 580 | Rose |
|  | +0.2 equiv. HCl | 490, 580 | Orange |
|  | +0.3 equiv. HCl | 490 | Yellow |
| **5-65** | Ethanol | 660 | Bluish green |
|  | +1.5 equiv. HCl | 540 | Rose |
|  | +3.0 equiv. $(CH_3)_2N$ | 660 | Bluish green |

A number of other spiro compounds have been examined.[146] The following are examples of the parent structures, together with the absorption wavelengths of the colored species formed by irradiation in dimethyl phthalate solution.

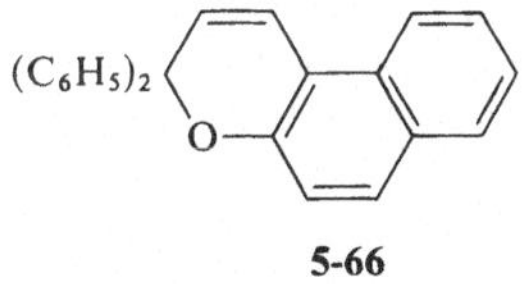

**5-66**

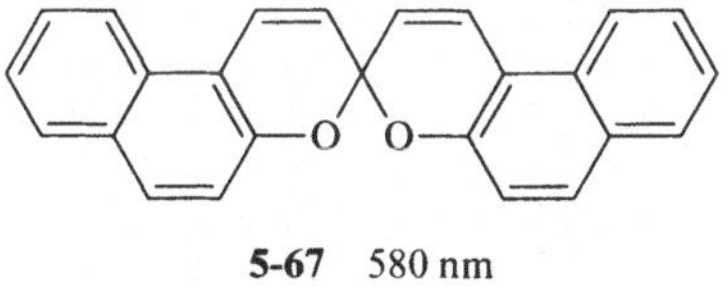

**5-67**   580 nm

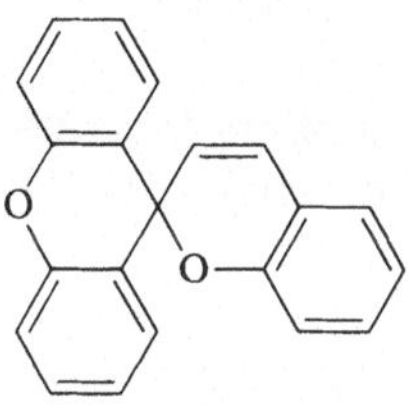

**5-68**

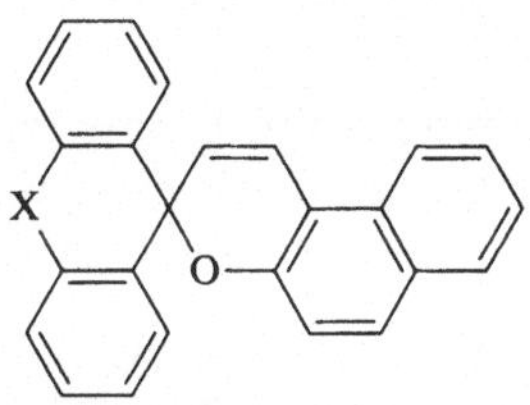

**5-69**   495 nm   X = O
**5-70**   590 nm   X = CH₃N

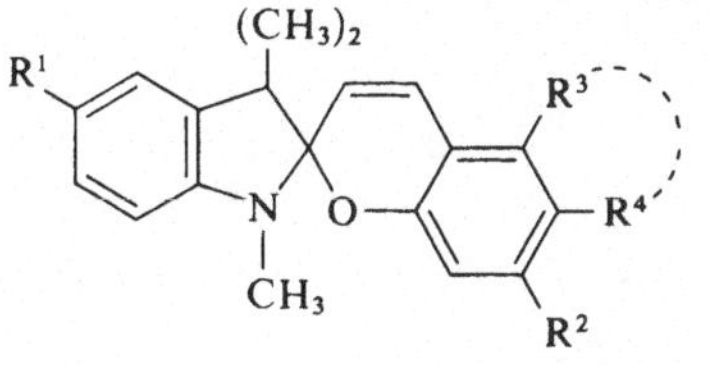

**5-71** $R^1 = R^2 = R^3 = R^4 = H$
**5-73** $R^1 = R^3 = R^4 = H, R^2 = CH_3O$
**5-75** $R^1 = R^2 = H, R^3—R^4 = C_4H_4$

**5-72** $R^1 = R^2 = CH_3O, R^3 = R^4 = H$
**5-74** $R^1 = CH_3O; R^2 = R^3 = R^4 = H$
**5-76** $R^1 = CH_3, R^2 = H, R^3—R^4 = C_4H_4$

*Table 5.3.* Rates of Conversion of **5-62b** into **5-62a** (R = H)

| $R^1$ | $k \times 10^{-5} (s^{-1})^a$ | Color of **5-62b** |
|---|---|---|
| 6'-NO$_2$ | 4.28 | Purple |
| 7'-NO$_2$ | 7.11 | Orange |
| 7'-Cl | 126.0 | Purple |
| 6',8'-Br$_2$ | 396.0 | Blue |
| 6'-NO$_2$-8'-CH$_2$CH:CH$_2$ | 21.0 | Purple |
| 6'-NO$_2$-8'-I | 0.633 | Purple |
| 6'-Cl-8'-NO$_2$ | 2.27 | Purple |
| 6'-Br-8'-NO$_2$ | 1.70 | Purple |
| 6'-CH$_3$O-8'-NO$_2$ | 1320.0 | Green |
| 6'-NO$_2$-8'-CH$_3$O | 55.3 | Blue |
| 5'-NO$_2$-8'-CH$_3$O | 25.2 | Green |
| 6'-NO$_2$-8'-Br | 0.367 | — |

$^a$ In ethanol solution at 6°C.

Berman *et al.*[147] have examined the effect of ring substituents on the thermal stability of the colored species (**5-62b**). They found that the thermal stability of the colored form can be increased by a suitable choice of ring substituents without adverse effect on the photochemical process. The thermal stabilities of derivatives of (**5-62**) as measured by the rate of conversion of (**5-62b**) into (**5-62a**) are given in Table 5.3.

### 5.2.2.10. Photochromic Anthrones and Sydnones

Anthrones (**5-77**) are polynuclear aromatic compounds with connecting double bonds.

In solution, and at low temperatures, the colorless form (**5-77a**) can be photochemically triggered to convert into the colored form (**5-77b**). The process is reversible under thermal conditions.[184,194]

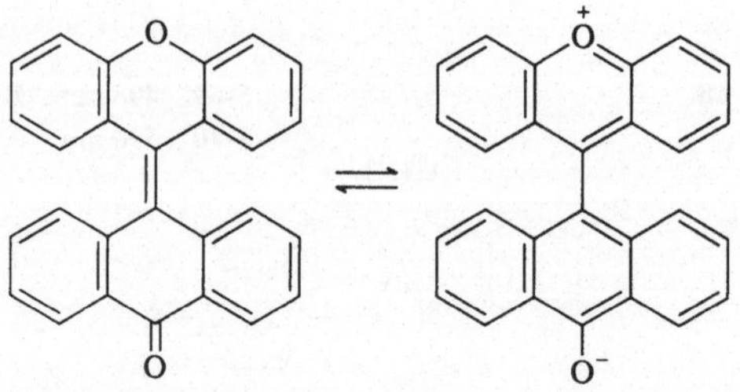

5-77a                                      5-77b

In general, the three related structures bianthrone (**5-78**), dixanthylene (**5-79**), and xanthylideneanthrone (**5-77**) show similar photochromic

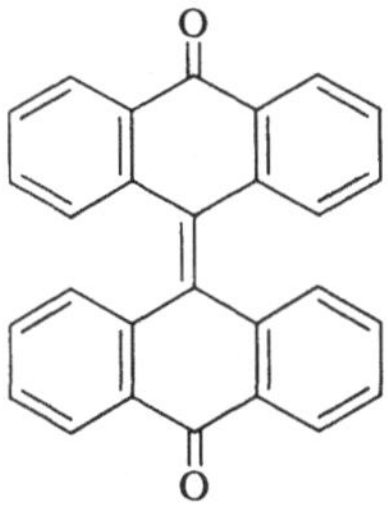

5-78 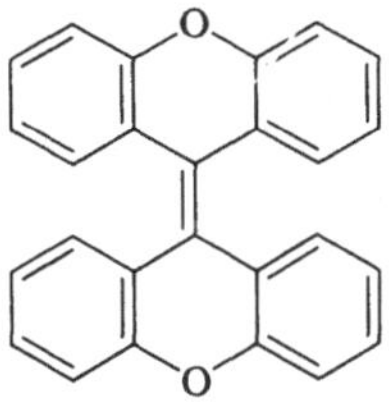                                  5-79

properties. Lewis and Lipkin[150] have found that the colored form of xanthyl-ideneanthrone (5-77b) has an increased dipole moment as compared to the colorless form (5-77a). This reinforced the assumption that the irradiated solution of this family of compounds contains an equilibrium shifted toward the structure (5-77b), while the colorless solution contains predominantly the reverse equilibrium. These compounds, because of their low-temperature photochromic properties, have been examined more for their thermochromic properties (see Section 5.2.1.2).

Sydnones, classified as dipolar compounds of the general structure (5-80), have been well known for their dipolar addition reactions, which were reviewed by Stewart[151] in 1964.

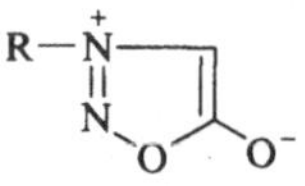

5-80

Their photochromism has been observed in the solid state, where the yellow-colorless crystals change to blue upon exposure to light. The most photochromic of these structures is $N$-(3-pyridyl)sydnone, which changes from yellow color to blue upon exposure to sunlight. The color change is reversible in the dark.[152,153]

Cohen and Schmidt[154] have suggested that the photochromism is due to dissociation of the molecule into a cation and anion of unspecified structure. Nespurek and Sorm[155] have reported that 4-isopropenyl-3-phenyl-sydnone as well as its dimer are also photochromic, while Greco and O'Reilly[156] prepared a range of 4-alkylene-3-phenylsydnones, three of which were photochromic.

## 5.2.2.11. Other Photochromic Compounds

Several other classes of compounds have been reported in the literature to be photochromic. These are less general and thus shall be surveyed in this section as a group rather than on an individual basis.

Derivatives of benzylamine,[157] for example *N*-(5-bromosalicyl-idene)benzylamine and *N*-(5-methylsalicylidene)benzylamine, have shown color changes from orange to red under sunlight.

Naphthalene derivatives[158,159] were also reported to show photochromic properties. For example, tetrachloro-1(2 or 4)-naphthalenone changes from colorless to amethyst, α-azoxynaphthalene[160] to deeper shades of yellow, and substituted 2-benzylidene-3-oxo-2,3-dihydrothionaphthalene[161] from colorless to red.

Benzoin derivatives,[162] for example, *o*- and *p*-nitrobenzylidenedesoxy-benzoin aminobenzene derivatives[163] and phenazine derivatives,[164] such as dimethylamino-2-methylphenazine and *o*-nitrobenzylideneisonicotinic acid hydrazide,[165] have shown a change from yellow to red.

Brucine salts[166] of bromo- and chloronitromethiomic acid have shown a color change from colorless to violet. Tetraphenyl derivatives of pyridine, triazine, pyran, and ethylene have also been reported to show photochromic properties. This group includes 2,4,4,6-tetraphenyl-1,4-dihydropyridine,[167] 2,4,4,6-tetraphenyl-3,5-dibenzoyltetrahydropyran,[167] tetrabenzoylethyl-ene,[168] and tetraphenyldihydrotriazine.[169]

Pyrrole derivatives[170] were also shown to be both photo- and thermo-chromic in the solid state. For example, 2-dichloromethyl-3,5-bisethoxycar-bonyl-4-methylpyrrole upon heating at 160°C converts from colorless to red. This color change is photochemically reversible.

Aromatic nitro compounds,[171] for example, phenylnitromethane and anhydrous quinoquinoline salts,[172] were also reported to show photo-chromic properties.

## 5.2.2.12. Photo-isomerization about Aromatic C,C-Double Bonds

The most studied *cis–trans* isomerizations of aromatic olefins fall within the stilbene family of compounds. An early review was published by Zechmeister.[173] The first report of *cis–trans* isomerization of stilbene (5-82) was by Smakula.[174] Irradiation of the *trans* isomer with light at 313 nm triggered complete isomerization into the *cis* form with a quantum efficiency between 0.7 and 1.0.

5-82                         5-81

Such isomerization was found to be general for other derivatives of stilbene such as *p*-nitro-, *p*-amino-*p'*-nitro-, and *p*-methoxy-*p'*-nitrostil-bene[175] as well as *sym*-di(9-phenanthryl)ethylene[176] (5-83) and 9,9'-phenan-throin[177] (5-84).

$$R^1R^2C:CR^1R^2$$

**5-83**  $R^1 =$ 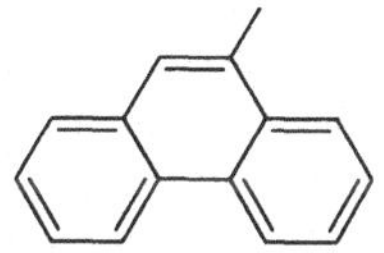 ,  $R^2 = H$

**5-84**  $R^1 =$ 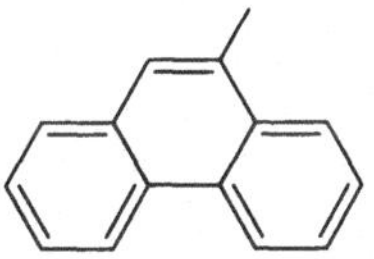 ,  $R^2 = OH$

Butadiene mono- and diphenyl derivatives, for example, 1-phenyl-1,3-butadiene[172] and 1,4-diphenylbutadiene,[179] have been reported[178] to isomerize readily under the action of sunlight. Similar *cis–trans* isomerization was also observed for ring-substituted derivatives of 1,4-diphenylbutadiene,[180] 1,6-diphenylhexatrienes,[181] and 1,8-diphenyloctatrienes.[182,183] $\alpha,\omega$-Diarylpolyenes containing seven, nine, and eleven conjugated double bonds showed similar photochemically triggered isomerizations.[184]

Photochemically induced *cis–trans* isomerizations of unsaturated aromatic carbonyl compounds, for example, benzalacetone, ethylstyryl ketone,[185] and cinnamonitrile,[186] have also been examined. Cinnamic acid and its derivatives[187,188] were found to undergo a photochemical *cis–trans* isomerization when irradiated at 313 nm and 253 nm. The direction of the attained equilibrium is a function of the irradiation and the specific cinnamic acid derivative.

Dyes which contain a double bond and a carbonyl group in their structure also undergo analogous isomerization. Examples are indigo and azo dyes (also see Section 5.2.2.6). Photochemically triggered *cis–trans* isomerizations of thioindigo dyes have been the most frequently studied.[189,190]

In solution, these dyes attain an equilibrium whose direction is dependent on solvent, temperature, and irradiation characteristics. Pechmann's dye (**5-85**), which shares common structural features with indigo dyes, also showed an analogous isomerization equilibrium in solution.[191]

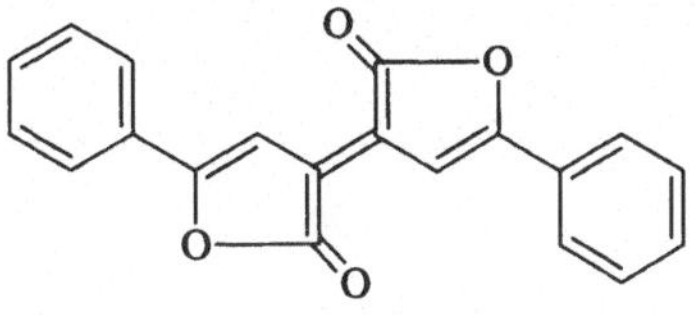

**5-85**

Cyanine dyes are also reported in the literature to show *cis–trans* isomerization.[192] Chromatographic analysis of the dye (**5-86**) showed three zones. When the zones were separated, their photochemical spectra were identical, due to a fast equilibrium attained by each isomer during the isolation. Other similar dyes include genacryl yellow (**5-87**), quinaldine red (**5-88**), and rhodamine derivatives (**5-89**).

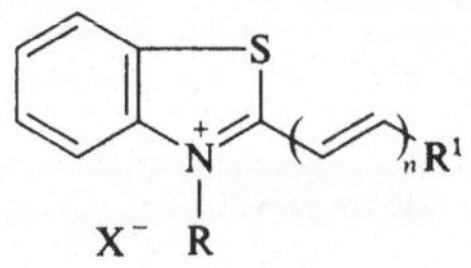

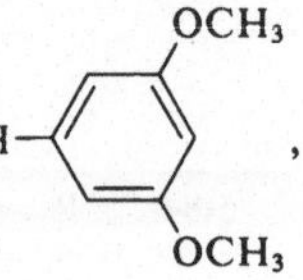

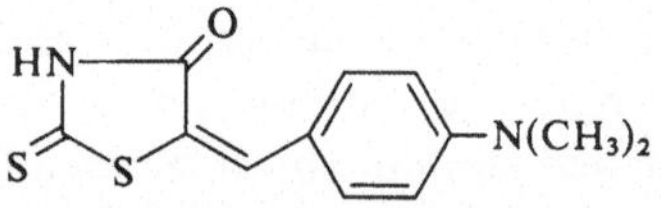

Analogous *cis–trans* isomerizations of merocyanine dyes were also observed during the photochemical irradiation of spiropyrans.[193-195]

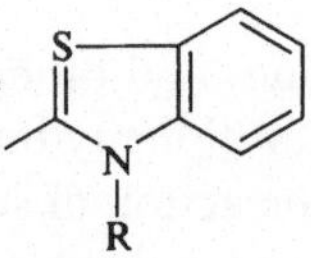

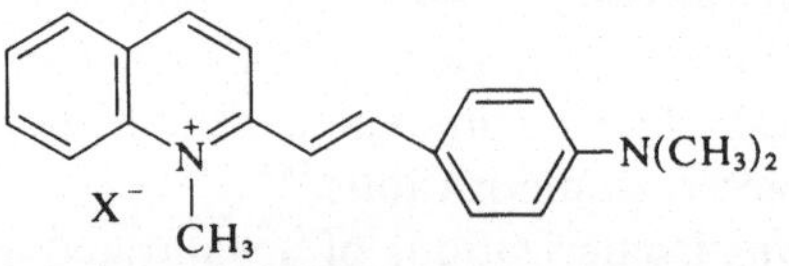

### 5.2.2.13. Photo-isomerization about C,N- and N,N-Double Bonds

*Cis–trans* isomerizations about C,N-double bond have been studied relatively little. Although compounds with C=N bonds in their structure are known to exist in both *cis* and *trans* forms, their equilibrium cannot be easily detected by dipole moment measurements or spectroscopic methods. Exceptions are the Schiff's bases (**5-90**) where one isomer can be stabilized by hydrogen bonding relative to the other isomer, and thus the equilibrium can be studied spectrophotometrically.[196]

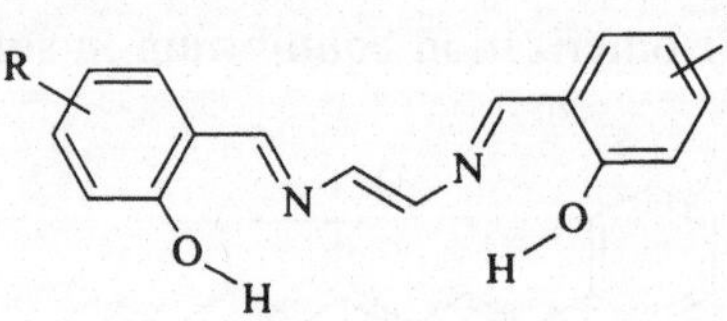

**5-90**

Compounds with an azo linkage have been more frequently studied than those with a C=N linkage. The invention[197] of a recording spectrophotometer for the ultraviolet and visible regions which utilizes a light beam of low intensity together with the invention of the rotating-shutter attachment for such a spectrophotometer allowed the periodic irradiation of an azo compound as well as the measurement of its corresponding spectrum. Such developments have permitted the study of the photochemical *cis-trans* isomerization of a number of aminoazo and hydroxyazo compounds.[198] These compounds were observed to undergo a drastic change during the photochemically triggered *cis-trans* isomerization. Such isomerization, which is reversible upon the removal of the triggering agent, is the basis of their photochromic properties. Hydrogen bonding between ortho hydroxyl groups and the azo group tends to stabilize the *trans* isomer.

### 5.2.2.14. Photochemically Irreversible Color Change

Photochemically triggered irreversible color changes are known in the chemical literature but to a lesser extent than the reversible changes. One example of such reactions is the polymerization of acetylene derivatives. Compounds possessing two or more conjugated triple bonds (**5-91**) have been reported to undergo irreversible polymerization under the action of UV light and in some cases with heat.[199,200,201] Such polymerization reactions yield conjugated polymers (**5-92**), which are also discussed in Chapter 4.

$$R^1-C{\equiv}C-C{\equiv}C-R^2 \rightarrow$$

**5-91**

**5-92**

| $R^1$ | $R^2$ | Absorption wavelength (nm) of the diacetylene monomer $(\varepsilon)^a$ |
|---|---|---|
| $(CH_3)_2CH$ | $(CH_3)_2CH$ | 229 (210), 241 (300), 256 (180) |
| $HOCH_2CH_2$ | $HOCH_2CH_2$ | 228 (540), 239 (540), 251 (360) |
| $CH_3CH(OH)CH_2$ | $CH_3CH(OH)CH_2$ | 226 (355), 241 (410), 264 (250) |
| $CH_3CH(OH)$ | $CH_3CH(OH)$ | 230 (1,930), 237 (1,800) |
| $C_3H_7CH(OH)$ | $C_3H_7CH(OH)$ | 232 (360), 242 (400), 255 (265) |
| 1-Cyclohexan-1-ol | 1-Cyclohexan-1-ol | 231 (293), 243 (347), 244 (332) |
| $C_2H_5C(CH_3)OH$ | $C_2H_5C(CH_3)OH$ | 230 (330), 242 (368), 255 (204) |
| $C_6H_5CH(OH)$ | $C_6H_5CH(OH)$ | 246 (1,159), 260 (785) |

$^a$ Ref. 202.

One characteristic of such acetylenic monomers is their low extinction coefficient as compared to conjugated dienes.[202] Their UV-triggered polymerization proceeds through a 1,4-addition reaction, as shown by a number of studies[210a] including an investigation by Raman spectroscopy.[203]

The solid-state reaction is a lattice-controlled polymerization where the monomer diacetylene molecule, aligned along its axis in a ladderlike fashion, adds to its neighboring molecule. The progress of such polymerizations can be monitored by visible spectroscopy, as a result of the color change which accompanies the polymerization reaction.

These highly conjugated polymers display colors varying from pale reds to deep blues and metallic gold, depending on the degree of polymerization and the substituents at the triple bonds. Such colors develop with the progress of the reaction from colorless monomers and are permanent.

A number of monomers of the general structure (5-93) have been prepared and polymerized into polymers of a variety of colors. The reaction is triggered by UV exposure of a dispersion of the monomer in a propanol-water medium. This reaction has been reviewed by Wagner[201a,204] and Melveger and Baughman.[203,205]

$$\text{R} \; \diagdown\!\!\!\!\!\bigcirc\!\!-C-C\equiv C-C-\bigcirc\!\!\!\!\!\diagup \; \text{R}$$

**5-93**

| R | Color of polymer | Percent yield of polymer[a] |
|---|---|---|
| $2\text{-}NO_2$ | Red | 2.0 |
| $3\text{-}NO_2$ | Red | 3.5 |
| $2\text{-}CH_3CONH$ | Red-violet | 19 |
| $3\text{-}CH_3CONH$ | Greenish blue | 25 |
| $2\text{-}C_6H_5CONH$ | Pale purple | 0.2 |
| $2\text{-}C_6H_5NHCONH$ | Violet | 3.0 |
| $3\text{-}C_6H_5NHCONH$ | Greenish blue | 15 |

[a] Ref. 204.

Certain dyes have been reported[206] to undergo irreversible photochemically triggered color change, e.g., styryl dyes of the general structure (5-94). The reported photochemical reaction took place in a polymeric film under a 150-W tungsten light exposure for about 15 seconds. A drop of 3–4 units in the dye density is generally observed with the light exposure.

These dyes can be prepared[206] under alkaline conditions through a nucleophilic addition of the heterocyclic moiety (5-99) onto a nitroaryl ring containing a good leaving group, such as a halogen.

Heterocyclic cationic dyes are also known to undergo photochemically triggered discoloration. Dyes with chromylium, pyrylium, and flavylium groups and their sulfur analogues have been reported[201] to undergo photochemical bleach reactions when exposed to light in presence of a sensitizer

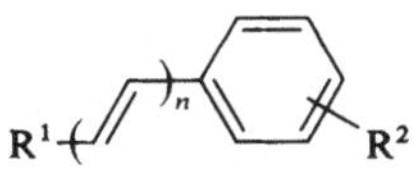

**5-94**

$R^1$ = heterocyclic group, $R^2$ = nitro group(s)

| $R^1$ | X | $n$ | $R^2$ | Absorption wavelength (nm) |
|---|---|---|---|---|
| 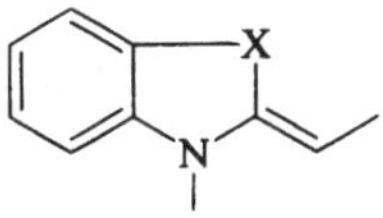 **5-95** | S | 0 | 2,4-$(NO_2)_2$ | 530 |
|  | S | 0 | 2,4,6-$(NO_2)_3$ | 565 |
|  | S | 1 | 2,4-$(NO_2)_2$ | 613 |
|  | Se | 0 | 2,4,6-$(NO_2)_3$ | 575 |
| 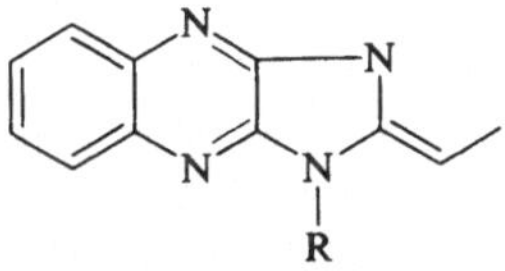 **5-96** | | 1 | 2,4-$(NO_2)_2$ | |
|  | R = allyl | | | 520 |
|  | R = ethyl | | | 530 |
| 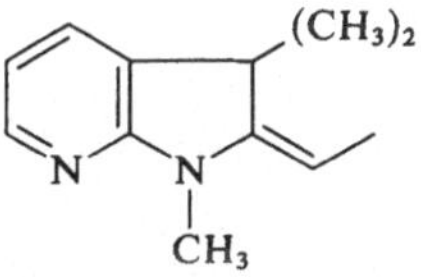 **5-97** | | 1 | 2,4-$(NO_2)_2$ | 480 |
| 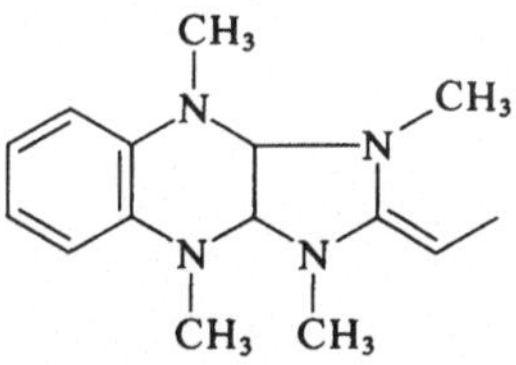 **5-98** | | 1 | 2,4-$(NO_2)_2$ | 530 |

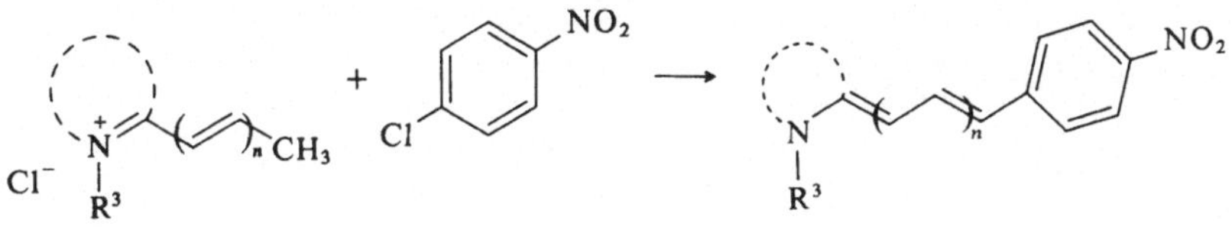

**5-99**

such as allylthiourea. Such dyes can be considered members of the cyanine family, with absorption wavelengths varying from 450 to 700 nm (yellow to green colors). The bleach reaction was observed in a polymeric film under a 500-W tungsten lamp.

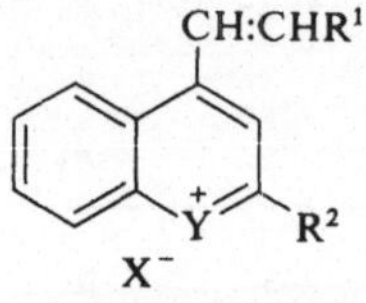

**5-100**

| Y | R$^1$ | R$^2$ | Absorption wavelength (nm) |
|---|---|---|---|
| S | C$_6$H$_5$ | C$_6$H$_5$ | 460 (yellow) |
| S | 4-CH$_3$OC$_6$H$_4$ | C$_4$H$_5$ | 540 (magenta) |
| O | C$_6$H$_4$ | C$_6$H$_4$ | 465 (yellow) |
| O | Thiophene | C$_6$H$_5$ | — (orange) |

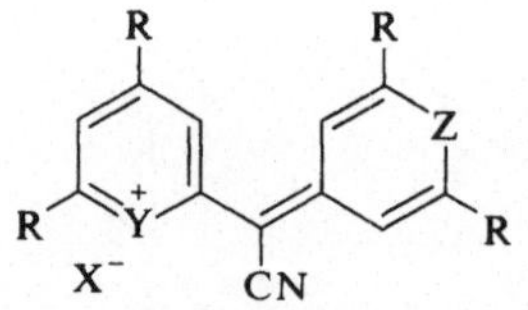

$R = C_6H_5$

**5-101**

| Y | Z | Absorption wavelength (nm) |
|---|---|---|
| O | S | 632 (cyan) |
| O | O | 565 (magenta) |

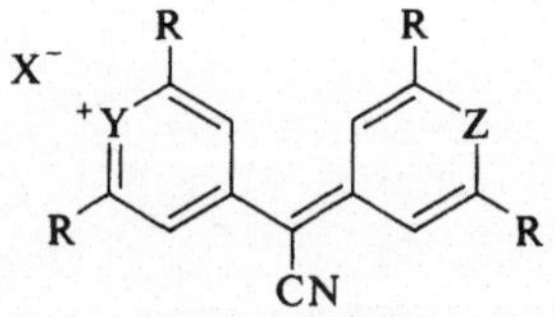

**5-102**   $R = C_6H_5$

| Y | Z | Absorption wavelength (nm) |
|---|---|---|
| S | S | 610 (greenish blue) |
| S | O | 550 (magenta) |
| O | O | 540 (magenta) |
| O | S | 575 (violet) |

The sensitizers reported useful in such a reaction include compounds which contain thiocarbonyl, mercaptan, carbonyl peroxide, or thioether groups, for example, 1-allyl-2-thiourea, $S$-diethylthiourea, and 3-allyl-1,1-diethyl-2-thiourea. Typical structures of these heterocyclic dyes are exemplified by (5-100), (5-101) and (5-102).

Formation of a dye by photolysis of a single compound has also been reported. Engler and Dorant[208] reported that 2-nitrochalcone (5-103), in the solid state and under UV exposure, undergoes a photo-induced rearrangement to an indigo dye structure (5-104).

5-103                                    5-104

A conceivable mechanism, suggested by Luwich[209] in 1968, involves a nucleophilic attack by the nitro group onto the double bond of the unsaturated olefin, followed by cyclization and coupling. The reaction provides a route to the formation of blue-colored compounds from chalcones which are commonly pale yellow. Substituents on the phenyl ring of the chalcone can provide a tool to control the reaction rate and the blue color of the dye formed.

## REFERENCES

1. (a) NCR Corp., British Pat. 1,109,554 (1968); (b) W. J. Becker and P. L. Foris, U.S. Pat. 3,364,023 (1968); (c) NCR Corp., British Pat. 1,073,999 (1976); (d) W. J. Becker and P. L. Foris, U.S. Pat. 3,359,103 (1967).
2. E. Berman and H. Schwab, U.S. Pat. 3,072,481 (1963).
3. (a) L. Hanel, U.S. Pat. 2,305,693 (1942); (b) O. Vierling, U.S. Pat. 2,305,799 (1942).
4. P. Waterworth and D. C. J. Reid, British Pat. 2,008,790A (1977).
5. D. J. Fry and B. R. D. Whitear, U.S. Pat. 3,589,898 (1971).
6. M. S. Agruss, U.S. Pat. 3,226,233 (1965).
7. R. Owen, U.S. Pat. 3,108,896 (1963).
8. W. Frankenburger and W. Zimmermann, U.S. Pat. 1,845,835 (1932).
9. L. Chalkley, U.S. Pat. 3,710,109 (1973).
10. J. G. Calvert and H. J. L. Rechen, *J. Am. Chem. Soc.* 74, 2101 (1952).
11. (a) L. Le Naour-Sene, U.S. Pat. 4,286,957 (1981); (b) G. S. Hammond, L. M. Stephenson and G. F. Vesley, *J. Chem. Educ.* 50, 529 (1973).
12. J. S. Splitter and M. Calvin, *J. Org. Chem.* 23, 651 (1958).
13. (a) V. M. Clark and A. Todd, *J. Org. Chem.* 24, 2012 (1959); (b) J. S. Splitter and M. Calvin, *J. Org. Chem.* 30, 3427 (1965); (c) W. D. Emmons, *J. Am. Chem. Soc.* 79, 5739 (1957).

14. J. Hammer and A. Macaluso, *Chem. Rev. 64*, 472 (1964).

15. G. R. Delpierre and M. Lamchen, *Quart. Rev.* (*London*) *19*, 329 (1965).

16. P. B. Gilman and R. G. Raleigh, U.S. Pat. 3,627,527 (1971).

17. (a) G. J. Sabongi, B. A. Lea, and S. S. C. Poon, European Pat. Appl. 0119831 (1984);
    (b) P. W. Jenkins, D. Heseltine, and J. D. Mee, U.S. Pat. 3,745,009 (1972).

18. W. L. Dilling, *Chem. Rev. 83*, 1 (1983).

19. W. Neugebauer and M. Tomanek, U.S. Pat. 3,070,443 (1962).

20. R. E. Buckles, *J. Am. Chem. Soc. 77*, 1040 (1951).

21. W. M. Moore, D. D. Morgan, and F. R. Stemitz, *J. Am. Chem. Soc. 85*, 829 (1963).

22. J. H. Day, *Chem. Rev. 63*, 65 (1963).

23. J. H. Day, *Chem. Rev. 68*, 649 (1968).

24. W. Theilacker, G. Kortum, and G. Friedheim, *Chem. Ber. 83*, 508 (1950).

25. A. Mustafa, *Chem. Rev. 43*, 509 (1948).

26. B. Pullman and A. Pullman, *Les Théories Electroniques de la Chimie Organique*, Masson et Cie, Paris (1952).

27. K. G. Lewis and C. E. Mulquiney, *Tetrahedron 33*, 463 (1977).

28. E. B. Knott, *J. Chem. Soc.*, 3038 (1951).

29. W. Dilthey and H. Wubken, *Ber. 61*, 963 (1928).

30. E. Bergman, R. E. Fox, and R. D. Thomson, *J. Am. Chem. Soc. 81*, 5605 (1959).

31. Y. Hirshberg, *J. Am. Chem. Soc. 78*, 2304 (1956).

32. R. Rim-Heiligman, Y. Hirshberg, and E. Fischer, *J. Chem. Soc.*, 156 (1961).

33. E. B. Knott, *J. Chem. Soc.*, 3038 (1951).

34. R. Wizing and H. Wenning, *Helv. Chim. Acta 23*, 247 (1940).

35. I. M. Heilborn, D. H. Hey, and A. Lowe, *J. Chem. Soc.*, 1330 (1936).

36. E. D. Bergmann, in: *Progress in Organic Chemistry* (J. W. Cook, ed.), Vol. 3, pp. 152-161, Butterworths Scientific Pub., London (1955).

37. A. Mustafa and M. E. Sobhy, *J. Am. Chem. Soc. 77*, 5124 (1955).

38. A. Schonberg, A. F. A. Ismail, and W. Asker, *J. Chem. Soc.*, 442 (1946).

39. Y. Hirshberg, E. Loewenthal, E. D. Bergmann, and B. Pullman, *Bull. Soc. Chim. Fr. 18*, 88 (1951).

40. A. Schonberg, A. E. K. Fateen, and A. E. M. A. Semour, *J. Am. Chem. Soc. 79*, 6020 (1957).

41. A. Schonberg and S. Nickel, *Ber. 64*, 2323 (1931).

42. E. D. Bergmann and H. Corte, *Ber. 66*, 39 (1933).

43. A. Schonberg and M. M. Sidky, *J. Am. Chem. Soc. 81*, 2259 (1959).

44. W. Theilacker, G. Kortum, and G. Friedheim, *Chem. Ber. 83*, 508 (1950).

45. A. Schonberg, A. Mustafa, and W. Asker, *J. Am. Chem. Soc. 76*, 4134 (1954).

46. H. P. Koch, *J. Chem. Soc.*, 387-394 (1949).

47. A. Mustafa and M. Kamel, *Science 118*, 411 (1953).

48. C. K. Ingold and P. G. Marshall, *J. Chem. Soc.*, 3080 (1926).

49. A. G. Cairns-Smith, *J. Chem. Soc.*, 182 (1961).

50. A. Senier and R. Clarke, *J. Chem. Soc.*, 2081 (1911).

51. L. Zechmeister, *Chem. Rev. 34*, 267 (1944).

52. L. Pauling, *Helv. Chim. Acta 32*, 2241 (1949).

53. L. Zechmeister, *Experientia 10*, 1 (1954).

54. M. P. Meunier, *Compt. Rend.*, *222*, 1528 (1946).

55. M. Calvin and H. W. Alter, *J. Chem. Phys. 19*, 768 (1951).

56. G. B. Kistiakowsky and W. R. Smith, *J. Am. Chem. Soc. 56*, 638 (1934).

57. G. B. Kistiakowsky and W. R. Smith, *J. Am. Chem. Soc. 58*, 2428 (1963).

58. P. E. Gagnon and L. P. Charette, *Can. J. Res. 19B*, 275 (1941).

59. Y. Hirshberg and E. Fischer, *J. Chem. Phys. 22*, 572 (1954).

60. P. J. W. Le Fevre and J. Northcroft, *J. Chem. Soc.*, 867 (1953).

61. K. E. Calderbank, R. J. W. Le Frevre, and J. Northcroft, *Chem. Ind.* 158 (1948).
62. P. Lunder and C. A. Winkler, *Can. J. Chem. 30*, 679 (1952).
63. (a) J. Stenhouse, *Liebigs Ann. 74*, 278 (1850); (b) J. Stenhouse, *Liebigs. Ann. 156*, 197 (1870).
64. (a) H. Schiff, *Liebigs Ann. 201*, 355 (1880); (b) H. Schiff, *Ber. Dtsch. Chem. Ges. 19*, 2153 (1886); (c) H. Schiff, *Liebigs Ann. 239*, 349 (1887).
65. T. Zincke and G. Muhlhausen, *Ber. Dtsch. Chem. Ges. 38*, 3824 (1905).
66. W. Borsche, H. Leditschtke, and K. Lange, *Ber. Dtsch. Chem. Ges. 38*, 3824 (1905).
67. (a) O. Fischer, *Liebigs Ann. 206*, 141 (1883); (b) R. R. Renshaw and N. M. Naylor, *J. Am. Chem. Soc. 44*, 862 (1922).
68. (a) W. Konig, *J. Prakt. Chem. 72*, 555 (1905); (b) W. Dieckmann and L. Beck, *Ber. Dtsch. Chem. Ges. 38*, 4123 (1905); (c) O. Aschan and A. Schwalbe, *Ber. Dtsch. Chem. Ges. 67*, 1830 (1934).
69. (a) A. P. Dunlop and F. N. Peters, *The Furans*, p. 622, Reinhold Pub. Co., New York (1953); (b) K. G. Lewis and C. E. Mulquiney, *Tetrahedron 33*, 463 (1977); (c) C. E. Mulquiney, Ph.D. Thesis, University of New England, Australia, 1971.
70. J. C. McGown, *J. Chem. Soc.*, 777 (1966).
71. C. F. Koelsch and J. J. Carney, *J. Am. Chem. Soc. 72*, 2280 (1950).
72. (a) E. N. Marvell, T. H. Li, and C. Paik, *Tetrahedron Lett.*, 2089 (1967); (b) E. N. Marvell, G. Caple, and I. Shahidi, *Tetrahedron Lett.*, 277 (1967); (c) E. N. Marvell, G. Caple, and I. Shahidi, *J. Am. Chem. Soc. 92*, 5641 (1970); (d) E. N. Marvell and I. Shahidi, *J. Am. Chem. Soc. 92*, 5646 (1970).
73. C-T. Chen, S-J. Yan, and J-C. Ho, *Bull. Inst. Chem. Acad. Sin. 13*, 49 (1967).
74. G. J. Sabongi, B. A. Lea, and S. S. C. Poon, British Pat. Appl. 8,307,023 A (1983).
75. (a) W. Konig, *J. Prakt. Chem.* 69, 105 (1904); (b) I. L. Knunyants *et al.*, *J. Gen. Chem. USSR 9*, 557 (1939); (c) H. E. Nikulajewski *et al.*, *Ber. 100*, 2616 (1967).
76. (a) T. Zincke, *Ann. 338*, 107 (1903); (b) T. Zincke, *Ann. 341*, 365 (1905); (c) T. Zincke, *Ann. 408*, 285 (1915).
77. (a) W. Konig, *J. Prakt. Chem. 72*, 555 (1905); (b) W. Konig, *J. Prakt. Chem. 88*, 193 (1913); (c) W. Konig, *Ber. 67*, 1274 (1934).
78. H. Stobbe, *Chem.-Ztg. 44*, 340 (1920).
79. L. Chalkley, *Chem. Rev. 6*, 217 (1929).
80. H. G. Brown and W. G. Shaw, *Rev. Pure Appl. Chem. 11*, 2 (1961).
81. S. S. Bhatnager, P. L. Kapur, and M. S. Hashmi, *J. Indian Chem. Soc. 15*, 573 (1939).
82. J. Van Overbeek, *Botan. Rev. 5*, 655 (1939).
83. R. Exelby and R. Grinter, *Chem. Rev. 64*, 247 (1964).
84. G. H. Dorion and A. F. Wiebe, *Photochromism*, The Focal Press, London (1970).
85. G. H. Brown, ed., *Photochromism*, in: *Techniques of Chemistry* (A. Weissberger, series ed.), Vol. 3, Wiley-Interscience, New York (1971).
86. M. D. Cohen and G. M. J. Schmidt, *J. Am. Chem. Soc. 66*, 2442 (1962).
87. A. Senier, F. G. Shepheard, and R. Clarke, *J. Chem. Soc. 101*, 1950 (1912).
88. A. Senier and F. G. Shepheard, *J. Chem. Soc. 95*, 441 (1909).
89. A. Senier and R. B. Forster, *J. Chem. Soc. 105*, 2462 (1914).
90. A. Aharoni, D. Dhkovski, E. H. Frei, and A. Tzalmova, *J. Phys. Chem. Solids 24*, 927 (1963).
91. G. Wattermark and L. Dogliotti, *J. Chem. Phys. 40*, 1486 (1964).
92. H. Blitz, *Ann. 305*, 170 (1899).
93. F. D. Cattaway, *J. Chem. Soc. 89*, 462 (1906).
94. C. V. Gheorghiu and B. Arrventieu, *Am. Sci., Univ. Jassy. 16*, 536 (1931).
95. H. Stobbe, *Ber. Verhandl. Sachs. Akad. Wiss. Leipzig 74*, 161 (1922).
96. M. Padoa and F. Graziani, *Atti. Accad. Lincei 18*, 269 (1909).
97. M. Padoa, *Atti. Accad. Lincei 18*, 694 (1909).

 98. M. Padoa and F. Graziani, *Atti. Accad. Lincei* 19, 489 (1910).
 99. H. Blitz, *Z. Physik. Chem.* 30, 527 (1899).
100. M. Padoa and L. Santi, *Atti. Accad. Lincei* 19, 302 (1910).
101. M. Padoa and L. Santi, *Atti. Accad. Lincei* 20, 675 (1911).
102. C. Gheorghiu, *Rev. Stiint. "V. Adamachi"* 32, 255 (1946).
103. W. O. Williamson, *Mineral Mag.* 25, 513 (1940).
104. R. Dickinson, I. M. Heilborn, and F. Irving, *J. Chem. Soc.*, 1888 (1927).
105. I. M. Heilbron, H. E. Hudson, and D. M. Huish, *J. Chem. Soc.* 123, 2273 (1923).
106. C. V. Gheorghiu, *Bull. Soc. Chim. Fr.* 1, 97 (1934).
107. H. Stobbe and H. Mallison, *Ber.* 64, 1226 (1913).
108. C. M. Bere and S. Smiles, *J. Chem. Soc.* 125, 2359 (1924).
109. R. Child and S. Smiles, *J. Chem. Soc.* 103, 2695 (1926).
110. H. Stobbe, *Ann.* 359, 1 (1908).
111. H. Stobbe, *Ann.* 380, 1 (1911).
112. H. Stobbe, *Ber. Verhandl. Sachs. Akad. Wiss. Leipzig* 74, 161 (1922).
113. L. Hänel, *Naturwiss.* 37, 91 (1950).
114. W. R. Brode, J. H. Gould, and G. M. Wyman, *J. Am. Chem. Soc.* 74, 4641 (1952).
115. W. R. Brode, J. H. Gould, and G. M. Wyman, *J. Am. Chem. Soc.* 75, 1856 (1953).
116. G. S. Hartley, *J. Chem. Soc.*, 633 (1938).
117. E. I. Stearns, *J. Opt. Sci. Am.* 32, 282 (1942).
118. W. R. Brode and G. M. Wyman, *J. Res. Nat. Bur. Stand.* 47, 170 (1951).
119. G. M. Wyman and W. R. Brode, *J. Am. Chem. Soc.* 73, 1487 (1951).
120. G. M. Wyman and A. F. Zenhausern, *J. Am. Chem. Soc.* 30, 2348 (1965).
121. H. Güsten, *Chem. Commun.*, 133 (1969).
122. D. L. Ross, J. Blanc, and F. J. Matticoli, *J. Am. Chem. Soc.* 92, 5750 (1970).
123. J. Blanc and D. L. Ross, *J. Phys. Chem.* 72, 2817 (1968).
124. C. R. Giullano, L. D. Hess, and J. D. Margetum, *J. Am. Chem. Soc.* 90, 587 (1968).
125. G. H. Brown, S. R. Odiseh, and E. J. Taylor, *J. Phys. Chem.* 66, 2426 (1962).
126. J. Lifschitz, *Ber.* 52, 1919 (1919).
127. J. Lifschitz, *Ber.* 58, 2434 (1925).
128. J. Lifschitz and L. C. Joffe, *Z. Physik. Chem.* 97, 426 (1921).
129. E. O. Holmes, *J. Am. Chem. Soc.* 44, 1002 (1922).
130. L. Harris, J. Kaminsky, and R. G. Simard, *J. Am. Chem. Soc.* 57, 1151 (1935).
131. L. Chalkley, U.S. Pat. 2,441,561 (1948).
132. B. K. Singh, *J. Am. Chem. Soc.* 43, 333 (1921).
133. B. K. Singh and B. Bhaduri, *Trans. Faraday Soc.* 27, 478 (1931).
134. M. Singh and T. R. Patt, *J. Indian Chem. Soc.* 19, 130 (1942).
135. W. C. Clark and C. F. Lothian, *Trans. Faraday Soc.* 54, 1790 (1958).
136. H. S. Gutowsky and R. L. Rutledge, *J. Chem. Phys.* 29, 1183 (1958).
137. A. J. Nunn and K. Schofield, *J. Chem. Soc.*, 583 (1952).
138. K. Schofield, *J. Chem. Soc.*, 2408 (1949).
139. R. Harwich, H. S. Mosher, and P. Passailaigue, *Trans. Faraday Soc.* 56, 44 (1960).
140. H. S. Mosher, C. Souers, and R. Hardwick, *J. Chem. Phys.* 32, 1888 (1960).
141. J. A. Sousa and J. Weinstein, *J. Org. Chem.* 27, 3155 (1962).
142. J. D. Margerum, L. J. Miller, E. Saito, M. S. Brown, H. S. Mosher, and R. Hardwick, *J. Phys. Chem.* 66, 2434 (1962).
143. A. L. Bluam, J. Weinstein and J. A. Sous, *J. Org. Chem.* 28, 1989 (1963).
144. O. Chaudé and P. Rumpf, *Compt. Rend.* 236, 697 (1953).
145. Y. Hirshberg, *J. Am. Chem. Soc.* 78, 2304 (1956).
146. Y. Hirshberg and E. Fischer, *J. Chem. Soc.*, 3129 (1954).
147. E. Berman, R. E. Fox, and F. D. Thomson, *J. Am. Chem. Soc.* 81, 5605 (1959).

148. Y. Hirshberg, *Compt. Rend. 231*, 903 (1950).
149. G. Kortum, W. Theilacker, and G. Littmann, *Naturwiss. 44*, 114 (1957).
150. G. N. Lewis and D. Lipkin, *J. Am. Chem. Soc. 64*, 2801 (1942).
151. F. H. C. Stewart, *Chem. Rev. 64*, 129 (1964).
152. M. J. Tien and I. M. Hunsberger, *Chem. Ind.*, 119 (1955).
153. M. J. Tien and I. M. Hunsberger, *J. Am. Chem. Soc. 77*, 6604 (1955).
154. M. D. Cohen and G. M. Schmidt, *Reactivity of Solids*, p. 556, Elsevier, Amsterdam (1961).
155. S. Nespurek and M. Sorm, *Czech. J. Phys. B25*, 1051 (1975).
156. C. V. Greco and P. J. O'Reilly, *J. Heterocyl. Chem. 7*, 761 (1970).
157. Z. Endo, *J. Chem. Soc. Japan 65*, 667 (1944).
158. G. Scheibe and F. Feichtmayer, *J. Phys. Chem. 66*, 2449 (1962).
159. H. Stobbe, *Chem.-Ztg. 44*, 340 (1920).
160. W. M. Cumming and K. J. Steel, *J. Chem. Soc., 123*, 2463 (1923).
161. M. A. Mostoslavskii and V. A. Izmailskii, *Zh. Obsch. Khim. 31*, 17 (1961).
162. H. Stobbe and F. J. Wilson, *Ann., 374*, 237 (1910).
163. L. Von Mechel and H. Stauffer, *Helv. Chim. Acta 24*, 151 E (1941).
164. P. Bartels, *Z. Physik. Chem. (Frankfurt) 9*, 74 (1956).
165. F. Mattu, R. Pirisi, and R. M. Manca, *Ann. Chim. (Rome) 42*, 632 (1952).
166. J. H. Backer, *Recl. Trav. Chim. Pays-Bas 55*, 915 (1936).
167. A. P. de Carvallio, *Compt. Rend. 200*, 6 (1935).
168. H. von Halban and H. Geigel, *Z. Physik. Chem. 96*, 233 (1920).
169. R. J. Walter, *J. Prakt. Chem., 67*, 445 (1903).
170. J. M. Brittain, R. A. Jones, R. O. Jones, and T. J. King, *J. Chem. Soc.*, 2656 (1981).
171. G. H. Dorion and K. O. Loeffer, U.S. Pat. 3,127,335 (1964).
172. W. Markwald, *Z. Physik. Chem. 30*, 140 (1899).
173. L. Zechmeister, *Chem. Rev. 34*, 335 (1944).
174. Z. Smakula, *Z. Physik. Chem. B25*, 90 (1934).
175. M. Calvin and H. W. Alter, *J. Chem. Phys. 19*, 768 (1951).
176. W. H. Martin, *Trans. Roy. Soc. (Canada), Sect. III, 34*, 35 (1940).
177. Y. Hirshberg and F. Bergmann, *J. Am. Chem. Soc. 72*, 5118 (1952).
178. O. Grummitt and F. J. Christoph, *J. Am. Chem. Soc. 73*, 3479 (1951).
179. A. Sandoval and L. Zechmeister, *J. Am. Chem. Soc. 69*, 553 (1947).
180. Y. Hirshberg, E. Bergman, and F. Bergman, *J. Am. Chem. Soc. 72*, 5120 (1950).
181. K. Lunde and L. Zechmeister, *J. Am. Chem. Soc. 76*, 2308 (1954).
182. L. Zechmeister and A. L. Le Rosen, *J. Am. Chem. Soc. 64*, 2755 (1942).
183. L. Zechmeister and A. L. Le Rosen, *Science 95*, 587 (1942).
184. C. F. Garbers, C. H. Engster, and P. Knarrer, *Helv. Chim. Acta 35*, 1850 (1952).
185. E. Baroni and H. Seifert, *Naturwiss. 29*, 560 (1941).
186. G. V. Bree, *Bull. Soc. Chim. Belg. 57*, 71 (1948).
187. B. K. Vaidya, *Proc. Roy. Soc. (London) A129*, 299 (1930).
188. A. R. Olson and F. L. Hudson, *J. Am. Chem. Soc. 55*, 1410 (1933).
189. W. R. Brode and G. M. Wyman, *J. Res. Nat. Bur. Stand. 47*, 170 (1951).
190. G. M. Wyman and W. R. Brode, *J. Am. Chem. Soc. 73*, 1487 (1951).
191. W. E. Brode, in: *Roger Adams Symposium Volume*, J. Wiley and Sons, New York (1955).
192. L. Zechmeister and J. H. Pinckard, *Experientia 9*, 16 (1953).
193. E. Fischer and Y. Hirshberg, *J. Chem. Soc.*, 4522 (1952).
194. Y. Hirshberg and E. Fischer, *J. Chem. Soc.*, 3129 (1954).
195. Y. Hirshberg and E. Fischer, *J. Chem. Soc.*, 297 (1954).
196. L. N. Ferguson and I. Kelly, *J. Am. Chem. Soc. 73*, 3707 (1951).
197. H. H. Cary, *Rev. Sci. Instrum. 17*, 558 (1946).
198. W. R. Brode, H. J. Gould, and G. M. Wyman, *J. Am. Chem. Soc. 74*, 4641 (1952).

199. G. Wagner, *J. Polym. Sci. 9*, 133 (1971).

200. G. Wagner, *Makromol. Chem. 145*, 85 (1971).

201. (a) G. Wagner, *Pure Appl. Chem. 49*, 433 (1977); (b) J. H. Krieger, *Chem. Eng. News*, Aug. 24, 24 (1980).

202. K. Bowden, I. Heilbron, E. R. Jones, and K. H. Sargent, *J. Chem. Soc.*, 1579 (1947).

203. A. J. Melveger and R. H. Baughman, *J. Polym. Sci., Polym. Phys. Ed. 11*, 603 (1973).

204. G. Wegner, *J. Polym. Sci. 9*, 133 (1971).

205. R. H. Baughman, *J. Appl. Phys. 43*, 4362 (1972).

206. D. M. Sturmer, British Pat. 1,399,751 (1972).

207. K. Drexhage, G. A. Reynolds, and J. J. Wrobel, *Res. Disclosure*, No. 17812 (1982).

208. C. Engler and K. Dorant, *Chem. Ber. 28*, 2497 (1895).

209. M. Luwich, Ph.D. Thesis, Weizmann Inst. of Science (1968).

# Index